F. MAYER, CYTOLOGY AND MORPHOGENESIS OF BACTERIA

HANDBUCH DER PFLANZENANATOMIE
ENCYCLOPEDIA OF PLANT ANATOMY
TRAITÉ D'ANATOMIE VÉGÉTALE

BEGRÜNDET VON K. LINSBAUER
FORTGEFÜHRT VON G. TISCHLER UND A. PASCHER

Herausgegeben von Professor Dr. H. J. BRAUN, Freiburg (Breisgau)
Professor Dr. S. CARLQUIST, Claremont (Calif./USA),
Professor Dr. P. OZENDA, Grenoble
Professor Dr. INGRID ROTH, Freiburg (Breisgau)

Spezieller Teil
Band VI, Teil 2

1986

GEBRÜDER BORNTRAEGER · BERLIN · STUTTGART

CYTOLOGY AND MORPHOGENESIS OF BACTERIA

by

Professor Dr. FRANK MAYER

Institute for Microbiology
University of Göttingen, FR Germany

With 234 figures

1986

GEBRÜDER BORNTRAEGER · BERLIN · STUTTGART

ISBN 3-443-14017-3

Printed in Germany

Preface

The mere description of cytological properties of bacteria in an inventory manner has not been the aim of this book. Cytological studies play a role in a complex system of investigations finally leading, as is hoped, to the comprehension of all aspects of bacteria. It is in this context that the results of cytological investigations should be considered to be of major importance. As compared to many other groups of organisms, bacteria have been more the subject for detailed studies on functional properties on the macromolecular level. To keep abreast with this development, cytological work has also to be extended to that level. In addition, physiological and genetic research involves experimental approaches often leading, as a side-effect, to changes in bacterial cell structure. A proper documentation of these changes improves the understanding of the experimental results. With this situation in mind, this book has been deviced such that the description of certain cytological facts usually is followed by a section dealing with selected aspects of related function and morphogenesis.

The recent development of new electron microscopic preparation, imaging, and image-evaluation procedures (low-temperature embedding, cryo-techniques, computer averaging) will not only provide tools suited for the correction of wrong views caused by the introduction of artifacts. These methods will also be used for the detection of so far unknown cytological properties of the bacterial cell. The potential of these techniques appears to be enormous. In this book, selected results recently obtained with some of these procedures are described and discussed.

I wish to express my thanks to Marcella Pieper for her very valuable help during the preparation of the figures. I thank my colleague D. G. Robinson who made helpful suggestions during the completion of the manuscript, my coworkers and colleagues F. Ebrahim-Nesbat, T. Jarchau, G.-W. Kohring, M. Lübben, H. Lünsdorf, M. Rohde and A. Weith who contributed original micrographs, and Zeiss, Oberkochen/FRG for light micrographs obtained with a Laser Scanning Microscope.

Göttingen, October 1985

Frank Mayer

Contents

1. Introduction

Bacteria are small ("micro"-) organisms. They share this property with small higher organisms such as certain algae, fungi, and protozoa. Most of the bacteria exhibit a simple shape, and they are much less well differentiated as compared to other organisms. The term "prokaryote" also used for the bacterial cell indicates that the lack of a distinct membrane-bound nucleus (such a nucleus is typical for all other organisms, the "eukaryotes") is one of the major structural aspects of bacteria. Other structural differences between prokaryotes and eukaryotes are evident such as the organization of the genetic material and the cell envelope, the differentiations developed for locomotion, the occurrence of pili and fimbriae for sexual interactions and adherence, respectively, and the mode of cell division. One of the important physiological properties of the eukaryotic cell is lacking in bacteria: they cannot perform endocytosis. Most of these and other characteristic features of bacteria are assumed to be "primitive". Nevertheless, bacteria obviously are very successful organisms, and they constitute a considerable share of the living matter on Earth. This may primarily be due to their small dimensions, high rate of reproduction, and physiological and genetic flexibility.

A thorough knowledge of bacterial ultrastructure has always been, and still is, one of the prerequisites for the understanding of many bacterial functions. Therefore, analysis and description of bacterial structures on cellular and molecular levels have not lost their importance despite the fact that major findings in modern microbiology regarding evolution, taxonomy, basic and applied genetics, and biotechnology have been obtained without instruments such as light or electron microscopes. Together with physiological, biochemical, chemical, and immunological approaches (Alberts et al. 1983, Stryer 1981), fine structure research (e. g., Armbruster et al. 1982, Bayer & Remsen 1970, Dubochet et al. 1983, Fuller & Lovelock 1976, Van Heel 1984, Holt & Beveridge 1982, Maunsbach 1984, Mayer 1978, Nanninga 1973, Nanninga et al. 1984, Rogers 1983, Unwin & Henderson 1975) has helped in elucidating bacterial locomotion and chemotaxis, photosynthesis, adaptation to extreme or other specific environments, cell nutrition, growth, division and differentiation, interactions with bacteriophages, sexual and other interactions between bacteria, colonization on and utilization of a wide variety of substrates, symbiotic as well as parasitic interrelationships with higher organisms, and aspects of pathogeneity in plants, animals and man, to name just a few problems.

The present state of knowledge on bacterial cytology, morphogenesis, and structure-function relationships cannot be covered in a book of this relatively moderate size. Therefore, only selected examples for each of the described topics have been chosen. For further details, the reader is referred to original papers, reviews and monographs presented in the bibliography. Additional information may be gleaned from the figures and figure legends supplementing the text.

For a better understanding of structural features, a very short introduction into relevant preparation and imaging techniques is given.

2. Methods of investigation

Bacterial structure and morphogenesis may be analyzed by a variety of light and electron microscopic preparation, imaging and image evaluation techniques. For a better comprehension of bacterial structures and molecular substructures, a cursory introduction into some aspects of the light and electron microscopy of bacteria follows below. Further references, besides those presented in the figure legends, for the detailed procedures are not given as various books on these topics are available.

2.1 Light microscopy

Prior to the introduction of electron microscopy light microscopy of bacteria was the only imaging technique available. Light microscopy has a long history and is still widely used in botany, zoology, microbiology and medicine. Thus a high standard in preparation and imaging techniques for biological samples has been achieved.

As with other biological samples, bacteria can be visualized by light microscopy as amplitude or phase contrast objects. Depending on the aspects to be investigated, various degrees of sophistication of preparation and imaging are needed in order to obtain adequate results. Usually, the examination of untreated samples gives information on presence, shape and dimensions of bacterial cells and the occurrence of bacterial spores. Further details (Gram type, distribution of nucleic acids or storage material in the cells, presence of flagella etc.) are only obtained when specific fixation and staining techniques are applied.

2.1.1 Unstained bacterial samples

Unstained living bacteria can be visualized by light microscopy (Fig. 1) of smear preparations. Microcolonies are used for the investigation of growth, cell division, spore formation or spore germination.

2.1.2 Fixation and staining procedures

Fixation of bacteria can be achieved by controlled heat treatment of air-dried smears. Structural artifacts caused by this procedure are negligible as they occur in dimensions beyond the resolving power of the light microscope.

Heat-fixed samples may be stained by different techniques depending on the purpose of the investigation. General staining (blue, red or violet) of bacterial cells is obtained by methylene blue (for cocci), carbol fuchsin (a mixture of basic fuchsin and phenol) (for rods and spirillum-like cells) or carbol gentian violet (for all bacterial types). A simple technique is the preparation of a "negatively stained" sample. A mixture of the bacterial suspension and indian ink is used for the preparation of a smear. After fixation by air drying, the bacteria appear light on a dark background. This technique is especially suited for the demonstration of slime or capsules.

Gram-staining

The heat-fixed smear is first stained with carbol gentian violet, rinsed with water, put in a solution of potassium iodide, followed immediately by treatment with ethanol, and again rinsed. Gram-

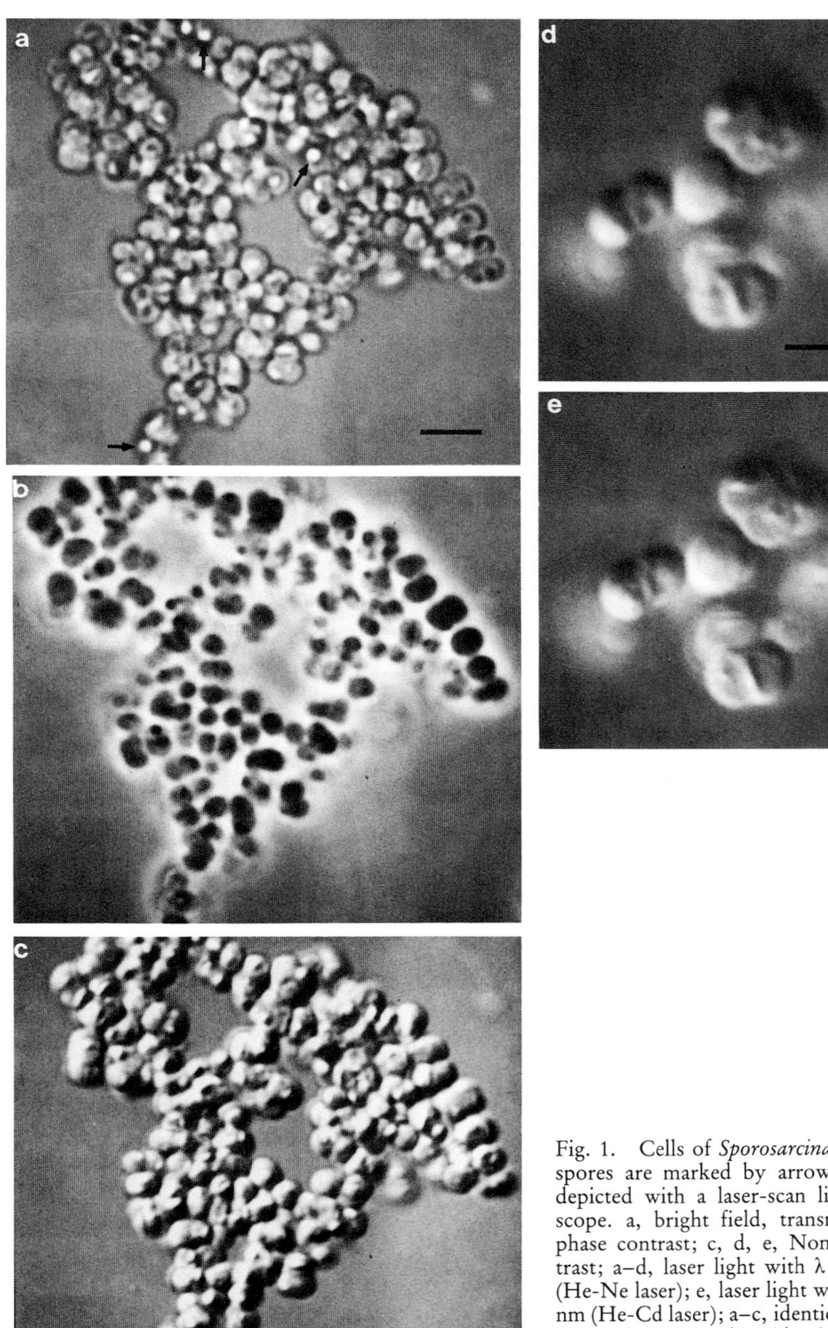

Fig. 1. Cells of *Sporosarcina halophila;* spores are marked by arrows. Samples depicted with a laser-scan light microscope. a, bright field, transmission; b, phase contrast; c, d, e, Nomarski contrast; a–d, laser light with λ = 633 nm (He-Ne laser); e, laser light with λ = 442 nm (He-Cd laser); a–c, identical magnification; bar: 5μm; d, e, identical magnification; bar: 5 μm. (Original micrographs by Zeiss, Oberkochen, FRG)

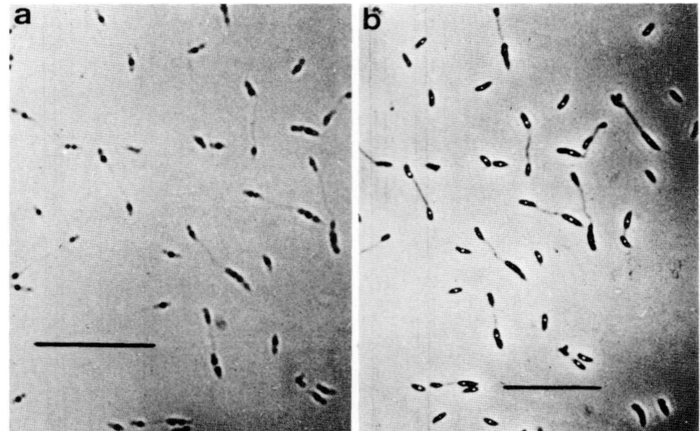

Fig. 2. Light micrographs of Giemsa-stained *Hyphomicrobium* strain *B-522* from an actively growing culture. a, bright field; b, phase contrast. Bar: 10 µm. (From Moore & Hirsch 1973 a)

negative bacteria can now be counterstained with dilute carbol fuchsin followed by rinsing with water. After air drying, Gram-positive bacteria are violet, Gram-negative cells are red. Bacterial cultures should not be older than 24 h in order to avoid misleading results.

Staining of nucleic acids

Nucleic acids in bacteria can be stained (the resulting colour is reddish violet) by the Giemsa staining procedure (Fig. 2). Samples with low numbers of bacteria can be investigated by treatment of formaldehyde-fixed cells (attached to a membrane filter) with acridine orange, followed by rinsing with water. The bacteria are detectable by their fluorescence (s. also below, Fluorescence light microscopy).

Staining of spores

Spores are stained, after heat fixation, CrO_3-treatment and water rinsing, with heated carbol fuchsin, followed by treatment with dilute H_2SO_4, rinsing with water and additional staining with methylene blue. The spores are red, whereas the vegetative cells appear blue.

Staining of flagella

Flagella or flagellar tufts can be visualized either by appropriate imaging techniques (phase contrast, dark field, s. below) without any staining, or by artificial "thickening" of the flagella by treatment of heat-fixed smears with tannic acid, $FeSO_4$ and basic fuchsin, followed by water rinsing and additional staining with carbol fuchsin. Flagella appear as fine red fibers.

Other staining procedures

Besides samples containing exclusively bacteria, preparations made from plants infected by bacteria (e.g., root nodules) can also be investigated by light microscopy of sections. The material is fixed in a solution of ethanol, formaldehyde and acetic acid containing $HgCl_2$. After rinsing in a solution

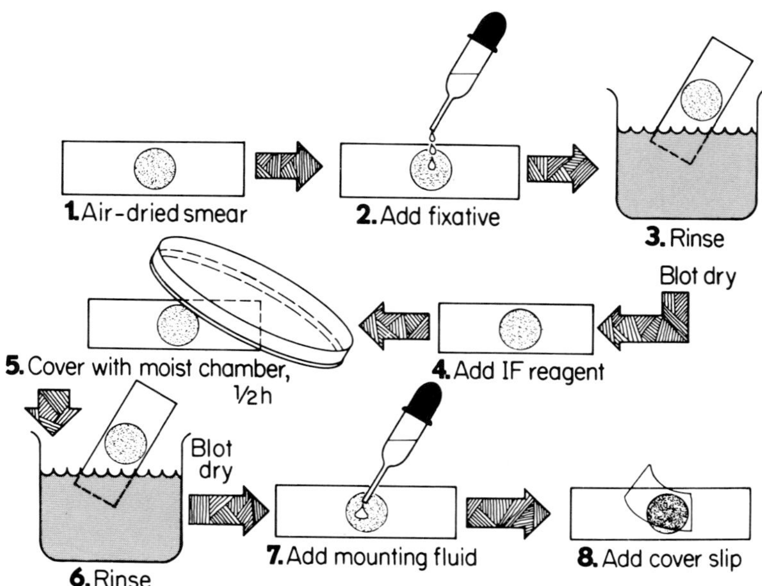

Fig. 3. Preparation of a slide for the fluorescent-antibody test (via direct immunofluorescence staining) to identify groups A and B streptococci. (From FACKLAM & WILKINSON 1981)

without $HgCl_2$ the material is paraffin embedded and sectioned. Finally, Gram staining is carried out on sections which have been freed from paraffin.

2.1.3 Immunolabelling techniques

One of the techniques usually applied is the FITC method (fluorescein isothiocyanate). Rhodamine or protein A-gold procedures can also be used. The direct technique (Fig. 3) is performed by binding of labelled antibodies to the respective antigens; the indirect technique involves antigen-specific (primary) antibodies which are the specific binding sites for the labelled antibody-specific (secondary) antibodies. The FITC method results in a bright green fluorescence. With protein A-gold, the resulting colour of a positive labelling reaction, in routine light microscopy, is purple.

2.1.4 Imaging techniques

The limitation of resolving power of the light microscope is especially evident when bacteria are to be analyzed.

Routine light microscopy procedures

The routine imaging techniques are bright field, dark field and phase contrast light microscopy. Nomarski interference contrast, when applied to bacterial samples, results in a distinct type of contrast of the cells which gives the impression of a cellular “relief” standing out against a flat background.

Recently, a scanning light microscope with a laser as the light source has become available. Besides having a high contrast and resolution (s. Fig. 1), this instrument combines the advantages of a light microscope with the typical properties of a scanning device, i.e., the potential for computerized image analysis.

Fluorescence light microscopy

Primary and secondary fluorescence in samples of bacteria can be observed with ultraviolet, blue or green light as excitation irradiation. In case of primary fluorescence, unfixed living cells should be analyzed (e.g., methanogenic bacteria). For secondary fluorescence, unfixed or fixed and adequately stained samples are prepared. Immunofluorescence is a technique of special interest for identification and localization purposes (s. above, Immunolabelling techniques).

2.2 Electron microscopy

The electron microscopy of bacteria and their components has been and still is the major tool for investigators working on cell structure and morphogenesis. Sample preparation plays a crucial role for reliability of the results. As in light microscopy, simple and more sophisticated preparation and imaging techniques are available. Some of them are described below.

2.2.1 Preparation techniques

Conventional sample preparation techniques involve some sort of pretreatment (air drying, chemical or physical fixation) and staining for contrast enhancement. In addition the analysis of intracellular structures usually requires thin specimens; this means that bacteria have to be broken open or sectioned. Danger of artificially altered structures is high at the cellular and macromolecular level. New developments, especially the introduction of cryo-techniques, seem very promising with respect to the reduction of such artifacts of preparation. One of the clearest examples for this is the structural concept of the "mesosome" (s. below).

Air drying

Air drying, i.e. the drying of cells or cell components onto a support film, is a simple preparation procedure. It is usually followed by metal shadowing (s. below) in order to provide sufficient contrast. The danger of artifact production is always present due to uncontrolled loss of water. This can be minimalized if, prior to drying, water is replaced by solvents with low surface tension.

Critical point drying, freeze drying

Controlled drying by the critical point drying procedure reduces considerably shrinking of the sample. This is achieved by the fact that the vapour of the solvent replacing the water (ethanol, acetone) and the solvent itself have the same volume, identical density and are coexistent. This brings the surface tension to zero. However, ethanol or acetone may themselves induce artifacts. Freeze drying is achieved by shock freezing followed by a controlled sublimation of the ice. Contrast is obtained by metal shadowing.

Metal shadowing

Metal shadowing is a technique for introducing contrast to specimens. For this metals like platinum, iridium, tantalum and others or their alloys are evaporated under high vacuum conditions. Samples can be pretreated by air drying, critical point drying or freeze drying (Figs. 4 and 5). Resolution is restricted, in the routine procedure, to 2 to 3 nm, depending on the type of apparatus and metal. Metal shadowing can be combined with carbon sublimation for higher resolution.

Negative staining

Bacterial cells or their components (membranes, macromolecules) mounted on a thin support film can be contrasted by application of specific metal salts dissolved in water (uranium or tungsten salts etc.). After air drying the samples appear light under the electron beam, surrounded by a dark background formed by the dried negative-staining salt (Figs. 6 and 8). Resolution is higher than with metal shadowing provided that the electron beam does not visibly interfere with the object or with the staining salt producing "burning".

Conventional ultrathin sectioning

After chemical fixation, dehydration, staining, embedding in liquid (monomeric) resin and polymerization of the resin, bacteria can be cut to give ultrathin sections of about 40 to 80 nm thickness.

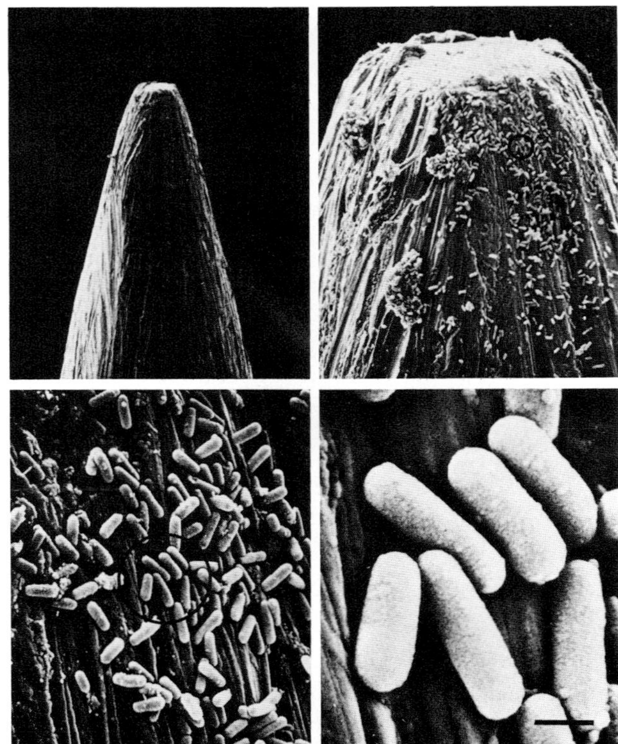

Fig. 4. Scanning electron micrographs, taken at progressively higher magnifications, show bacterial cells on the point of a pin. Bar: 1 μm. (From ALBERTS et al. 1983, after BRAIN)

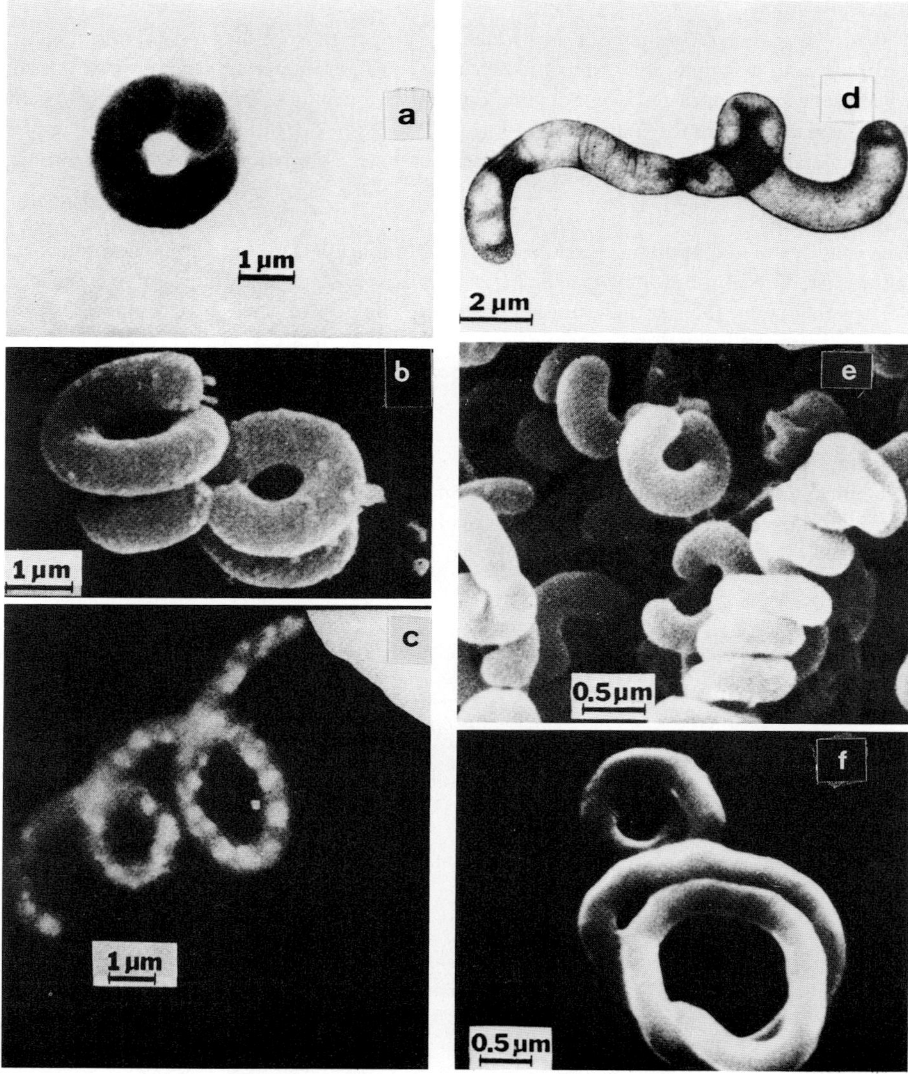

Fig. 5. Bacterial cell shapes. Normal morphological forms of *Microcyclus* species. a, b, c, *Microcyclus flavus* (after RAJ); d, *Microcyclus major* (after LARKIN); e, f, *Microcyclus marinus* (after RAJ). (From RAJ 1981)

Post-staining with metal salt solutions further enhances their contrast. Resolution is limited to values around 2.5 to 3 nm depending on the quality of fixation, type of staining and section thickness. Under certain conditions, sections can be used for immunocytochemical investigations (s. below).

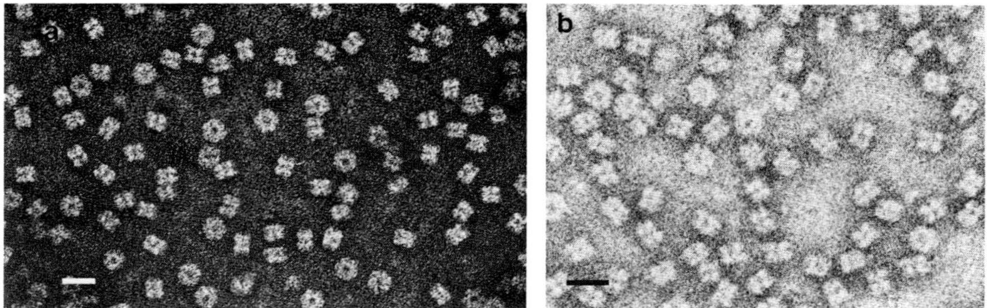

Fig. 6. Negatively stained samples. a, b, isolated enzyme glutamine synthetase (a, deep stain; b, shallow stain). Bars: 20 nm. (Original micrographs F. MAYER)

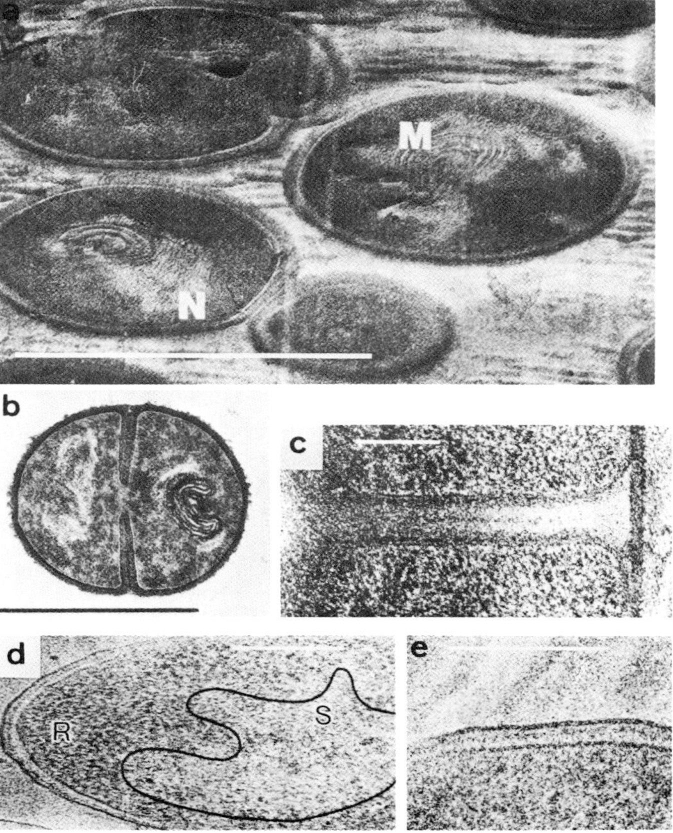

Fig. 7. Electron microscopy of frozen-hydrated bacteria. a, amorphous frozen-hydrated section of *Staphylococcus aureus* fixed overnight in OsO_4 before freezing. Nucleoids and "mesosomes" are marked N and M, respectively; b, a conventionally prepared *S. aureus* cell; c, septum of *S. aureus;* d, *E. coli*; view showing the smooth (S) central and rough (R) external regions of the cytoplasm. A tentative borderline has been drawn between these two regions; e, view showing the typical aspect of the *E. coli* cell envelope. Bars: 1 µm in a and b, 0.1 µm in c, and 0.2 µm in d and e. (From DUBOCHET et al. 1983)

Low temperature embedding

Bacterial samples can be prepared for ultrathin sectioning by low temperature embedding techniques (e. g. Lowicryl technique). This is performed by starting with a conventional or modified aldehyde fixation followed by dehydration at low temperature, infiltration with resin at low temperature and polymerization of the resin by ultraviolet light at low temperature. The ultrathin sections (cut at room temperature) are especially suited for cytochemical and immunocytochemical labelling procedures. Modifications have been published where "low temperature" resins have been used at room temperature, only UV polymerization being carried out at 4°C. Polar or nonpolar resins are available. Polar (hydrophilic) embedding medium allows one to keep the sample, during infiltration, in a partially hydrated state. With nonpolar (hydrophobic) embedding medium, freeze substitution methods can be applied. In this case, infiltration and polymerization temperatures need not be raised above −35°C.

Cryo ultramicrotomy

Cryo ultramicrotomy has been developed in order to avoid artifacts caused by chemical fixation. The samples are sectioned in the frozen hydrated state (Fig. 7) and can either be viewed in the conventional transmission electron microscope or in the cryo electron microscope. This fixation and sectioning technique is especially suited for immunocytochemical work. A typical artifact is compression of the section.

Freeze fracturing and freeze etching

These methods start with a shock freezing of the sample thus avoiding formation of destructive ice crystals. The frozen sample is then cleaved or "sectioned" under high vacuum conditions and metal shadowed after or without "etching" of the exposed sample surface through sublimation of ice. Carbon is sublimated to produce a stabilizing additional layer on top of the metal replica. The sample is then taken out, and the replica is freed from organic material by a chemical cleaning procedure. In a cleaved sample, both fracture faces may be obtained under appropriate conditions. Fracturing occurs preferentially within biological membranes and along other structural inhomogeneities. The replica give an impression of the three-dimensional inner structure of the sample. Artifacts cannot always be avoided under routine conditions.

2.2.2 Imaging techniques

Several electron microscopic imaging techniques are in use for bacteria and their components: conventional transmission mode (CTEM), scanning mode (SEM), scanning transmission mode (STEM). CTEM and STEM can be used under bright and darkfield conditions. Image manipulation and evaluation are simplified by the scanning modes, as well as microanalysis of the composition of the sample. Analysis of the distribution of light and medium heavy elements in the transmission mode is possible by use of a special prism-mirror-prism electron spectrometer, resulting in high resolution elemental maps with a high sensitivity of detection (element masses down to 10^{-21} g). Another variation of the CTEM mode is cryo electron microscopy, i. e. viewing of the frozen sample in the cooled microscope stage. Recent studies have shown that one of the limiting factors of electron microscopy of biological samples is beam damage. As a result, low or minimum dose procedures have been developed for viewing and micrography.

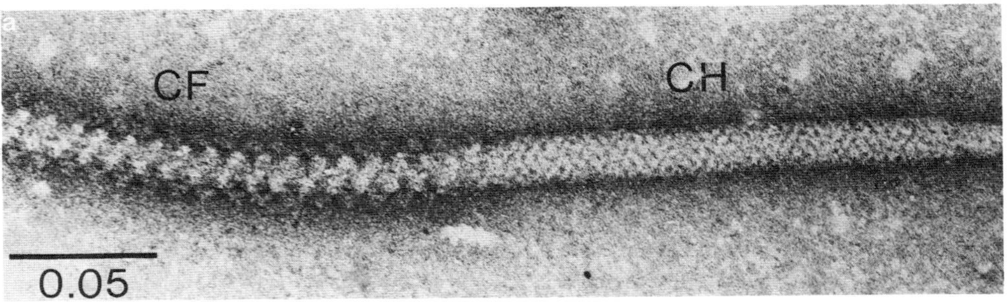

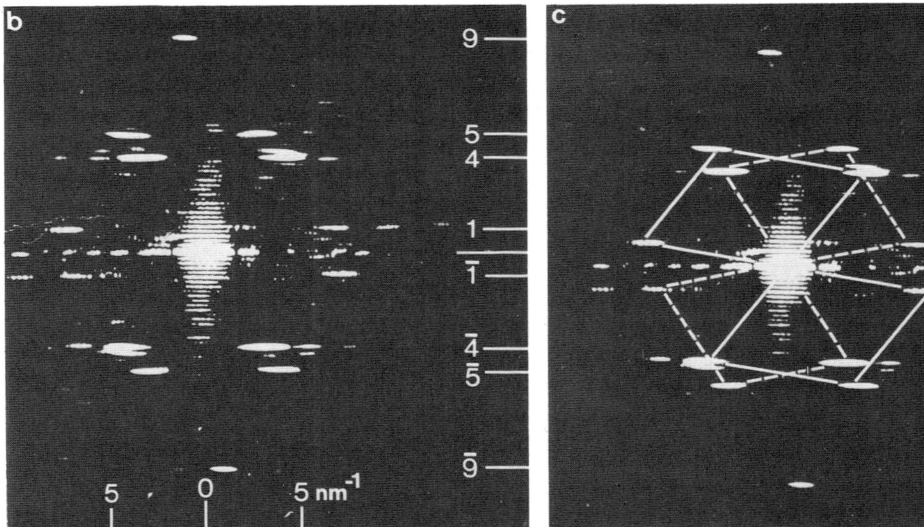

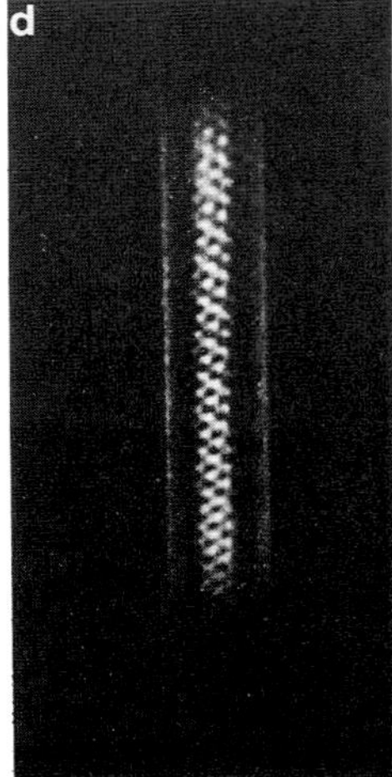

Fig. 8. a, hook (CH) and filament (CF) of the complex flagellum of *Pseudomonas rhodos 9-6* after negative staining. The globular substructure of the hook is clearly distinct from the helical outer structure of the filament. Bar represents length in µm; b, light-optical diffraction pattern of the hook with layer lines indicated; c, light-optical diffraction pattern of the hook with reciprocal lattice superimposed. The lattices for the two sides of the hook are distinguished by continuous and broken lines; d, two-sided filtered image of the hook obtained by employing the maxima of both reciprocal lattices together. (From RASKA et al. 1976)

2.2.3 Image evaluation

In the early days of electron microscopy, micrographs were "evaluated" by looking at them or by measuring details with a simple ruler. Since then, techniques have been developed which make use of lasers (light optical diffraction [Fig. 8], filtering and image reconstruction), microdensitometers, sophisticated electronic devices and computer programs (Fig. 9). Great progress has been made in the evaluation of natural or artificial ordered arrays depicted under differing viewing angles; this technique can be used for the three-dimensional representation of macromulecules or aggregates thereof (e. g. surface layers of bacteria, purple membrane). A resolution around 0.4 nm or better has been obtained for certain biological samples under ideal preparation, imaging and image evaluation conditions.

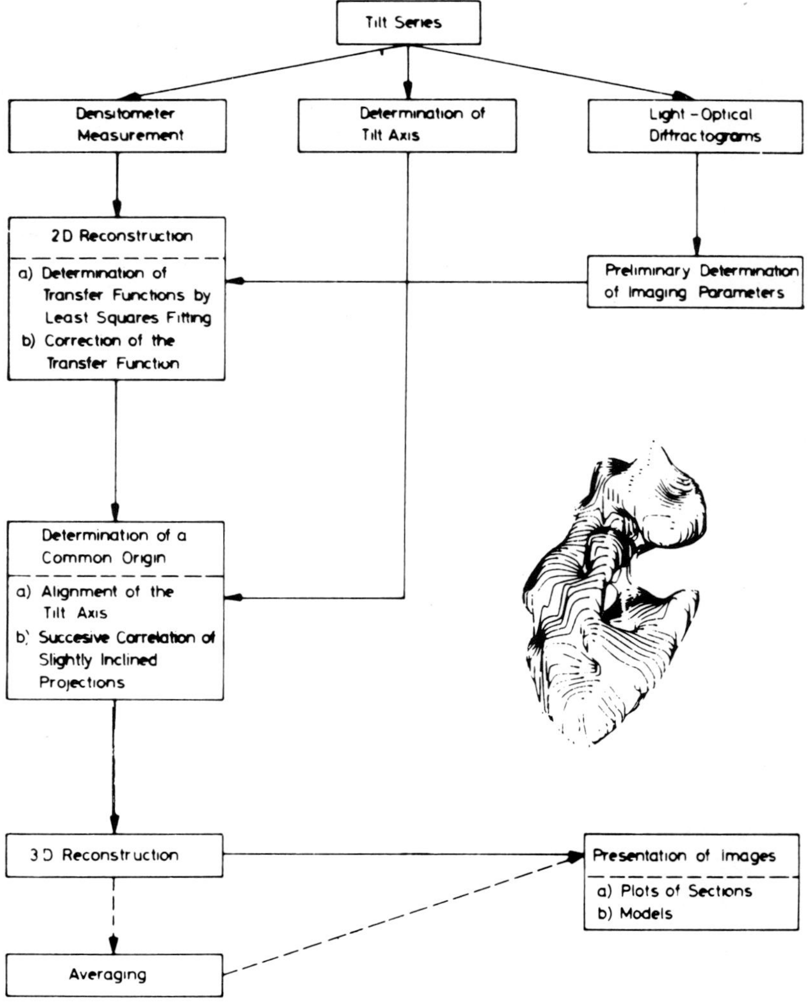

Fig. 9. Principal pathway for 3D reconstruction. (From HEGERL et al. 1984)

2.2.4 Cytochemical techniques

Combinations of structural data and chemical composition, both recorded by electron microscopy, have helped improve our comprehension of structure-function relationships. Data on the identification, chemical composition and location of bacterial cell components in dimensions within and beyond the resolving power of the light microscope have been obtained by various electron microscopic cytochemical techniques.

Gram staining for electron microscopy

The Gram reaction is a routine technique in light microscopy (s. above). Using potassium trichlor (η^2-ethylene) platinum (II) (TPt) instead of $KJ-J_2$ for the staining of *Bacillus subtilis* and *Escherichia coli,* it was shown that TPt produced an anion in aqueous solution which was compatible with crystal violet (CV) of the Gram stain. Thus, the insoluble complex formed by the interaction of TPt with CV contained platinum as an electron dense substance for electron microscopy and allowed

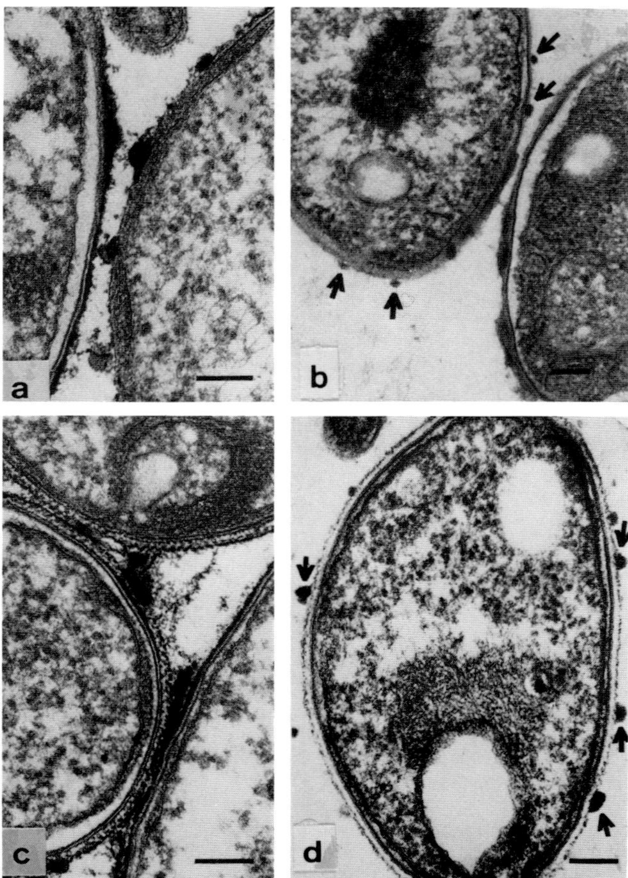

Fig. 10. *Pedomicrobium*-like budding bacteria. a, b, strain *869;* c, d, strain *868;* both strains were grown for 3 days and then fixed for electron microscopy in the presence (a, c) and absence (b, d) of ruthenium red. Wall surface polymers (arrows) appear dark Bars: 0.1 µm. (From Ghiorse & Hirsch 1979)

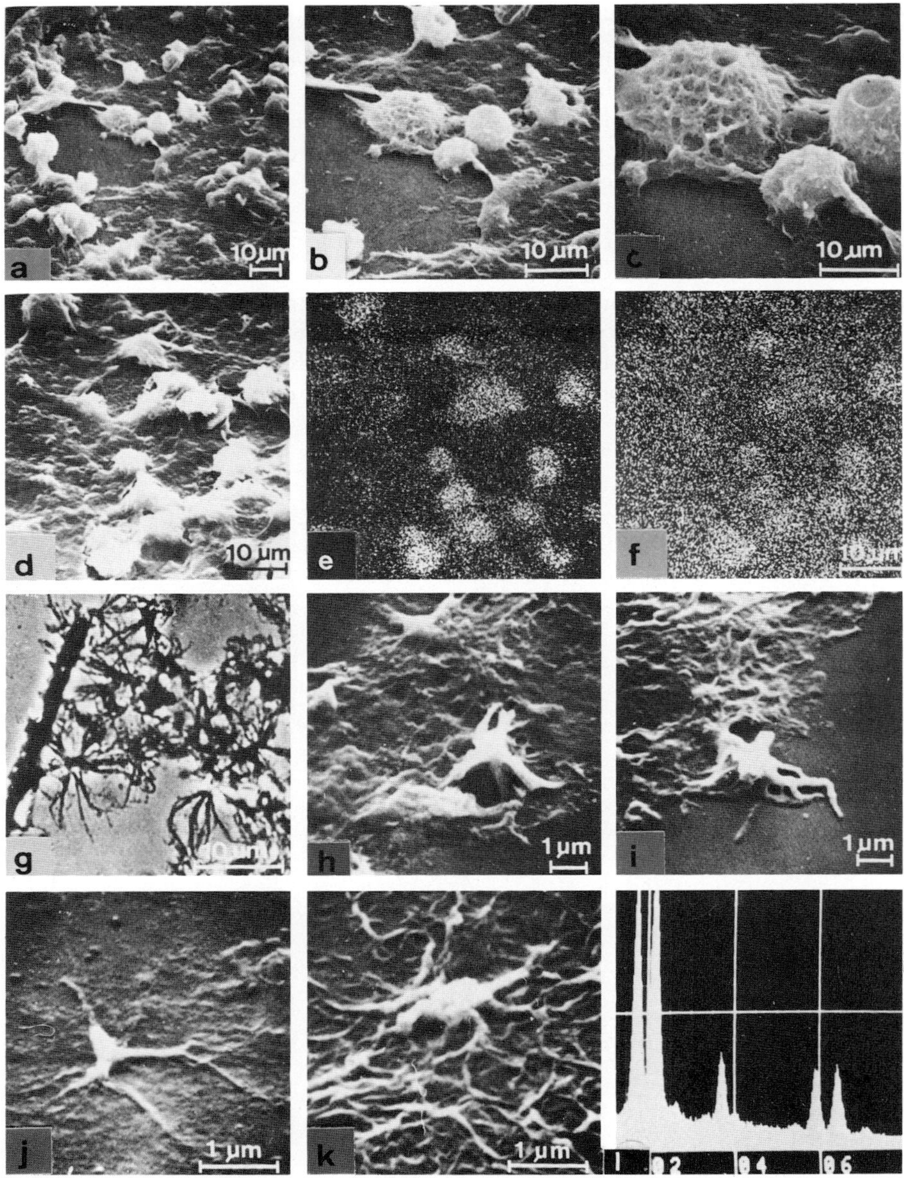

Fig. 11. Capsules of bacteria; iron and manganese containing structures. a–f, *Siderocapsa geminata;* g–e, *Metallogenium personatum;* a–d, capsules closed or with a hole; e, MnKα; f, FeKα; e and f show simultaneous presence of manganese and iron; g, light microscopy; h–k, young organisms; l, simultaneous manganese and iron. (From Hanert 1981)

the Gram staining mechanism to be applied at the electron microscopical level. It was shown that a CV–TPt complex was both within the cells and at the cell surface. In *B. subtilis*, the decolorization step dissolved the precipitate from the cell surface, but the internal complex remained within the cell. This was not the case for *E. coli*. The decolorization step removed both surface bound and cellular CV–TPt. During its removal, the outer membrane was sloughed off, only the murein sacculus and plasma membrane remained. Probably, the plasma membrane was also perturbed, but it was retained within the cell by the murein sacculus.

Ruthenium red staining

Extracellular polyanions and acidic mucosubstances can be stained with ruthenium red in a simple procedure (Fig. 10). Thus, bacterial surface layers containing these compounds can be viewed in ultrathin sections as a result of their electron dense appearance.

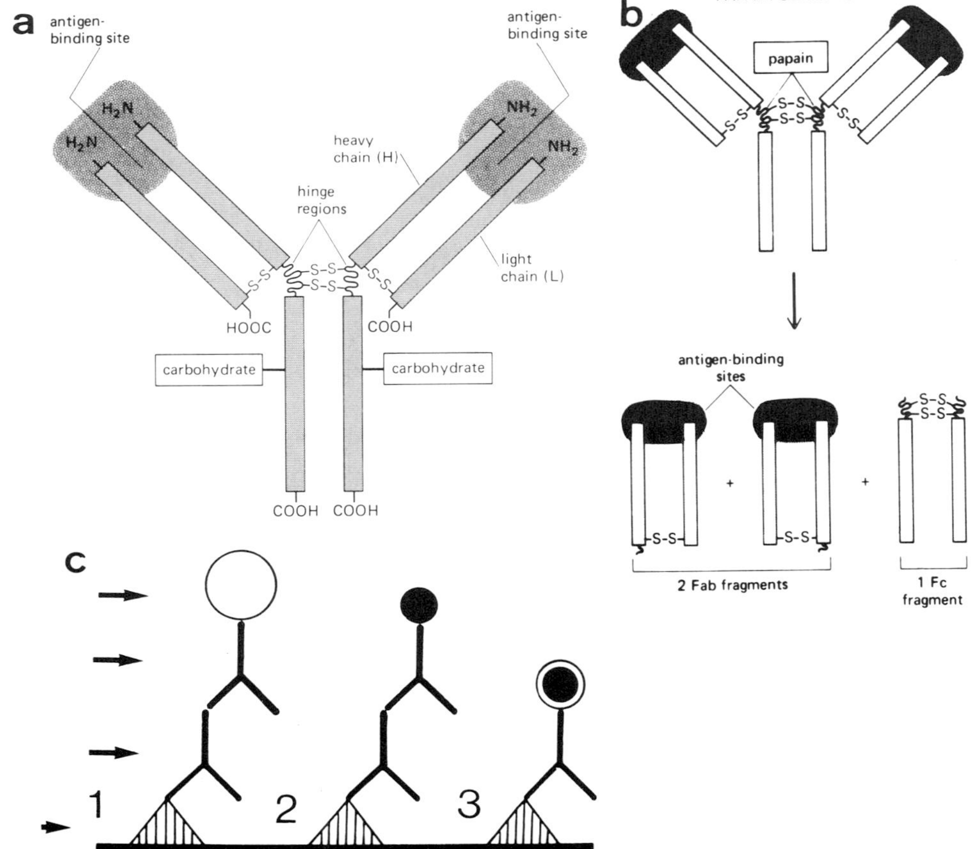

Fig. 12. a, Schematic drawing of a typical antibody molecule; b, fragments produced when antibody molecules are cleaved with papain. (From Alberts et al. 1983). c, labelling techniques using IgG antibodies. 1 = ferritin technique; 2 = gold technique; 3 = protein A-gold technique. (From Kohring et al. 1985)

Other techniques

Identification and localization of specific enzymes in bacterial cells can be achieved by procedures resulting in formation of electron dense precipitates due to specific reactions catalyzed by the enzymes in situ. Usually, ultrathin sectioning is employed for sample preparation. Metals and certain other components can be identified and localized by micro-analysis (Fig. 11).

2.2.5 Autoradiography

Bacteria can be grown in the presence of radioactive compounds. Depending on the types of compounds, they can be transported into the cell and used as building material for different cell components, or converted into simpler or complex intermediates or products of cell metabolism. In ultrathin sections, the positions of the radioactivity at the moment of cell fixation can be detected by exposing and "developing" a radiation-sensitive negative "film" covering the section. The density distribution on the film will then, together with the electron micrograph of the section, allow the localization of the radioactivity with respect to the cell structure. The resolution is limited by the geometry of the sample, by type and local concentration of the radioactive compound and by the "photographic" process.

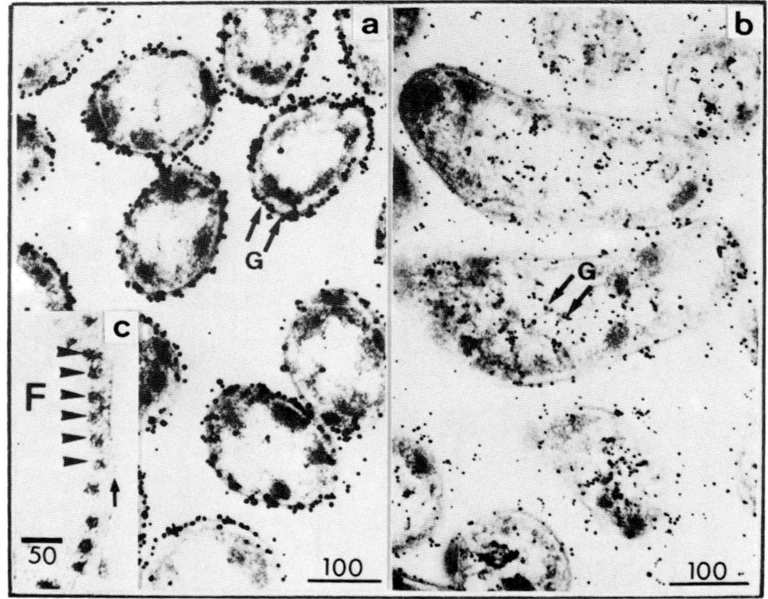

Fig. 13. Localization of the enzyme CO dehydrogenase in *Pseudomonas carboxydovorans*. a, the enzyme is located at the cell periphery in the exponential growth phase; b, in the stationary growth phase, the enzyme is also found in the cytoplasm; c, in the exponential growth phase, the enzyme is attached to the inner (cytoplasmic) aspect of the cytoplasmic membrane. a, b, immunogold technique, postembedding labelling; c, immunoferritin technique, pre-embedding labelling. F = ferritin; G = gold. Dimensions are given in nm. (From Rohde et al. 1984)

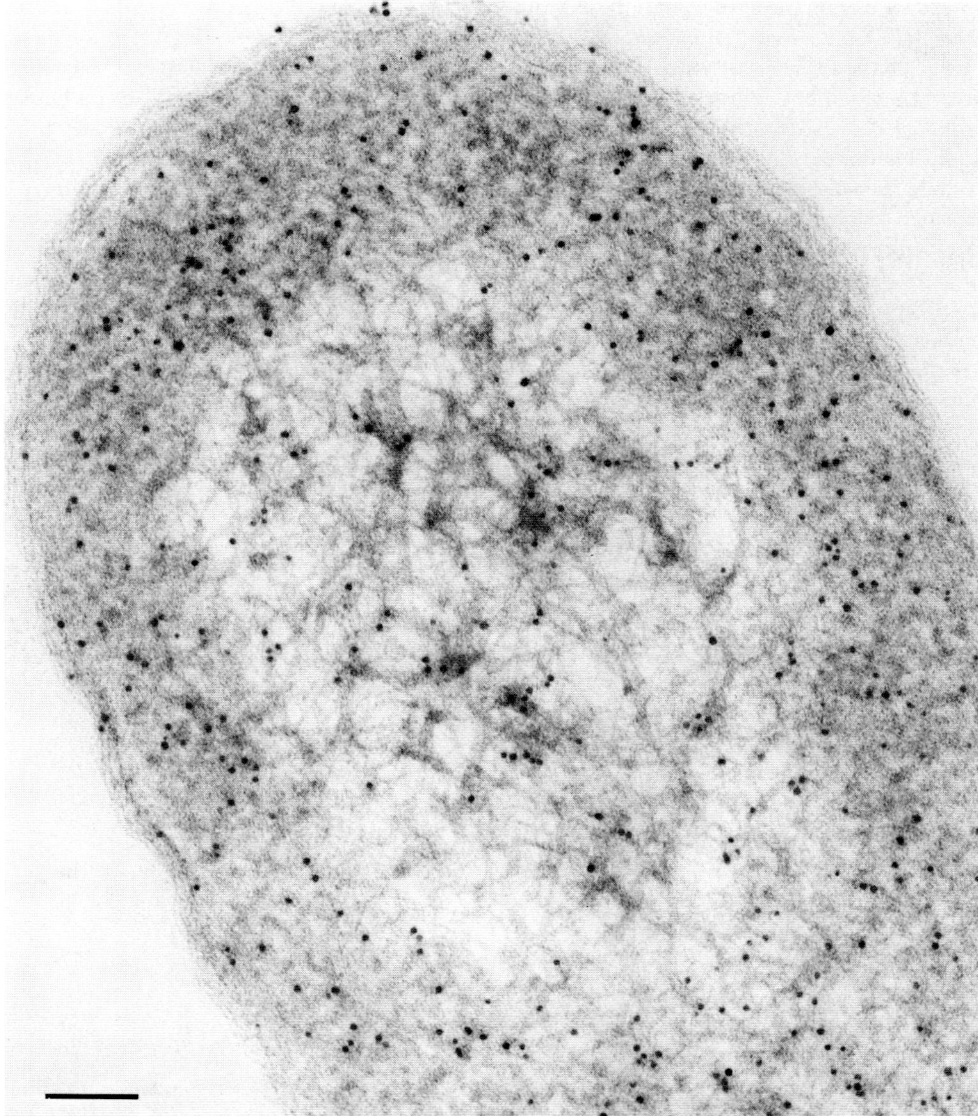

Fig. 14. Ultrathin section through a cell of *Alcaligenes eutrophus* embedded in a low-temperature resin (Lowicryl); immunogold labelling of the soluble hydrogenase. Bar: 0.1 µm. (Original micrograph M. ROHDE)

2.2.6 Immunolabelling techniques

Identification and localization of bacterial components can be achieved by various immunolabelling procedures (Fig. 12). The routine techniques have in common a requirement for isolation of the respective component (the antigen), the production of antiserum directed against it, and the labelling of the antigen-antibody complexes. Usually, either polyclonal, monovalent or monoclonal IgG antibodies are used. Labelling with ferritin or gold is obtained either by labelling a second IgG antibody directed against the first type of (antigen specific) antibody, or by binding colloidal gold (with or without a covering layer of protein A around the gold particles) to the specific antibody. Both ferritin and gold can easily be detected under the electron microscope. The techniques can be applied to intact, sectioned (Figs. 13 and 14) or freeze-fractured cells or to isolated cell components. Chemical fixation as well as low temperature embedding or cryo techniques are used for sample preparation. In many cases, cryo techniques show distinct advantages over chemical fixation with respect to preservation of antigeneity. Another procedure for localization of antigenic sites, especially at the macromolecular level, is the direct observation of antigen-antibody or antigen-IgG Fab fragment complexes (e. g., enzyme molecules with IgGs or IgG Fab fragments bound to a component in the active center) by negative staining (Fig. 15). X-ray studies have revealed the size of a peptide that can be accomodated by the antibody-combining site, a pocket or crevice of 1.5 by 0.6 nm, with an approximate depth of 0.6 nm. This crevice can accomodate a peptide of 5 (β-pleated sheet conformation) to 10 (α-helix conformation) amino acids (Poljak et al. 1976). Hopp & Woods (1981) suggested that antigenic determinants are usually located in, or immediately adjacent to, regions of greatest local hydrophilicity.

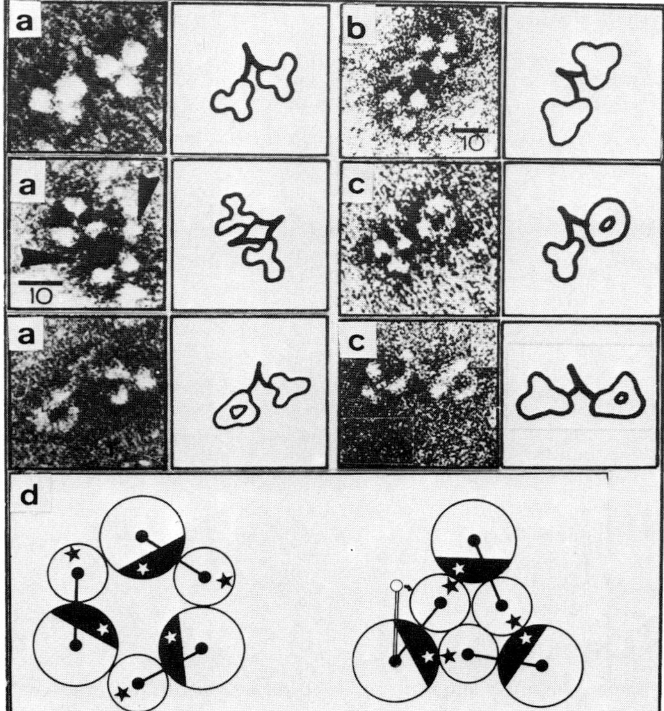

Fig. 15. Structure of citrate lyase isolated from *Rhodopseudomonas gelatinosa.* a, anti-subunit L-IgG-enzyme complexes; the small Y-shaped structures are IgG antibodies; the enzyme occurs as "star" or as "ring"; b, anti-subunit M-IgG-enzyme complex; c, anti-subunit S-IgG-enzyme complexes; d, proposed structure-function model indicating variations in the relative positions of the enzyme subunits. (From Zimmermann et al. 1982)

3. The structural concept of the bacterial cell

Bacteria show considerable variations in size, shape and fine structure depending on species, strain and even growth phase. However, there are structural principles common to all bacteria (Figs. 16 and 17). They may occur as single cells or aggregates of low complexity. Usually, the bacterial cell is enveloped by a peptidoglycan layer covered by additional surface layers of more or less complex composition. The innermost part of the envelope, the cytoplasmic membrane, encloses the cytoplasm containing, without obvious compartmentation, nucleic acids, ribosomes, enzymes, storage material and other cell components within the cytosol. Often, differentiations of the cytoplasmic membrane occur, increasing the membrane surface by forming intracytoplasmic membranes bearing components for biochemical or biophysical reactions. Besides surface layers, bacteria may possess fimbriae and pili for interaction with the environment. For locomotion, many cells exhibit flagella or other systems bringing about cell movement.

Variations on this basic structure are mainly found with regard to the cell envelope; some bacteria lack a typical peptidoglycan layer, or they are wall-deficient. Other variations have been developed

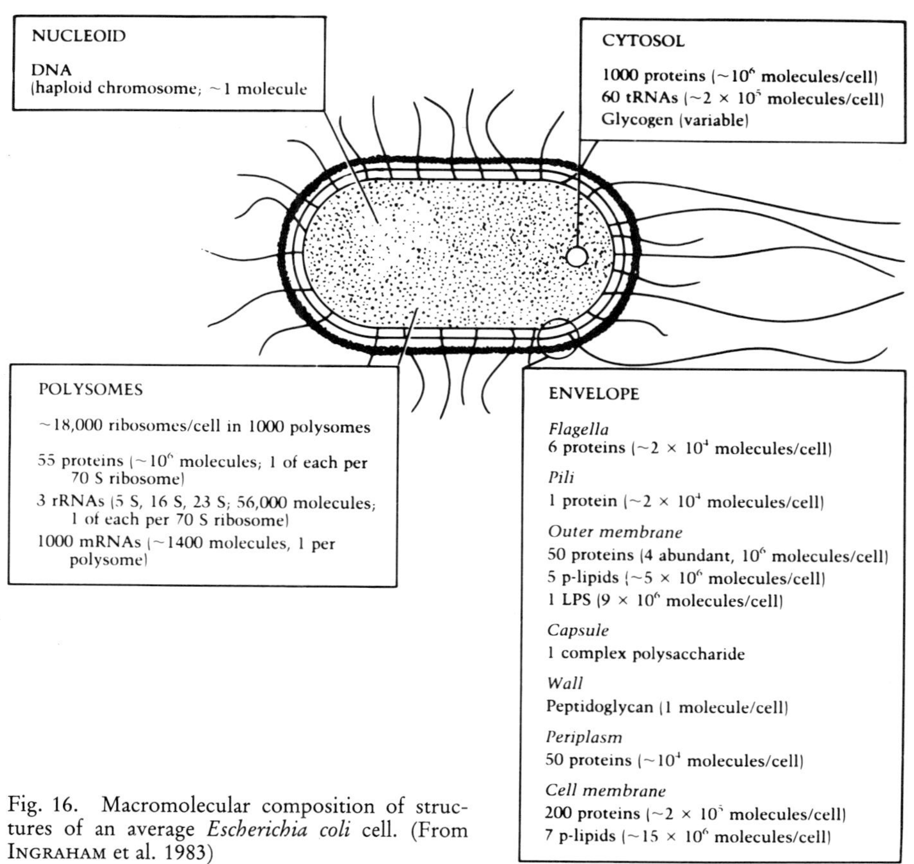

Fig. 16. Macromolecular composition of structures of an average *Escherichia coli* cell. (From INGRAHAM et al. 1983)

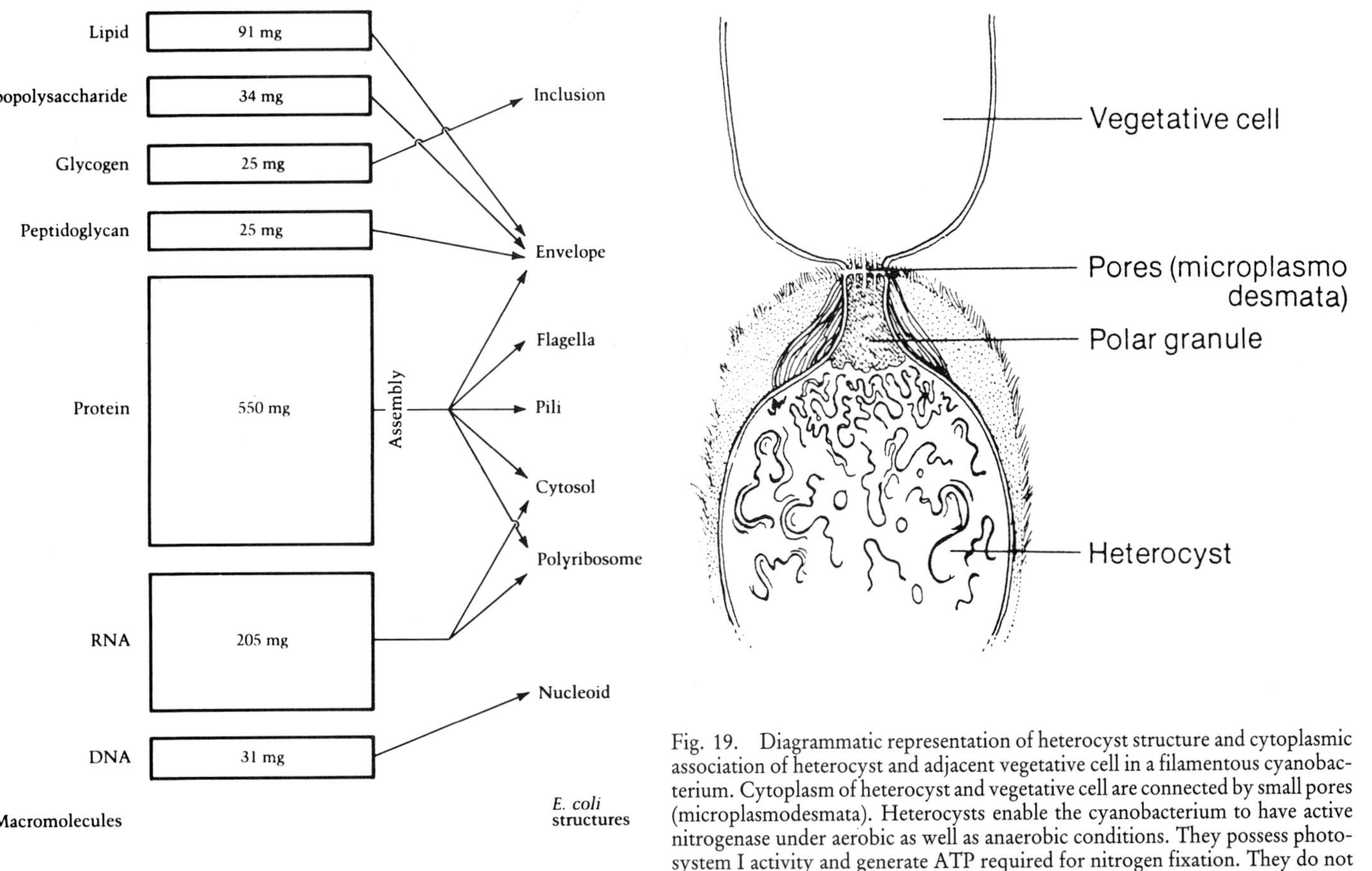

Fig. 17. Macromolecules needed to produce 10^{12} cells (1 g, dry weight) of *E. coli B/r*. The capsule has been omitted. (From Ingraham et al. 1983)

Fig. 19. Diagrammatic representation of heterocyst structure and cytoplasmic association of heterocyst and adjacent vegetative cell in a filamentous cyanobacterium. Cytoplasm of heterocyst and vegetative cell are connected by small pores (microplasmodesmata). Heterocysts enable the cyanobacterium to have active nitrogenase under aerobic as well as anaerobic conditions. They possess photosystem I activity and generate ATP required for nitrogen fixation. They do not contain photosystem II activity, which generates oxygen. When nitrogenous compounds such as NH_3 and NO_3 fall below a critical level, the cyanobacterium responds by producing heterocysts. The vegetative cells, which lie on either side of the heterocyst, supply nutrients to the heterocyst. The glycolipid envelope of the heterocyst provides a barrier to the entrance of oxygen. (From Boyd 1984)

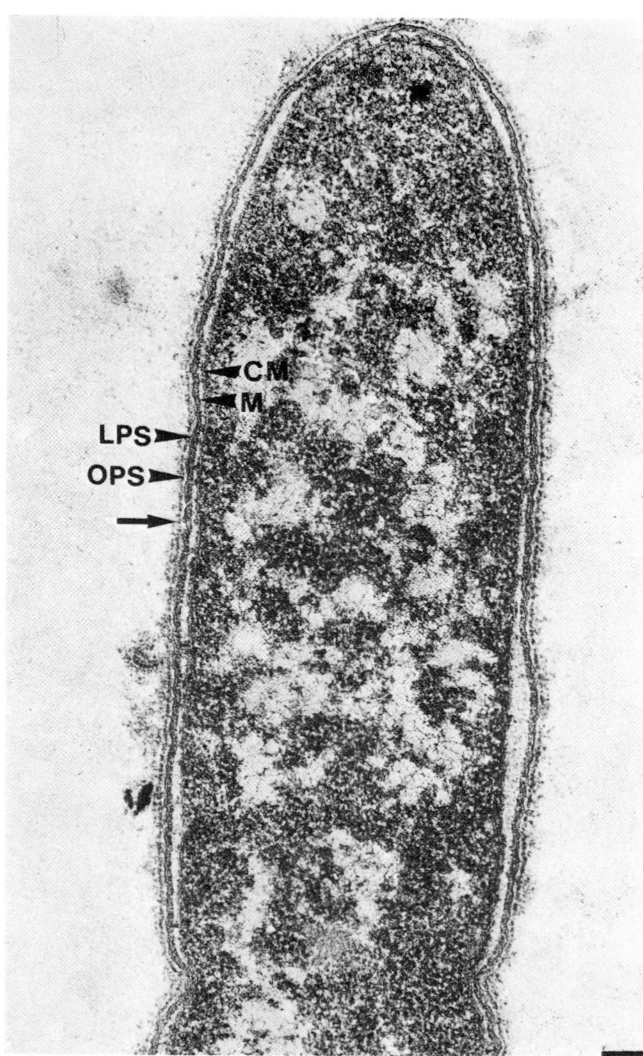

Fig. 18. Thin section of a cell of *Yersinia enterocolitica* strain *Ye 75 S* during cell division after incubation with anti-*Ye 75 S*-serum. Between the "double track" of the outer membrane and the sera depositions (arrow) an electron lucent layer (O-specific polysaccharides) of the LPS (OPS) can be seen. – CM = cytoplasmic membrane; M = murein layer; LPS lipopolysaccharide; OPS = O-specific polysaccharide. Bar: 0.1 μm. (From Acker 1977)

in order to encounter unfavourable environmental conditions; spores and other persistent cell forms may be part of a life cycle consisting of several phases.

Cell duplication usually takes place by a simple mode of division (binary fission) resulting in two equal daughter cells (Fig. 18). Nevertheless, cell division comprises a sequence of highly ordered and regulated steps. Variations of cell division include modifications of the direction of division planes, occasionally even leading to true branching, and creation of specialized cells, such as heterocysts in cyanobacteria (Fig. 19), as parts of filaments otherwise consisting of normal cells.

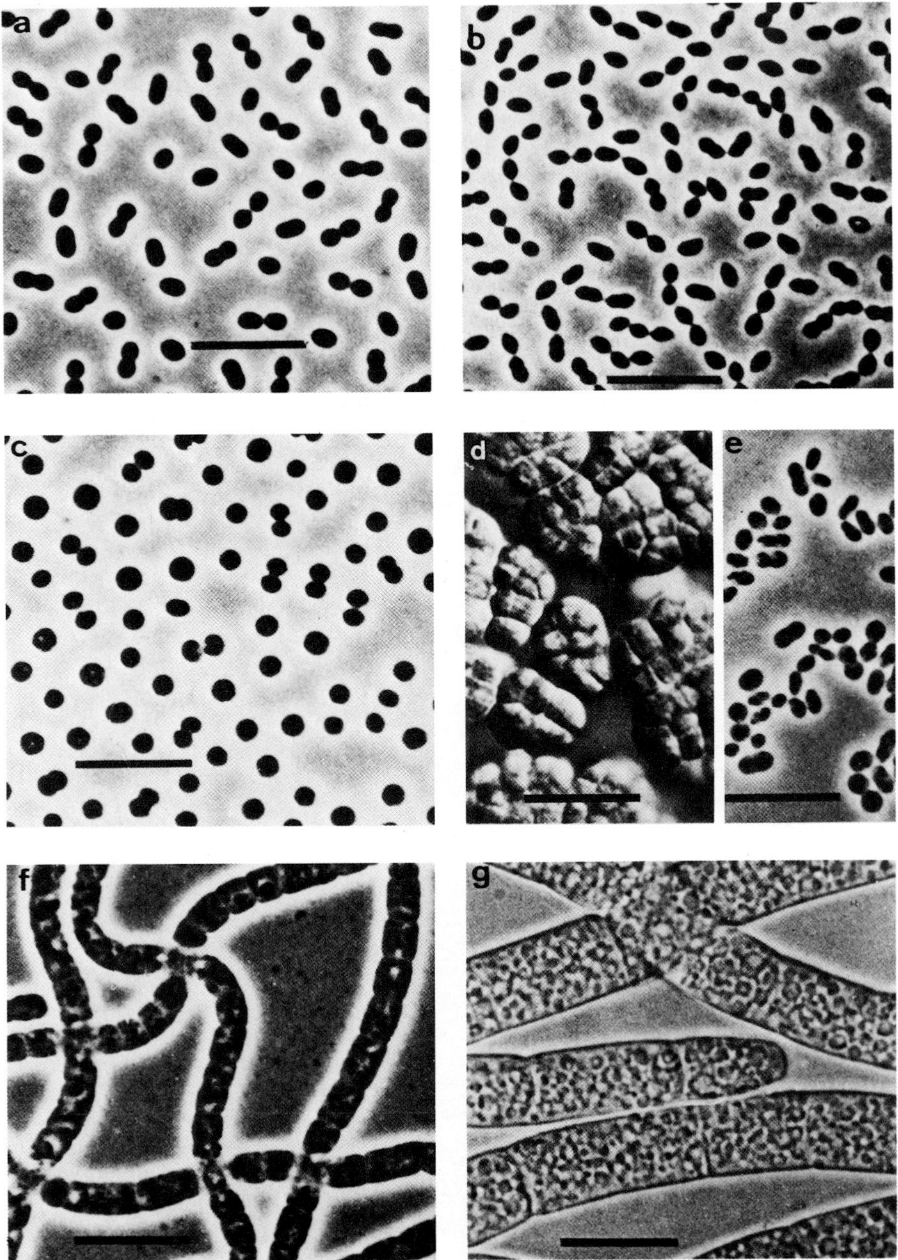

Fig. 20. Shapes of bacterial cells; sulfate-reducing bacteria. a, *Desulfobacter postgatei;* b, *Desulfobulbus propionicus;* c, *Desulfococcus multivorans;* d, e, *Desulfosarcina variabilis;* f, *Desulfonema limicola;* g, *D. magnum.* Bars: 10 μm. (From Pfennig et al. 1981)

4. Determination of cell shape

Various different cell shapes (Figs. 20–28) occur in microorganisms: spheres or coccoid bacteria, rods, helically grown cells (spirillum-type organisms), comma-shaped (vibrios), box- and plate-shaped types (JAVOR et al. 1982; STOECKENIUS 1981, WALSBY 1980), filaments, bacteria with stalks and prosthecae, and branched systems. Manifestation of cell shape occurs during assembly of the

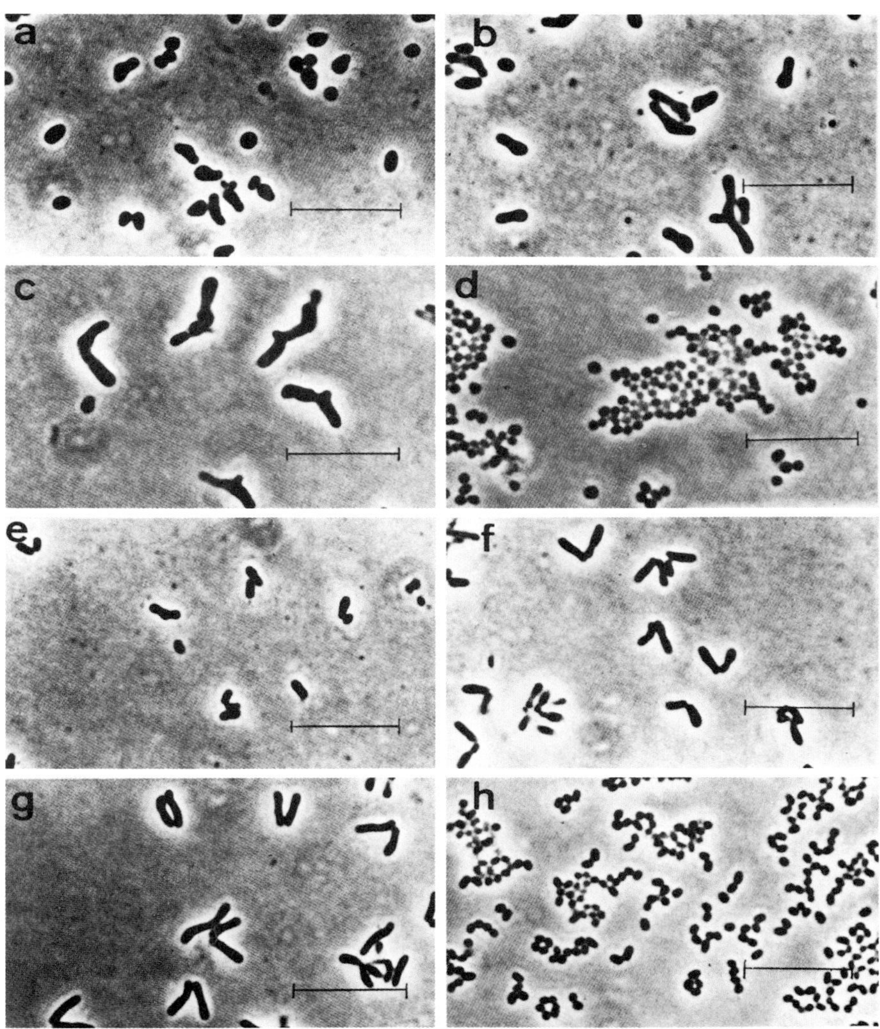

Fig. 21. Shape of bacterial cells; coryneform bacteria. a–d, *Arthrobacter globiformis;* a, after 6 h; b, after 12 h; c, after 24 h; d, after 3 days; e–h, *Brevibacterium linens;* e, after 6 h; f, after 12 h; g, after 24 h; h, after 3 days. Bars: 10 μm. (From KEDDIE & JONES 1981)

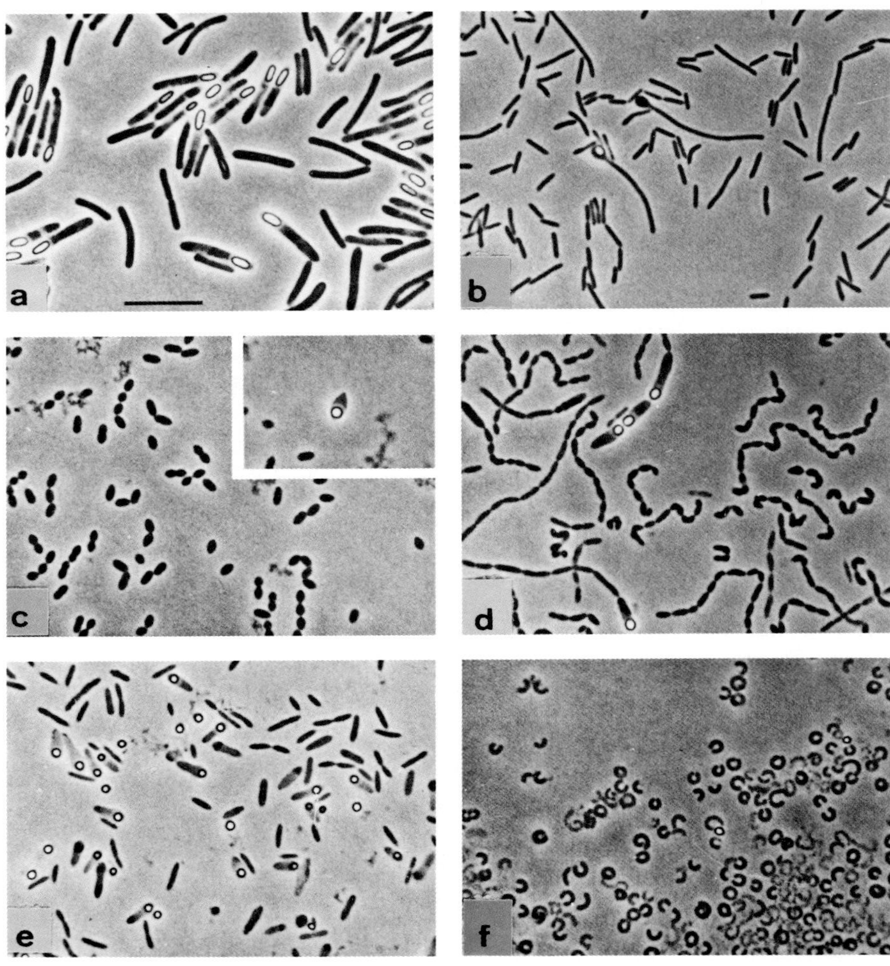

Fig. 22. Shapes of bacterial cells; the genus *Clostridium;* all micrographs same magnification as given in a; bar: 10 µm. a, *C. butyricum;* b, *C. thermocellum;* c, *C. coccoides;* d, *C. oroticum;* e, *C. indolis;* f, *C. cocleatum;*

envelope, i. e. the cytoplasmic membrane and the wall. Usually, the cell shape of a certain bacterial strain is either fixed or it is variable within a certain, defined range (Trueba et al. 1982). Thus, asking for the cause of cell shape means to ask how genetic information is translated into morphology. Cell shape, as Henning & Schwarz (1973) have put it, is a multivalent term with several aspects. One of them is geometry (cylinder, sphere etc.) and cell dimension, one is dynamic and refers to variations in dimensions of a given geometry (s. above) that occur as a result of growth and division. As also pointed out by Henning & Schwarz (1973), the two known mechanisms responsible for the formation of structures possessing a defined morphology are self-assembly and sequential assembly of polypeptides. These two mechanisms would allow one to build biological

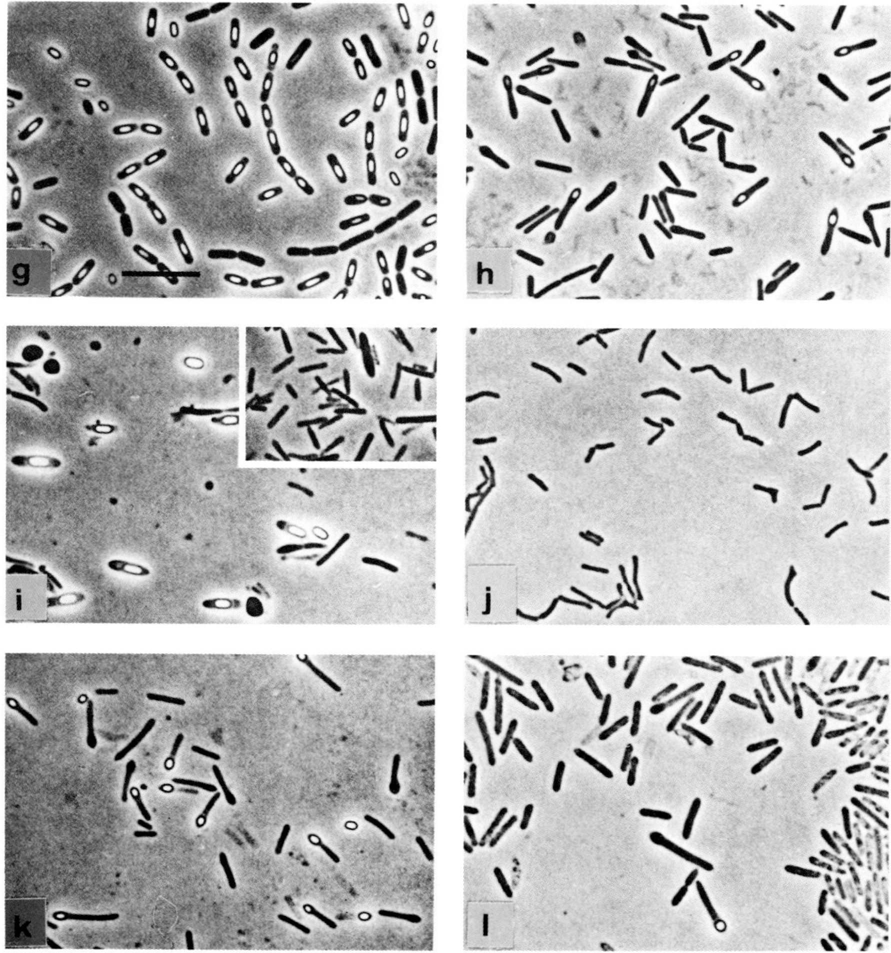

Fig. 22. g, *C. bifermentans;* h, *C. sporogenes;* i, *C. acetobutylicum;* j, *C. ramosum;* k, *C. cadaveris;* l, *C. tetanomorphum.* (From GOTTSCHALK et al. 1981)

structures of almost any morphology, but they cannot effect any variability in composition and structure of a final product. Moreover, it has been shown that even in self-assembly other factors ("epigenetic control") than the amino acid sequence may be responsible for directing subunit assembly.

From the components making up the cell envelope, the murein was shown to play a major role in the maintenance of cell shape. However, a clear shape-determining or shape-specifying role for this component has not yet been documented (s. Bacteria lacking peptidoglycan). The other main constituents of cell envelopes are lipids (mainly phospholipids), proteins, and additional compounds such as polysaccharides, lipopolysaccharides, or teichoic acids. Evidence has been obtained

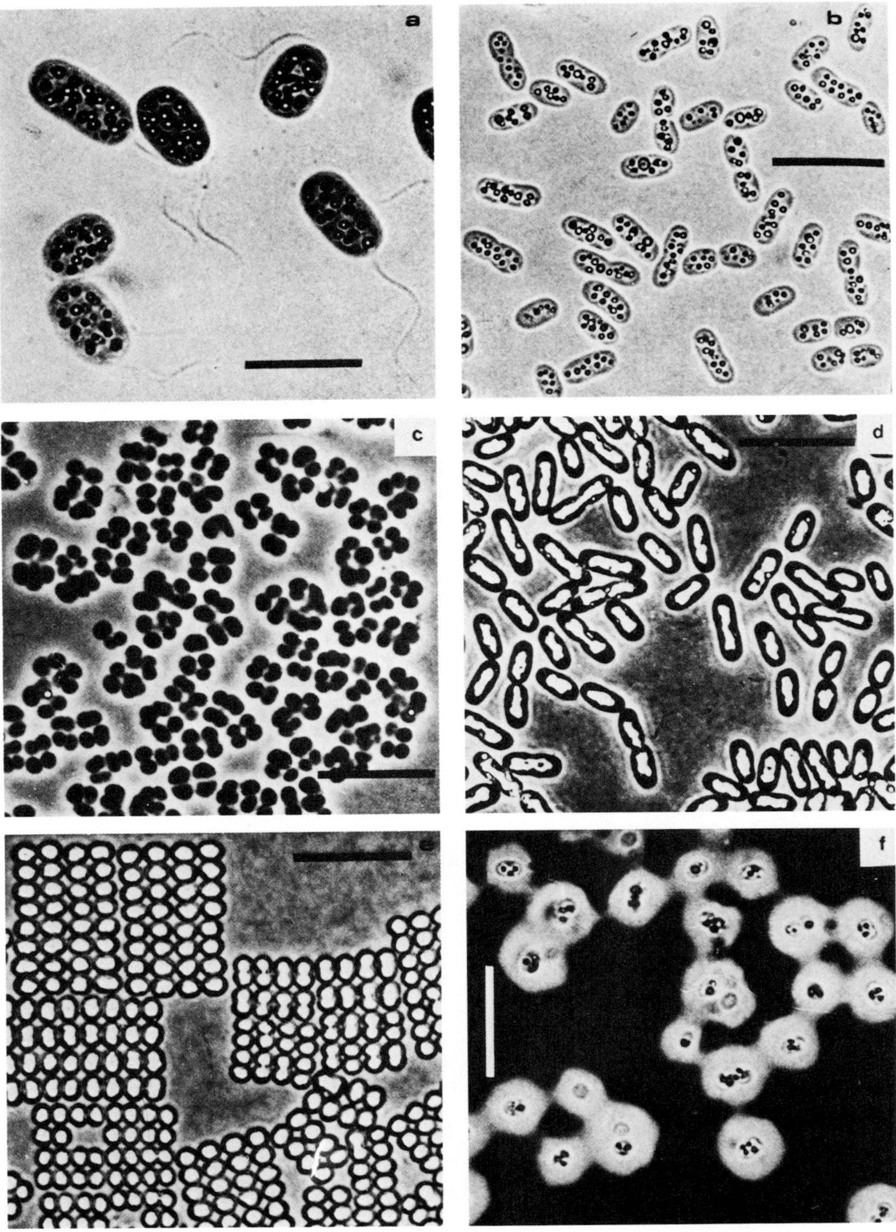

Fig. 23. Morphology of Chromatiaceae. a, *Chromatium okenii;* b, *C. vinosum;* c, *Thiocapsa roseopersicina;* d, *Thiodictyon elegans;* e, *Thiopedia rosea;* f, *Amoebobacter pendens.* Bars: 10 μm. (From TRUEPER & PFENNIG 1981)

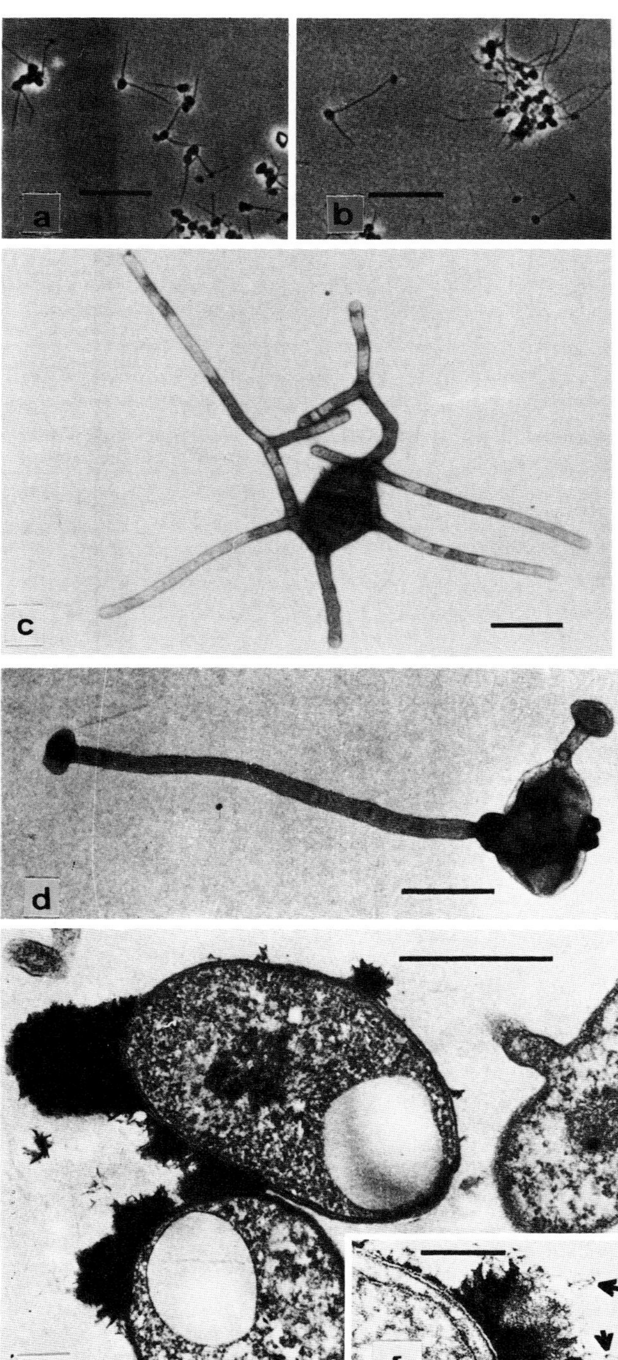

Fig. 24. *Pedomicrobium*-like budding bacteria. a, strain *869*; b, strain *868;* both strains grown for several days and prefixed with 0.05% glutaraldehyde. Phase contrast; bars: 10 µm; c, strain *869;* prefixed with 0.05% glutaraldehyde; negative staining. Bar: 1µm; d, strain *869* grown in medium containing traces of iron and manganese. Dense particles on the mother cell are thought to be manganese oxides. Negative staining. Bar: 1 µm;
e, f, thin sections of strain *869.* e, dense accumulations on cell surfaces contain manganese but no iron; f, at higher magnification particles on cells fixed in the presence of ruthenium red are more textured, and polymer filaments (arrows) are seen radiating outward from their surfaces. Bars: 1 µm in e, 0.1 µm in f. (From GHIORSE & HIRSCH 1979)

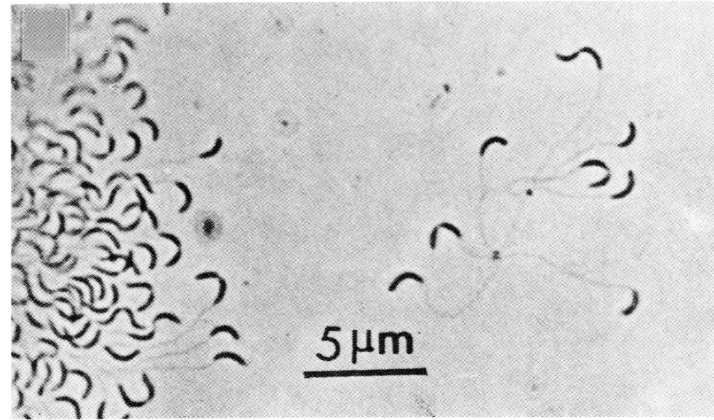

Fig. 25. Prosthecate bacteria; rosettes of *Caulobacter crescentus*. (From SCHMIDT 1981)

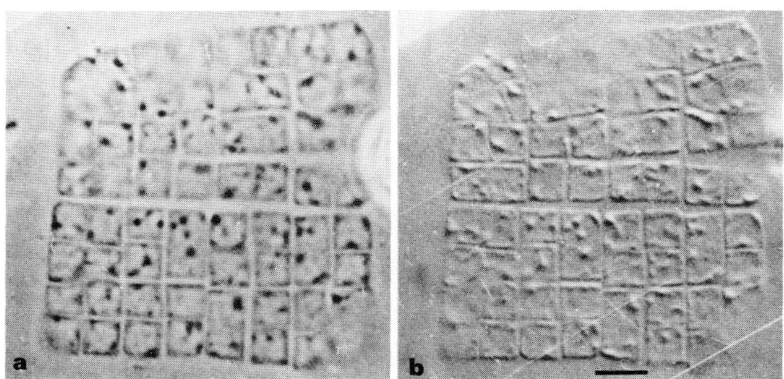

Fig. 26. Light micrographs (a, phase contrast; b, Nomarski interference) of a group of square bacteria. The dark bodies seen in phase contrast are cellular inclusions bulging against the cell wall. Bar: 5 µm. (From KESSEL & COHEN 1982)

that lipids may contribute to shape determination, although they do not possess the necessary shape-specifying properties. The same may be true for lipopolysaccharides, polysaccharides and teichoic acids. Therefore, it has become fairly clear that, at least primarily, the specificity must reside in proteins, be it as structural proteins or enzymes. Indications in favour of this conclusion are growing especially from work on reconstitution of cell shapes, and from studies on cell surface layers in a wide variety of bacteria including archaebacteria. It should be added that the shape of a bacterial cell is the result of a combined action of shape-determining components and the internal cell pressure (turgor) brought about by the solutes (KOCH 1984, KOCH et al. 1981).

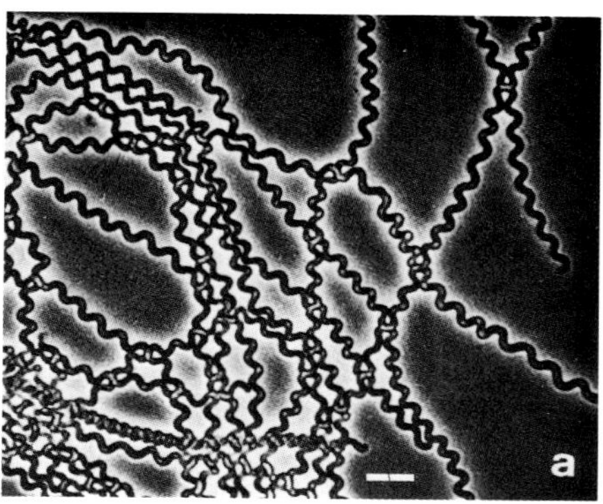

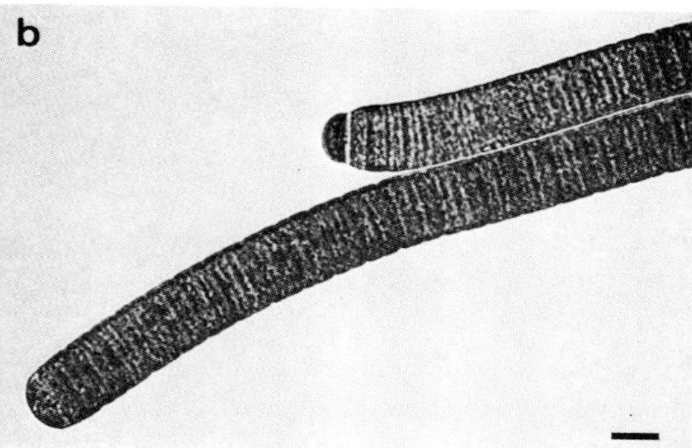

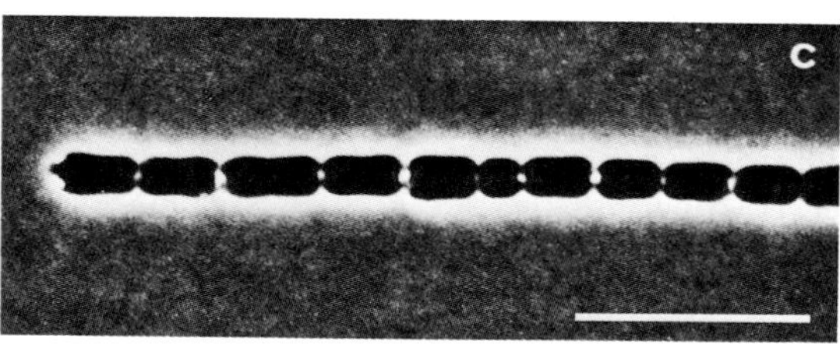

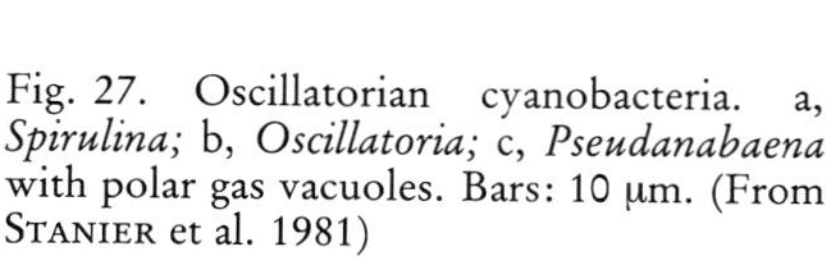

Fig. 27. Oscillatorian cyanobacteria. a, *Spirulina;* b, *Oscillatoria;* c, *Pseudanabaena* with polar gas vacuoles. Bars: 10 μm. (From STANIER et al. 1981)

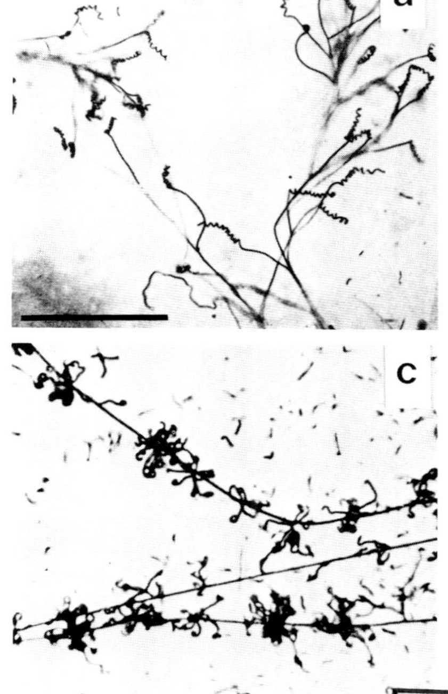

Fig. 28. Morphology of aerial mycelium of three streptomycetes.
a, *Streptomyces* sp., spore chains form helical structures with up to 10 turns;
b, *Streptoverticillium* sp.; ends of spore chains sometimes hook-like;
c, *Streptomyces pallidus.* Bar: 0.1 mm. (From KUTZNER 1981)

5. Cytology of the bacterial cell

5.1 The cytosol

In contrast to structurally defined components of the bacterial cell, the cytosol as seen in the electron microscope is not organized. As discussed by INGRAHAM et al. (1983), it is a matter of some debate whether the cytosol, in vivo, is just a disordered solution (a "bag of enzymes") or indeed highly organized (Fig. 29). Considering its constituents (mostly protein; tRNA, storage polymers, small quantities of a large number of metabolites, building blocks, vitamins, cofactors, ATP, inorganic ions) and the fact that all but a small fraction of the catabolic and biosynthetic reactions of cell metabolism occur here, the organization of the cytosol should be of great interest. It is a concentrated solution. As a consequence, protein-protein interactions might provide some degree of molecular order or organization in vivo. Enzymes with cooperative or sequential functions might be in physical contact, forming clusters. In fact, enzymes for several pathways (e. g., the arginine biosynthetic pathway) were found to be mutually attracted and aggregated. INGRAHAM et al. (1983) assume that whatever order exists in the cytosol, it is probably brought about by self-assembly of protein aggregates. This process might be supported by miniature chemical gradients, e. g. those created by ATP generation at the cell membrane and ATP consumption in other regions of the cell. All remaining assembly reactions of the cytosol may be assumed to be macromolecule-processing reactions.

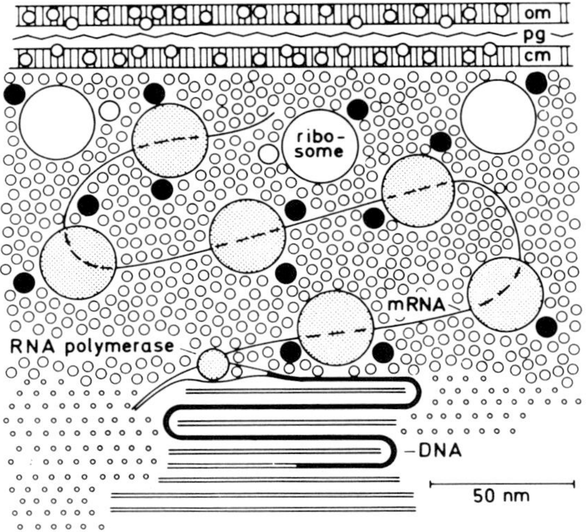

Fig. 29. Schematic representation of a 200 nm^2 area of *E. coli.* The small circles represent average-sized proteins. The large black circles represent the elongation factor EF-TU. In the envelope some proteins with a mol wt of 40 Kdal are drawn. – cm = cytoplasmic membrane, om = outer membrane, pg = peptidoglycan. (From NANNINGA et al. 1984)

5.2 Cell envelope

5.2.1 General principles

The cell envelope, i. e. all peripheral layers including the cytoplasmic membrane of a bacterium (Figs. 29–31), varies considerably, depending on species, strain or even physiological state of the cell (AMAKO & UMEDA 1977, BEVERIDGE 1981, COSTERTON et al. 1974, DE PETRIS 1967, DE VOE et al. 1971, DREWS et al. 1978, GLAUERT & THORNLEY 1969, GOLECKI 1977, GOLECKI & DREWS 1974, KELLENBERGER & RYTER 1958, LUGTENBERG & VAN ALPHEN 1983, NANNINGA 1970, RASTOGI et al. 1984, SALTON 1964, SALTON & OWEN 1976, WEIDEL & PELZER 1964). Whereas the cytoplasmic membrane (s. below, treated in an extra section) is one of the sites of location of cell components controlling or facilitating specific physiological and biochemical properties such as chemoreception, passive and active transport, synthesis of biomacromolecules and a diversity of other metabolic properties, the outer parts of the cell envelope are qualitatively different in their functions (ANDERSON 1949, ARCHIBALD 1980, CHENG et al. 1970, 1971, INGRAHAM et al. 1983, INOUYE & BECKWITH 1977, PARKER & MUNN 1984, RASTOGI et al. 1981, RAUVALA 1983, SHUR & ROTH 1975, WETZEL et al. 1970). Besides a role in maintaining cell shape and turgor and mechanically protecting and supporting the cytoplasmic membrane and cell content, an important function

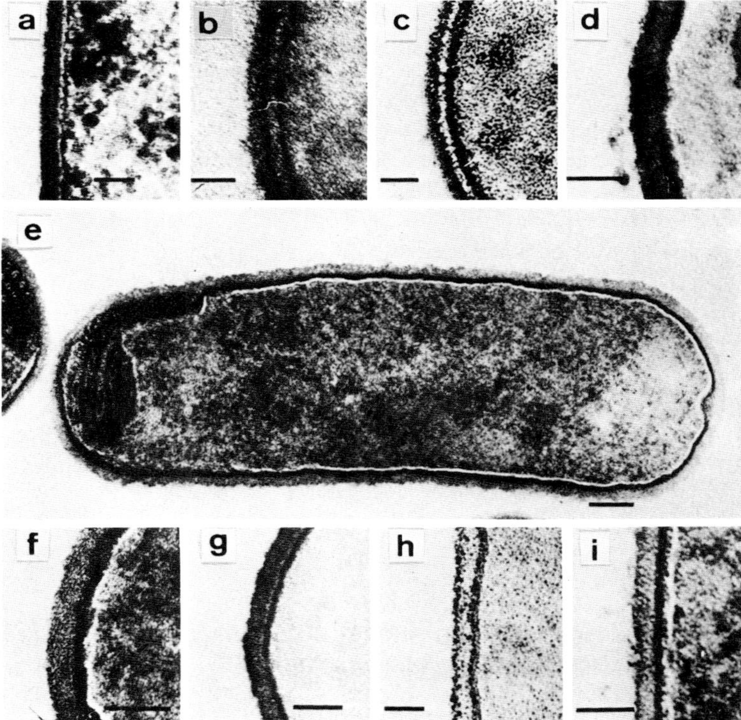

Fig. 30. Ultrastructure of the cell envelope of *Microbacterium flavum* (a–d) and *Lactobacillus* (e–i) after application of different staining procedures: a, e, f, lead citrate; b, g, phosphotungstic acid, c, h, Thiéry; d, i, ruthenium red. Bars: 50 nm in a–d and f–i, and 0.1 µm in e. (From RASTOGI et al. 1984)

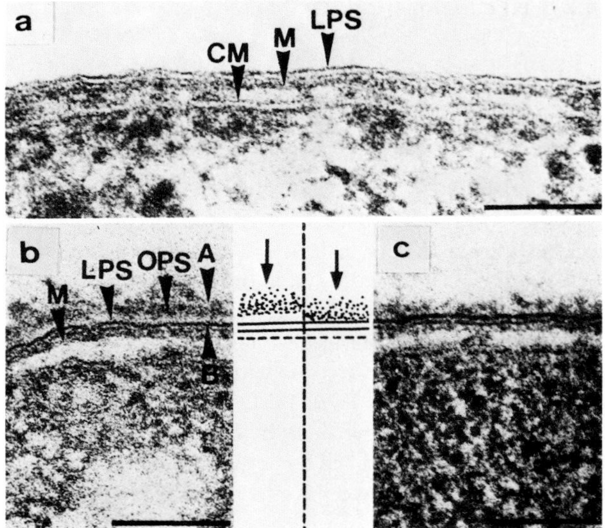

Fig. 31. *Yersinia enterocolitica* strain *Ye 75 S*. a, lipopolysaccharides barely visible; b, *Ye 75 S;* c, *Ye 75 R*. Incubation with their antisera. In the sketch, sera depositions at *Ye 75 S* and *Ye 75 R* cells are represented by dots (arrows). *Ye 75 R* cells (c) lack the electron-lucent layer (OPS) of *Ye 75 S* cells (b). Bars: 0.1 μm. CM = cytoplasmic membrane; M = murein layer; LPS = lipopolysaccharide; OPS = O-specific polysaccharide. Arrows A and B in b mark a distance of 20 nm. (From Acker 1977)

is interaction with the environment. The spectrum of interactions is large, ranging from contact with bacteria of the same species, other bacteria, eukaryotic cells or solid surfaces to bacteriophage adsorption. These interactions are made possible by special chemical and structural features of the cell envelope components.

Analysis of structure-function relationships of cell envelopes has a long history and is especially important in research on bacterial pathogeneity.

Depending on the result of a staining reaction (s. Staining procedure, and Gram staining for electron microscopy), bacteria may be subdivided into Gram-positive and Gram-negative cells (Bartholomew & Mittwer 1952, Beveridge & Davis 1983, Davis et al. 1983, Salton 1963, Scherrer 1963). With rare exceptions (methanogens, halobacteria, mycoplasmas and L-forms), staining and the structure of the cell are correlated. Gram-positive cells exhibit a thick inner wall layer consisting mainly of peptidoglycan and teichoic acids, whereas the peptidoglycan network in Gram-negative cells is thin. A typical Gram-negative cell contains an “outer membrane” surrounding the peptidoglycan layer.

In the following chapter, these envelope layers are described, together with additional bacterial surface structures mediating specific cell properties.

5.2.2 Structures and components in the wall of Gram-positive bacteria as revealed by electron microscopy

Gram-positive bacteria are known to contain, in their walls, peptidoglycan and accessory polymers (s. below). Ultrathin sections through Gram-positive cells (Fig. 30) prepared for electron microscopy according to established procedures (fixation and staining) (Rogers et al. 1980; further references therein) show a thick wall layer, occasionally with a trilaminar appearance. Wall thickness as revealed by electron microscopy depends on the mode of sample preparation, but also on species and growth phase. Common values for wall thickness range between 20 nm and 30 nm, but values of up to 100 nm have been reported. Considerable variations have been found when samples are compared which were obtained by ultrathin sectioning of whole cells, or of isolated walls, after different modes of fixation and staining, or by freeze fracturing. In addition unequivocal, naturally occurring variations in wall thickness have been demonstrated. Wall synthesis is coupled to protein and nucleic acid synthesis through formation of wall biosynthetic enzymes and the supply of a limited amount of amino acids and nucleotides. If protein synthesis is stopped by chloramphenicol, wall synthesis may still proceed. In *Staphylococcus aureus* cells treated in this way, wall thickness increased from 30 nm to 100 nm. In the tryptophan auxotroph *Bacillus subtilis trp*, progressive wall thickening was observed during incubation in the presence of wall amino acids and glucose, but only in the absence of tryptophan. During the thickening process, movement of an innermost darkly staining band of the original wall towards the center of the wall could be followed. This observation might suggest a physical or chemical difference between the original innermost wall layer and the central wall region. Ultrathin sections of glutaraldehyde fixed unstained *Bacillus subtilis* cells exhibit a double or triple banded wall structure supporting the view that the accessory polymer teichoic acid is concentrated in the outer layer of the wall (s. below). However, other detailed studies have led to other opinions. It has, for example, been suggested that banded wall structures might more likely be due to a variable packing of wall material than to the distribution of teichoic acids. The assumption of a scattered distribution of teichoic acid in the wall was supported by results of studies performed on *Staphylococcus aureus* with different staining procedures (lead citrate, uranyl acetate, ruthenium red) applied to osmium-fixed cells. In addition, micrographs of sections were obtained after treatment with gold-coupled concanavalin A and by a concanavalin A-peroxidase technique, allowing a similar interpretation. Extraction of the accessory polymers from walls by different reagents (acids, formamide) occasionally resulted in a reduced measurable wall thickness; however, the trilaminar appearance persisted. It was concluded that the peptidoglycan component, at least in streptococcal walls, can by itself cause the trilaminar aspect. Antibodies raised against peptidoglycan were found to react specifically with both sides of the wall, indicating the distribution of this polymer in different wall areas. In conclusion, one should not be dogmatic about relationships between wall appearance and the distribution of different polymers within the wall. Nevertheless, the assumption that some infra-structure in walls of Gram-positive bacteria is present, seems to be reasonable (Rogers et al. 1980) (further details s. Molecular composition and structure of peptidoglycan, and Features specific for Gram-positive cell walls).

5.2.3 Structures and components in the wall of Gram-negative bacteria as revealed by electron microscopy

Morphologically, several layers making up the outer structure of Gram-negative cells (Figs. 32–35) can be distinguished (Rogers et al. 1980; further references therein). These are the cytoplasmic membrane (by definition part of the cell envelope but not part of the cell wall), the periplasmic space, the peptidoglycan layer, a more or less structureless zone (the "intermediate zone"), the outer membrane, and lipopolysaccharides (Acker 1977). The dimensions reported for the thickness of these layers vary considerably, depending on bacterial strains and methods of preparation for electron microscopy.

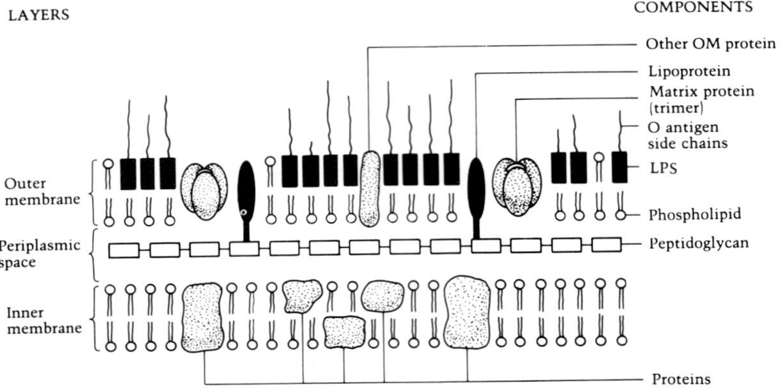

Fig. 32. Diagram of a Gram-negative cell envelope. – LPS = lipopolysaccharide; OM = outer membrane. (From INGRAHAM et al. 1983)

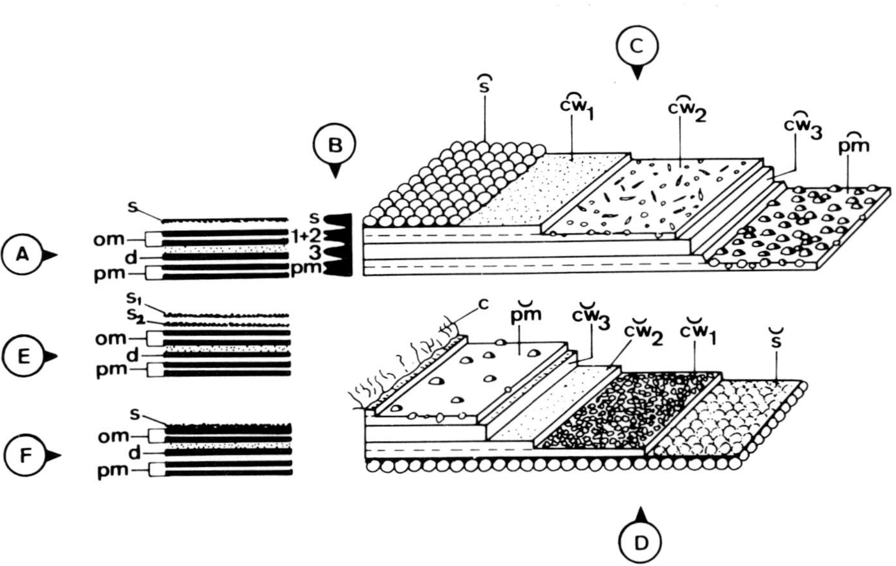

Fig. 33. Diagrams showing the relationship between the layers revealed in thin sections (A) and in freeze-etched preparations (B, C, and D) of the cell envelope of a typical Gram-negative bacterium which has an additional S-layer. E, two or more separate surface layers have been found in some Gram-negative bacteria; F, S-layer in close contact with the outer dense layer of the outer membrane. – c = cytoplasm; cw = cell wall layer; d = dense layer; om = outer membrane; pm = cytoplasmic membrane; S = S-layer. (From SLEYTR 1978)

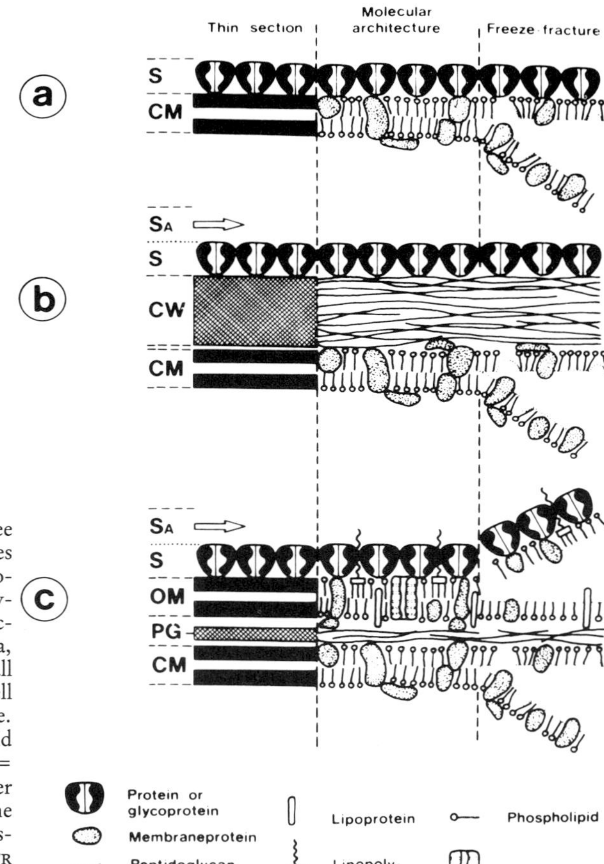

Fig. 34. Schematic diagram of the three main categories of bacterial cell envelopes containing S-layers. Left, thin section profiles; middle, molecular architecture showing major components; right, freeze-fracture behavior of cell envelopes. a, Archaebacteria, which lack a rigid cell wall component; b, Gram-positive cell envelope; c, Gram-negative cell envelope. CM = cytoplasmic membrane; CW = rigid wall layer (Gram-positive bacteria); OM = outer membrane; PG = peptidoglycan layer (Gram-negative bacteria); S = crystalline S-layer. Arrows indicate locations for possible additional S-layers (S_A). (From SLEYTR & MESSNER 1983)

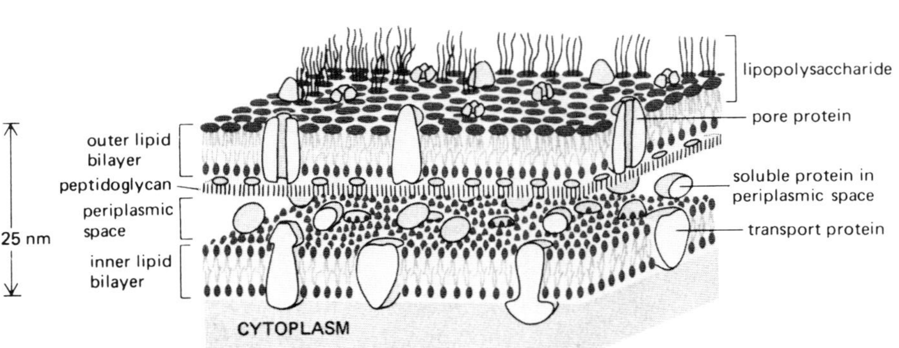

Fig. 35. Schematic view of a section of the envelope of an *Escherichia coli* bacterium. (From ALBERTS et al. 1983)

5.2.4 Peptidoglycan: Molecular composition, structure, and implications for taxonomy

Peptidoglycan (murein, mucopeptide) is the basic component of the wall of most bacteria (exceptions s. below) (Formanek et al. 1974, Ghuysen & Shockman 1973, Höltje & Schwarz 1985, Murray et al. 1965, Weidel et al. 1960). Its constituents are glycan chains with peptide substituents which are cross-linked to form a network enveloping the cell. Peptidoglycan seems to play a role in the maintenance of cell shape in Gram-positive and, presumably to a considerable extent, in Gram-negative cells (exceptions s. below). Degradation of peptidoglycan by lysozyme (Birdsell & Cota-Robles 1967, Webb 1948, Weibull 1953) causes the loss of the cell's original shape. The cell is transformed into a spherical body, a spheroblast, which is only stable in hypertonic medium.

The analysis of the composition and structure of peptidoglycan of Gram-positive and Gram-negative bacteria has a long history. One of the first findings from the early electron microscopy of

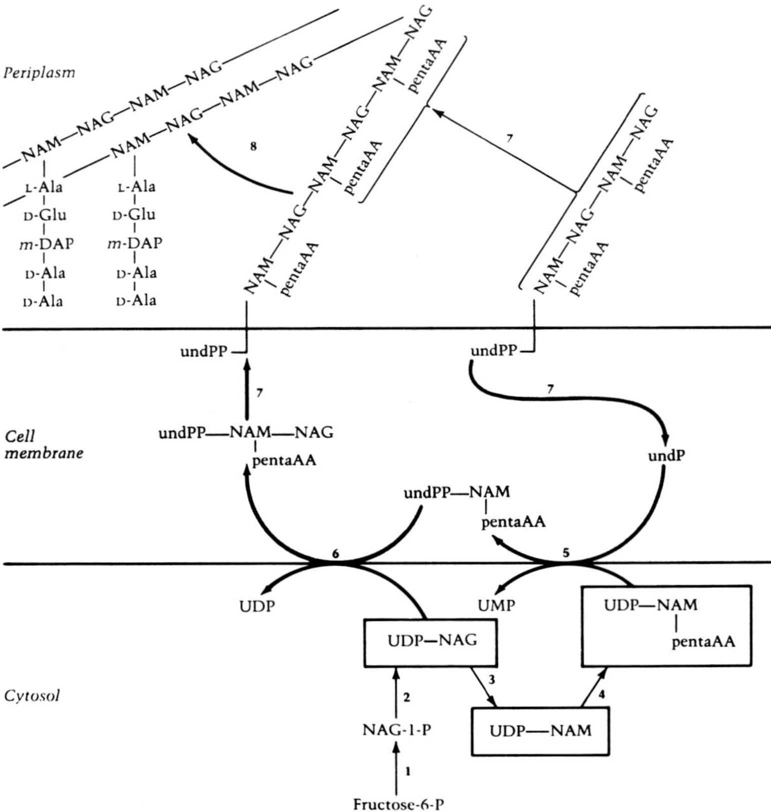

Fig. 36. Polymerization of peptidoglycan. – NAG = N-acetylglucosamine; NAM = N-acetylmuramic acid; und, undP, undPP = undecaprenol and its phosphate and pryophosphate derivatives; pentaAA = the pentapeptide L-alanine-D-glutamate-m-diamino-pimelate-D-alanine-D-alanine; UMP = uridine monophosphate; UDP = uridine diphosphate. (From Ingraham et al. 1983)

bacteria was that Gram-positive cells revealed a thick peptidoglycan layer whereas Gram-negative cells exhibited only a very delicate envelope structure. At that time, this indicated that the peptidoglycan of Gram-negative bacteria might consist of a few or even only one layer whereas it presumably formed a multilayered envelope in Gram-positive bacteria (s. above). More recent electron microscopy has revealed further structural details. Fig. 37 presents a model of the arrangement of the glycan chains in the *E. coli* sacculus as deduced from the effect of endopeptidase digestion. The results obtained from those experiments suggest that the glycan chains run more or less perpendicular to the long axis of the cell.

It was only by chemical and physical analyses that a further insight into the composition and architecture of peptidoglycan could be obtained (FREHEL & RYTER 1979, GUINAND et al. 1979, HOYLE & BEVERIDGE 1984, MILLER et al. 1966, SCHLEIFER & KANDLER 1972). In addition, the synthesis and structure of this wall polymer (Fig. 36) and many of its variations as well as relationships to chitin and cellulose are now well unterstood. In Gram-negative bacteria there are no bridging amino acids at all; in Gram-positive bacteria, a peptide of up to four different types of amino acids may form the bridge giving rise to the complex nature of this type of peptidoglycan.

According to the chemical nature of peptidoglycan, several classifications of this wall polymer have been proposed. In general, the types of cross-linking and bridging have been used in these systems of classification as the most important characteristic features for classification.

Variations in peptidoglycan structure and composition appeared to have correlations regarding relationships between bacteria. Taxonomic implications have become evident. SCHLEIFER & KANDLER (1972) deduced from their investigations four basic statements which have been summarized by ROGERS et al. (1983) as follows:

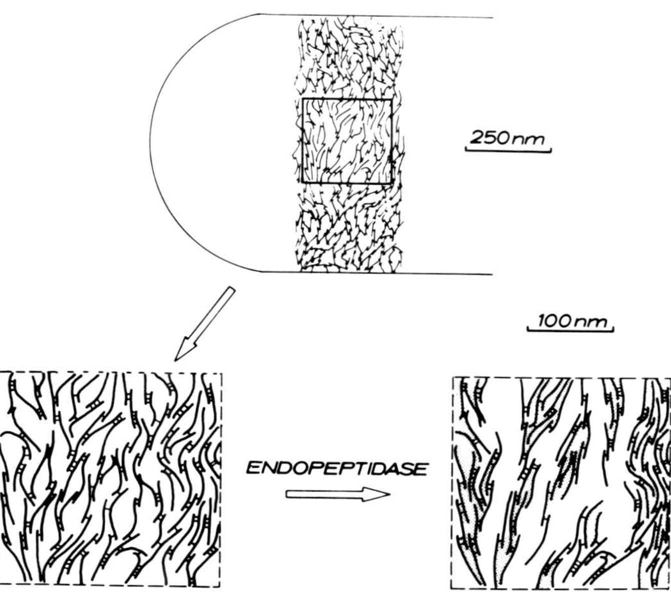

Fig. 37. Model of the arrangement of the glycan chains in the *E. coli* sacculus as deduced from the effect of endopeptidase digestion. (From NANNINGA et al. 1984, after VERWER et al.)

(1) Widespread distribution. Peptidoglycan is a characteristic constituent of almost all prokaryotic cells. Higher organisms such as fungi show a more simplified wall structure, in which the muramic acid has lost its 3-O-lactyl ether group (and hence its peptides) so that the resulting glycan becomes converted to chitin. In higher plants even the 2-acetamide groups of chitin are lost, so that cellulose results.

(2) Multiple gene involvement. The formation of a peptidoglycan structure clearly requires the action of many genes, since at least 20 different enzymes are involved in its biosynthesis. The diverse peptidoglycan types differ from each other by more than one gene, so that given peptidoglycan structures tend to be stable, and are unlikely to have arisen by convergence.

(3) Possibility of change without loss of viability. Although the peptidoglycan structure is stable during periods of investigation, Gram-positive species have achieved numerous solutions to the basic structural problem, all of which are compatible with survival.

(4) Changes should be recognizable as a tendency towards a higher development; in other words, an evolutionary direction should be deducible. Although this is naturally the most debatable criterion (ROGERS, 1980), SCHLEIFER & KANDLER (1972) have proposed a possible evolutionary sequence in which the multilayered peptidoglycans of Gram-positive bacteria are taken to be the most primitive, and to have the greatest variation in cross-links, whereas the most developed are the peptidoglycans of Gram-negative bacteria and spirochetes, in which bridging amino acids have disappeared from the cross-links (s. above).

5.2.5 Structural and biochemical features specific for Gram-positive walls

5.2.5.1 Teichoic and teichuronic acids

Polymers other than peptidoglycan are often found in Gram-positive cell walls as accessory components (Fig. 38). They have been characterized as consisting of polymerized polyol phosphates. Often, they contain side-chains of saccharide units and ester linked D-alanine residues (BIRDSELL et al. 1975, ROGERS et al. 1980). The significance of these wall polymers called teichoic acids was previously somewhat misjudged by the finding that similar compounds were shown to be extractable from the cytoplasmic membrane as well, even from the same organism ("intracellular", "membrane" teichoic acids). This latter type of polymeric acids has now been identified as lipoteichoic acids (ROGERS et al. 1980; further references therein), a name referring to their composition, i. e. a glycerol teichoic acid linked to a glycolipid, the latter component enabling the molecule to be attached with one end to the cytoplasmic membrane, with the hydrophilic part of the molecule reaching into the wall. Teichoic acids have been found to be covalently linked to peptidoglycan. As a general rule, this is achieved by means of a link consisting of three glycerol phosphates attached to N-acetyl-glucose-amine which is bound to a phosphate on the C_6 of muramic acid. Teichoic acids can be easily extracted with cold trichloroacetic acid due to the presence of a sugar-1-phosphate.

Several functions of teichoic acids have been discussed (ROGERS et al. 1980). Their most likely function appears to be as mediators for the access of divalent cations, especially Mg^{2+}, to the cell. This affinity might be caused by the many equally spaced phosphate groups. If cells of *Bacillus subtilis* var. *niger* were Mg^{2+}-limited, then the cell wall contained teichoic acids. Under phosphate limitation, however, teichuronic acid (s. below) replaced teichoic acid. As a second function, the effects of teichoic acid on an autolytic enzyme have been described. In *Streptococcus pneumoniae*, the lytic enzyme requires the presence of choline on the teichoic acid. However, for other bacteria

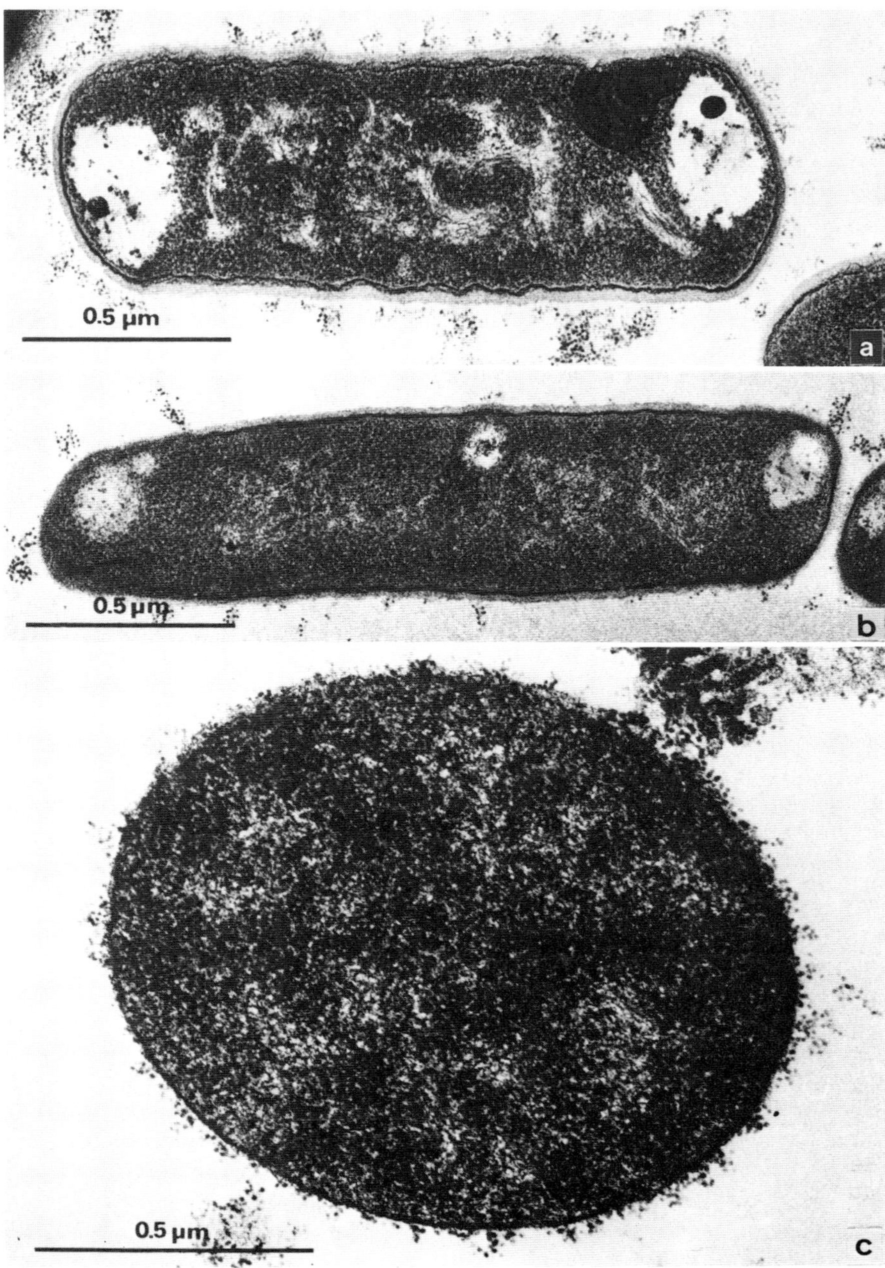

Fig. 38. Cellular location of the lipoteichoic acids treated with specific antiserum and ferritin-conjugated antibodies directed against the specific antiserum. a, *Lactobacillus fermenti;* intact cell; b, *L. casei;* intact cell; c, protoplast of *L. fermenti.* (From Van DRIEL 1973)

also having lytic enzymes strongly bound to the cell wall, there are no indications that teichoic acids are involved. A third function of teichoic acids is their effect on the relationships of bacteria with their environment, e. g. immunological properties. In lactobacilli (VAN DRIEL et al. 1973), they play an important role in serological classification; a similar situation is observed for some streptococci and staphylococci. Teichoic acids are known to form part of the receptor for many of the bacteriophages that infect Gram-positive bacteria. Two different phages (*SP 50* and *∅ 25*) specific for *Bacillus subtilis* require different receptors (GIVAN et al. 1982). Together with high proportions of teichoic acids glucosyl substituents on teichoic acids are necessary for binding phage *∅ 25*. In contrast the binding of phage *SP 50* is not highly dependent on the presence of glucosyl substituents. Such differences could be of value in the use of phages as probes for the localization and distribution of teichoic acids in the wall and for studies on the growth of the cell wall. A further function of wall teichoic acids is associated with their charge. Maintenance of repulsive charges may be desirable enabling cell populations to disperse in the medium and thereby utilize nutrients effectively (MIORNER et al. 1983).

In teichuronic acids, the acidic nature of the molecule is derived from carboxyl groups of glucuronic acid residues which alternate with N-acetylgalactosamine. These compounds have been detected in the walls of a variety of Gram-positive bacteria (ROGERS et al. 1980; further references therein). The linkage of teichuronic acids has been studied in detail on *Micrococcus lysodeikticus (luteus)*. It has been shown that, during synthesis of this polymer by membrane preparations, a polyprenol phosphate lipid intermediate is first linked to N-acetylglucosamine, and then to two residues of N-acetylmannosaminuronic acid. Thereafter glucose and N-acetylmannosaminuronic acid are added alternately. Finally, the polymer is transferred to the peptidoglycan where it seems to be linked by a phosphodiester group to the C_6 hydroxyl group of a muramic acid residue. In addition to a function in divalent cation economy (s. above), teichuronic acids appear to be related somehow to cell separation. The presence of this polymer on the surface of wildtype cells of *Micrococcus luteus* promoted cell separation as compared to a mutant strain which showed growth in large cell aggregates and lacked teichuronic acids, at the same time requiring a much higher concentration of Mg^{2+} for growth than the wildtype. Investigation of a novobiocin-resistant mutant of *Bacillus licheniformis* deficient in teichuronic acid revealed that activity and binding of autolysine depended on teichuronic acid but not teichoic acid.

5.2.5.2 Lipids and waxes

In mycobacteria, nocardia and corynebacteria, the walls contain up to 30% of their weight as lipids extractable by organic solvents. This situation is only given for these groups, but not for other Gram-positive bacteria. In some mycobacteria a fraction containing a tetrasaccharide of peptidoglycan joined to esterified arabogalactan was found by extraction of whole cells. It is a major component of the “wax D” fraction (ROGERS et al. 1980, SALTON 1964; further references therein). It is assumed not to represent a truely soluble peptidoglycan but rather a product of autolysis. Such fragments were shown to have great importance as immunoadjuvants.

5.2.5.3 Proteins

Only few proteins have been found in the walls of Gram-positive bacteria (Fig. 39) (LAMED et al. 1983, ROGERS et al. 1980, SALTON 1964). Instead of teichoic acid, the walls of a mutant of *Bacillus subtilis 168* contain a high molecular weight (255 000 D) protein consisting of subunits (105 000 D and 155000 D). Protein A of *Staphylococcus aureus* is another example (FORSGREN & SJÖQUIST 1966, GODING 1968). It has an extended shape, a molecular weight of 42000 D and is sometimes covalently linked to peptidoglycan, sometimes excreted into the medium. It is believed to contain repetitive regions placed outside the wall which bind the Fc-fragments of immunoglobulins; the other part of the molecule is assumed to be wall-bound. Protein A is used in immunocytochemical experiments for binding gold particles to immunoglobulins (s. above, Methods of investigation). Type M proteins of group *A* streptococci have been found to be associated with the wall. They play an important role in virulence, protecting the bacteria from phagocytosis. Molecular weights between 32000 D and 180000 D have been reported, and multiple subunits are assumed. T and R proteins have also been isolated from walls of streptococci. Wall proteins have not only been observed in group *A* streptococci but also in groups *B, C* and *G* (HUIS in 'T VELD & LINSSEN 1973).

5.2.5.4 Polysaccharides

Many Gram-positive bacteria contain neutral polysaccharides covalently linked to the peptidoglycan of their walls (FREHEL et al. 1982, FREHEL & RYTER 1982, HEIDELBERGER et al. 1936, ROGERS et al. 1980; further references therein). Wall polysaccharides in streptococci play an important role as antigens (WILKINSON 1975, WILKINSON & JONES 1976). Group *A* strains, group *A* variant strains and group *C* strains each have specific neutral polysaccharides as deduced by immunological evidence. Group *A* and group *A* variant strains contain polysaccharides consisting of 1,3-linked

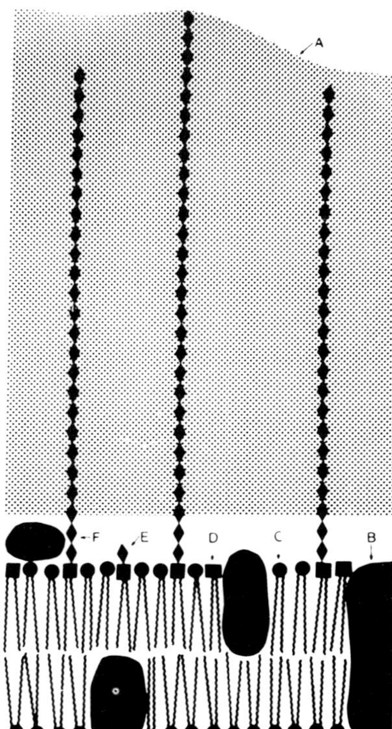

Fig. 39. Diagrammatic representation of a cell wall (A) and membrane profile of a lactobacillus. The membrane components shown are protein (B), phospholipid (C), glycolipid (D), phosphatidyl glycolipid (E), and lipoteichoic acid (F). The lipoteichoic acid molecules are depicted with their glycolipid moieties as part of the membrane lipid bilayer and the polar polyglycerophosphate chains extending into the cell wall. (From VAN DRIEL et al. 1973)

rhamnose residues, and in group *A* these have additional non-reducing N-acetylglucosamine terminals. Group *C* wall polysaccharides contain, as the immunodominant sugar, N-acetylgalactosamine instead of N-acetylglucosamine. The polysaccharides appear to be linked to the peptidoglycan by phosphodiester bonds to the C_6 of muramic acid, thus being similarly connected to the major wall polymer as are teichoic and teichuronic acids. This type of linkage seems to be restricted to Gram-positive bacteria.

Walls of mycobacteria (ROGERS 1983, ROGERS et al. 1980) contain a polysaccharide which is a complex made up of arabinogalactan esterified with long-chain fatty acids, the mycolic acids. The arabinogalactan is covalently linked to N-glycollylmuramic acid residues in the glycan chains of peptidoglycan.

Not all neutral polysaccharides are covalently linked to peptidoglycan. They may occur as capsules, or they may even be set free into the medium (s. Capsules, sheaths and slime). An encapsulated strain of *Lactobacillus casei* var. *rhamnosus* has been found to contain a rhamnose-rich polysaccharide in the capsule that other strains have incorporated into their wall. Such polysaccharides may play a role in pathogeneity as examplified by the pathogenic corynebacteria.

5.2.6 Structural and biochemical features specific for Gram-negative walls

5.2.6.1 The periplasmic space or "periplasmic gel"

Application of classical preparation techniques usually results in the electron microscopic visualization of an electron transparent gap between the cytoplasmic membrane and the peptidoglycan layer about 3 to 4 nm wide. Its existence in vivo is uncertain because any degree of plasmolysis might produce such a gap. Modified fixation procedures resulted in sections that did not exhibit such a gap. A new concept for bacterial envelope structure, especially the peptidoglycan layer and the "periplasmic space" has been developed by HOBOT et al. (1984) (Fig. 40). They have employed the "progressive lowering of temperature embedding technique" and freeze-substitution for the preparation of ultrathin sections for electron microscopy. A comparison of their results with standard embedding procedures reveals a new aspect of cell envelope structure in *E. coli* cells. The envelope was delimited by an electron-dense layer, beneath which a uniform, "matter-containing" layer is seen located between the outer and the cytoplasmic membrane. There was no "empty" periplasmic space. The peptidoglycan appeared to be more hydrated than has been previously assumed; it is gel-like and possibly fills the entire space between the outer and the cytoplasmic membrane. The peptidoglycan is probably more cross-linked in that part underlaying the outer membrane and less so towards the cytoplasmic membrane, with which it forms a gel layer which is looser than the cross-linked layer mentioned above. The hydration even of the cross-linked layer was found to be much higher than is usually assumed. The authors introduced therefore the term "periplasmic gel". Periplasmic proteins (HEPPEL 1971, JOHANNSSEN et al. 1984, KOHRING & MAYER 1985) would be freely diffusable within this gel. By osmotic shock, it would be compressed, and the periplasmic enzymes would be released through the outer membrane (HEPPEL 1969, NEU & HEPPEL 1965). MACALISTER et al. (1977) have localized alkaline phosphatase at the cell periphery. They discussed the possibility that this enzyme, as other periplasmic enzymes, is synthesized in association with the cytoplasmic membrane and vectorially transported to the periplasmic area, where it assumes its tertiary and quaternary structure and acquires its enzymatic activity (s. also below, Cell growth). Experiments performed in order to determine the site where the restriction endonuclease EcoRI is located in the *E. coli* cell (Fig. 41) revealed that this enzyme is mainly found trapped between the

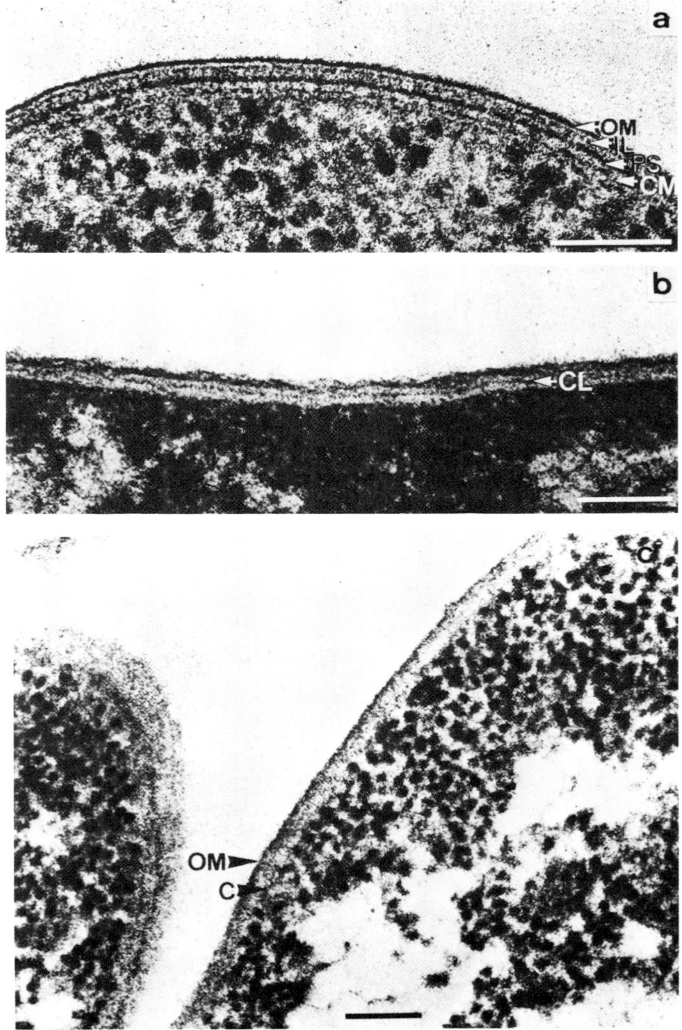

Fig. 40. Structure of the envelope of *E. coli*. a, freeze-substituted *E. coli B*. The cell envelope has a constant, uniform width and is divided into the outer membrane (OM), intermediate layer (IL), periplasmic space (PS), and cytoplasmic membrane (CM). There is a slight patterned appearance to the intermediate layer. The periplasmic space is not empty, i. e., it is electron-opaque. Section stained with uranyl acetate and lead citrate; b, glutaraldehyde-fixed *E. coli B* with low-temperature dehydration (−35°C) and embedded in Lowicryl K4M. The envelope appears to be uniform between the outer membrane and the cytoplasmic boundary. Sometimes a stained central line (CL) is seen. No periplasmic space is visible. Section stained with uranyl acetate and lead citrate; c, as b, but observed in STEM with no section staining. Matter is contained between the outer membrane (OM) an the cytoplasmic boundary (C). No periplasmic space is visible. Bars: 0.1 µm. (From Hobot et al. 1984)

cytoplasmic membrane and the outer membrane. A similar location has been suggested by Mayer & Reichenbach (1978) for a number of restriction endonucleases in gliding bacteria.

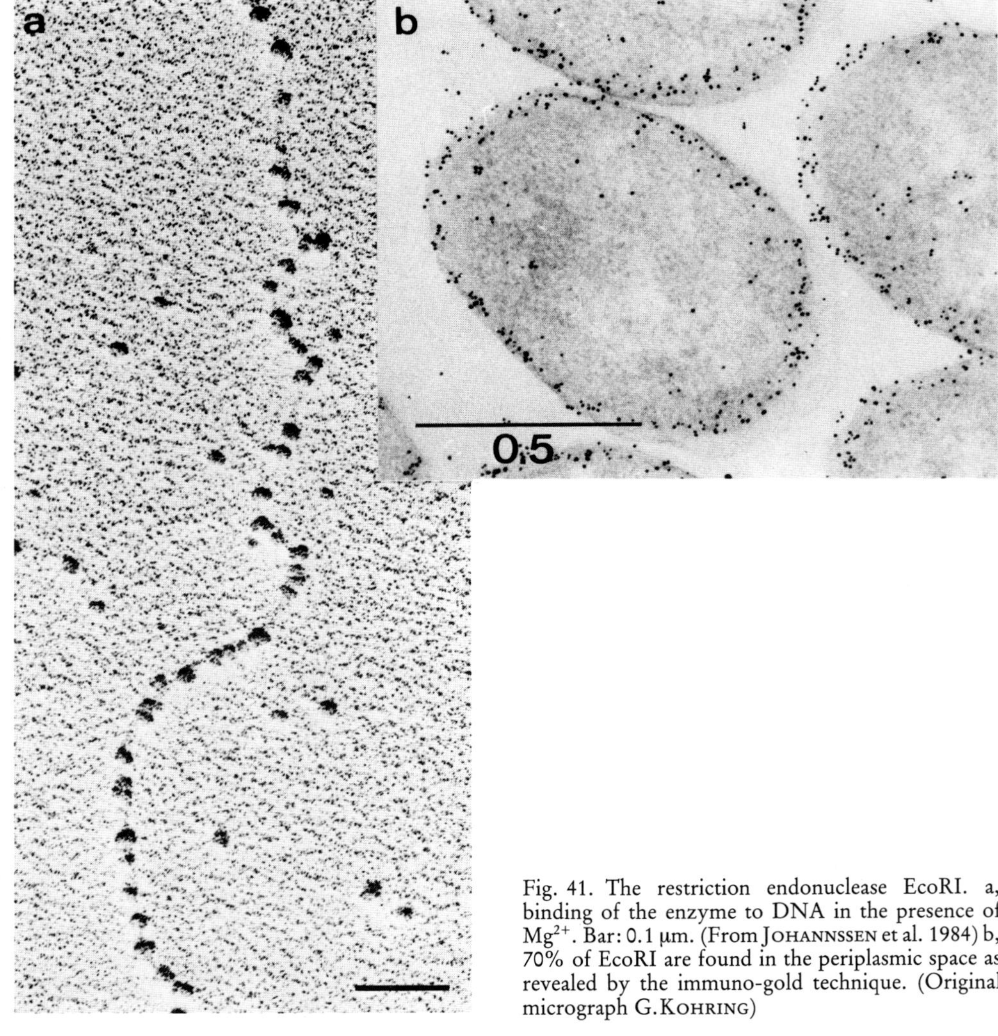

Fig. 41. The restriction endonuclease EcoRI. a, binding of the enzyme to DNA in the presence of Mg^{2+}. Bar: 0.1 µm. (From JOHANNSSEN et al. 1984) b, 70% of EcoRI are found in the periplasmic space as revealed by the immuno-gold technique. (Original micrograph G. KOHRING)

5.2.6.2 The "intermediate layer"

Located outside the peptidoglycan, the "intermediate layer" is defined as a gap between peptidoglycan and outer membrane. According to the results obtained by HOBOT et al. (1984) (s. above) it might not exist; instead, the "periplasmic gel" might also fill this space. Proteolytic enzymes were shown to remove protein from this area, assumed to be material connecting peptidoglycan to the outer membrane. Whether or not all of this material is chemically identical with a free part of a lipoprotein inserted into the outer membrane has not yet been determined.

5.2.6.3 Lipoprotein

In *E. coli*, a specific lipoprotein has been found (BRAUN 1975) the major part of which appears to be integrated into the outer membrane, but with about one third of it being covalently linked to peptidoglycan. Its lipophilic properties are attributed to a masked N-terminal cysteine, with fatty acid residues present on the amino group and on a diglyceride. The protein exhibits a highly ordered α-helical structure of about 20 residues, a β-loop, an α-helical part almost extending to the N-terminal cysteine, and a non-helical C-terminal sequence about 10 residues long. Separation of cytoplasmic and outer membranes by lysozyme treatment caused the lipoprotein to move with the outer membrane, with peptidoglycan attached to it. This finding indicates that, in *E. coli*, the peptidoglycan is actually linked to the outer membrane (except during peptidoglycan synthesis where it is attached to the cytoplasmic membrane; s. below). It was concluded that this lipoprotein helps maintain the structural integrity of the cell envelope. A biosynthetic precursor, the prolipoprotein, has been described. A similar type of lipoprotein has been found in *Serratia marcescens*. However, several *Proteus* species and *Pseudomonas fluorescens* appear to lack this lipoprotein.

5.2.6.4 The outer membrane

Most Gram-negative bacteria (exceptions s. below) contain an outer membrane (Fig. 42) (MIZUSHIMA 1985; further references therein) easily visible in electron micrographs of conventionally prepared ultrathin sections as two dense lines separated by a transparent one, i. e. exhibiting an aspect identical to a "unit membrane" or typical "bilayer". In such samples and in negatively stained preparations, the outer membrane appears to be wavy or wrinkled; occasionally "bleb" formation is observed. However, it may be concluded from freeze-etched samples and from sections visualized in the frozen-hydrated state that these structures are artificially produced by shrinkage of the cell.

Several major functions have been attributed to the outer membrane (ROGERS et al. 1980, INGRAHAM et al. 1983; further references therein). One of them is a selective permeability barrier,

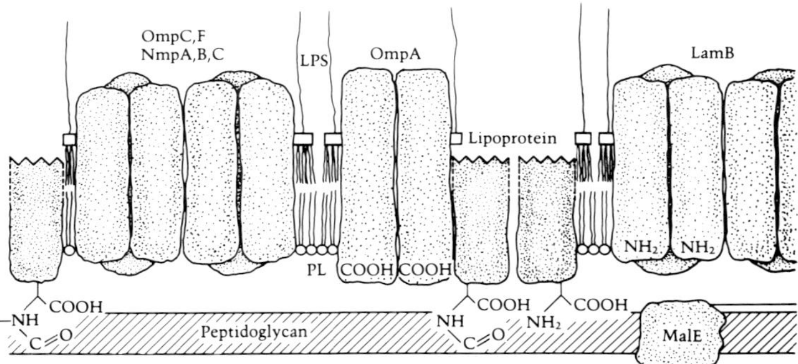

Fig. 42. Schematic representation of the molecular organization of the major proteins in the outer membrane of Gram-negative bacteria. The diagram shows the amide linkage of lipoprotein to the peptidoglycan, the transmembrane nature of OmpA and the porins (OmpC, OmpF, NmpA, NmpB, and NmpC). The LamB protein is an outer membrane protein that functions as a porin for the transport of maltose and maltodextrins; it is also a receptor for bacteriophage λ; it is shown in proximity to the MalE protein, a periplasmic maltose-binding protein. – LPS = lipopolysaccharide; PL = phospholipid. (From INGRAHAM et al. 1983, after OSBORN)

keeping in the periplasmic space or "intermediate layer" (or the "periplasmic gel", s. above) active proteins, and keeping out hydrophilic compounds above a certain size, and a wide range of hydrophobic substances. Secondly, the outer membrane controls the specific uptake of certain metabolites. Thirdly, it is a site for the adsorption of bacteriophages (Fig. 43). Finally, as the lipid A-ends of the lipopolysaccharides are integrated into this membrane, its function as a carrier for O-antigens is obvious.

The molecular composition of the outer membrane enabling it to perform these functions has been studied. One major component is protein. Compared with the cytoplasmic membrane, the outer membrane possesses only a very limited number of different types of proteins, indicating a relative lack of enzymic activities. However, the ratio of total protein to phospholipids in the outer membrane is higher than that in the cytoplasmic membrane (2.2:1 as compared to a value of 1.5:1) (ROGERS et al. 1980). Some outer membrane proteins occur in high copy numbers. These are the matrix proteins or porins (DI RENZO et al. 1978, GARTEN et al. 1975, HINDENNACH & HENNING 1975, LEE et al. 1979, ROSENBUSCH 1974, ROSENBUSCH et al. 1980, SMIT & NIKAIDO 1978). They were found to form ordered patterns (hexagonal lattices with spacing 7.7 nm) on isolated peptidoglycan layers with bound lipoprotein. Protein 1 a has a molecular weight of 36000 D.

The relative amounts of proteins 1 a and 1 b are influenced by the type of medium used to grow the bacteria. The fact that both proteins exist in the extended β-conformation might influence their function. It hase been suggested that they span the outer membrane, forming pores through the membrane in a triplet association. Recent investigations have provided further insight into the structural organization of these complexes. Another type of protein found in the outer membranes is the *tolG* protein. Varying molecular weights have been determined for it, ranging from 27000 D to 48000 D, and its molecular weight has been found to be "heat modifiable" (ROGERS et al. 1980). As mentioned above, a further protein of the outer membrane is the lipoprotein described by BRAUN (1975). Experiments with mutants of *E. coli* indicated that the proteins in the outer

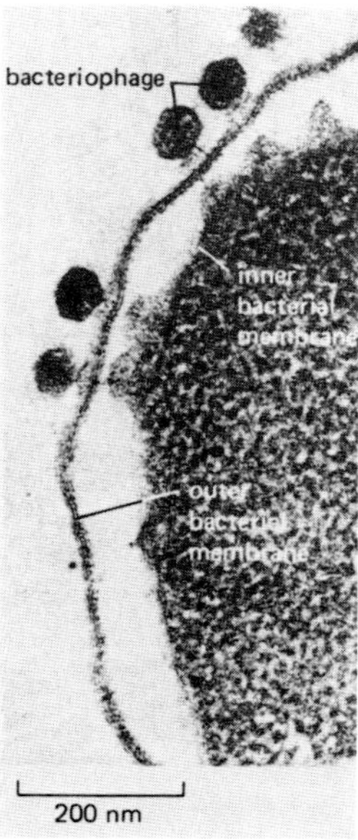

Fig. 43. Bacteriophages attached to adhesion sites in *E. coli*. The virus appears to inject its DNA into the cell interior at these sites. (From ALBERTS 1983, after BAYER et al.)

membrane might be involved in processes such as growth (s. below) and cell lysis, and in the determination of cell shape (s. below).

Bacteriophage adsorption (BRAUN et al. 1973, BRAUN & WOLFF 1973) and specific mechanisms for the penetration of hydrophilic substances may also be mediated by proteins of the outer membrane. In *E. coli*, induction of the *lam B* gene by maltose switches on the passage of maltodextrins through the outer membrane. At the same time, adsorption of bacteriophage λ to the cell is mediated. This indicates that the *lam B* gene product (VOS-SCHEPERKEUTER et al. 1984) acts as the phage receptor as well. There are numerous other examples for interrelationships of bacteriophage receptors and permeability proteins, e. g. interactions between uptake of nucleosides and *T6* adsorption (the protein responsible for this feature is not one of the recognized porins), or porins acting as receptors for the bacteriophages *T2* and *T4*.

The second major component of the outer membrane encompasses phospholipids. In general, the same phospholipids are present in the outer membrane and in the cytoplasmic membrane (ROGERS et al. 1980, INGRAHAM et al. 1983). However, the contents of phosphatidylglycerol and cardiolipin are lower in the outer membrane whereas the ratio of phosphatidylethanolamine is increased. The amount of phospholipid per unit mass of membrane in the outer membrane has been found to be lower than in the cytoplasmic membrane. Most phospholipids of the outer membrane are assumed to be located in the inner leaflet whereas the outer leaflet contains predominantly the lipid A-end of the lipopolysaccharides.

The third major component entails these lipopolysaccharides (INGRAHAM et al. 1983, KAUFFMANN 1944, SCHMIDT 1972, SCHMIDT et al. 1970). They form the O-somatic antigens and are responsible for the endotoxicity of Gram-negative bacteria. Besides the well known antigens of enterobacteria (Kauffmann-White scheme for *Salmonella*, Kauffmann-Perch scheme for *Proteus*), the "Enterobacterial Common Antigen" (ECA), being specific for the whole family, has been analyzed (KUHN et al. 1984). The lipopolysaccharides are composed of the lipid A constituent integrated into the outer leaflet of the outer membrane (s. above), and a second part, the polysaccharide moiety, being composed of the core and the O-specific side chains. R mutants usually have relatively large proportions of lipid A, rendering the lipopolysaccharides highly lipophilic. In contrast, other lipopolysaccharides have been found to be dissolved in the water phase of phenol extractions. The core portion of the molecules contains ketodeoxyoctonic acid (KDO), heptose, hexoses and amino sugars. The O-specific side chains are made up of repeating units specific for particular serotypes. The structure of lipid A has been worked out in detail (ROGERS et al. 1980; further references therein). The lipophilic property is caused by fatty acids attached to β-1,6-linked disaccharides of D-glucosamine.

The importance of lipopolysaccharides in the structure of Gram-negative bacteria is shown by the fact that stable mutants lacking these entities have not yet been isolated.

The endotoxicity of lipopolysaccharides is generated by the interaction of the lipid A-part of the molecule with complement, thus producing a loss of haemolytic activity and other symptoms such as pyrogenicity (ROGERS et al. 1980).

5.2.6.5 Interconnections between envelope layers

Possible interconnections between layers of the outer bacterial structures have been observed (BAYER 1968a, 1968b, 1975, BAYER et al. 1982) as bridges running between the outer side of the cytoplasmic membrane and the inner side of the wall of Gram-positive bacteria. In Gram-negative cells, e. g. *E. coli*, areas of contact between wall layers and cytoplasmic membrane (Fig. 43) are

maintained after plasmolysis. The same areas were found to be attachment sites for bacteriophages and sites of synthesis of lipopolysaccharides (INGRAHAM et al. 1983). Such interconnections may be indicators for regulatory events taking place during cell growth and division.

5.2.7 Regular surface layers present on bacterial walls

Gram-positive and Gram-negative bacteria as well as bacteria lacking peptidoglycan in their wall may have a cell envelope containing regular cell surface layers (S-layers) (SLEYTR & MESSNER 1983). Ultrathin sectioning (Fig. 44), freeze-etching (Fig. 45), metal-shadowing (Fig. 46), and negative staining (Fig. 47) have been the techniques for sample preparation in investigations of these regular wall layers. Numerous studies have revealed considerable variations in structure (BAUMEISTER & KÜBLER 1978, BURLEY & MURRAY 1983, KAWATA et al. 1974, KESSEL 1978, MASUDA & KAWATA 1979, 1981, 1983, TAYLOR et al. 1982, THOMPSON & MURRAY 1982, VAARA 1982). Nevertheless, the observed cell wall profiles may be classified into three categories (Fig. 34). Although not a universal feature of all bacteria, these regular patterns of proteins or glycoproteins have been found in a wide range of species. S-layers are often lost from wild type bacteria when the cells are grown in the laboratory.

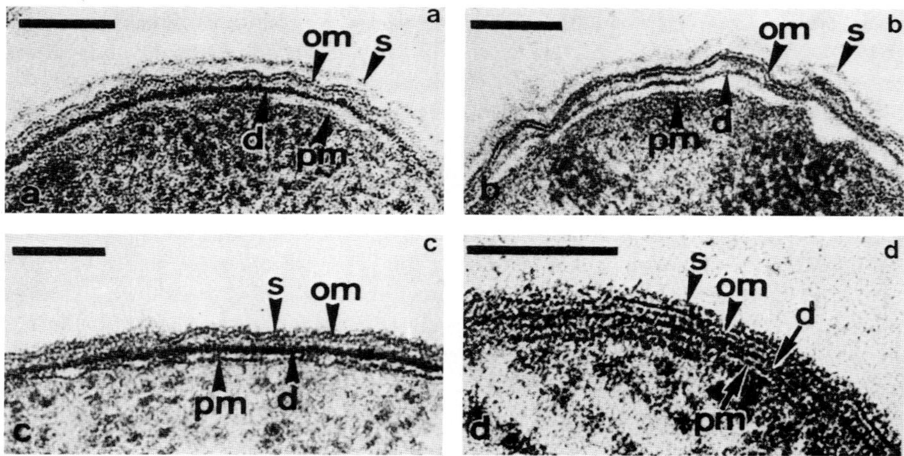

Fig. 44. Ultrathin sections of envelopes of Gram-negative bacteria with regularly arranged S-layers. a, *Acinetobacter* sp. strain *MJT/F5/5;* b, as a, but heat-treated; c, *A.* sp. strain MJT/F5/199A; d, *Spirillum serpens.* – d = dense layer; om = outer membrane; pm = cytoplasmic membrane; s = S-layer. Bars: 0.1 μm. (From SLEYTR 1978)

5.2.7.1 Ultrastructure of S-layers

In addition to the analysis of S-layers attached to whole cells, most high resolution studies have been performed on isolated S-layers. Most regular arrays were observed to be hexagonal or tetragonal, with a center-to-center spacing between adjacent morphological units within a range of 5 to 20 nm. Averaged images showed tetrameric morphological units in the tetragonal lattice and hexameric units in the hexagonal lattice. Image processing procedures such as linear integration, optical filtering and digital filtering with computers have been applied to in vitro S-layer self-assembly products, achieving resolutions between 1.5 and 2.5 nm (GROSS et al. 1984, RASCH et al. 1984). In contrast to the cylindrical part of the cell where the S-layer patterns are generally uniform over large areas, an accumulation of faults can be observed occurring at the cell poles, giving the impression of a mosaic composed of small crystallites. It has been found that S-layer crystallites may rearrange and fuse on the cell surface to form larger areas of regular patterns. Isolated S-layer protomers can be reattached to the surface of the type of cell wall from which they have been originally removed (MASUDA & KAWATA 1980). Moreover, when mixtures of two different S-layer protomers from two suitable bacteria were used, both types of lattices were formed on one cell (Fig. 48). This has been shown for *Clostridium thermohydrosulfuricum* and *C. thermosaccharoly-*

Fig. 45. Surface layers of bacteria. a, freeze-etched preparation of *Desulfotomaculum nigrificans;* b, *Clostridium thermohydrosulfuricum;* c, *Bacillus stearothermophilus;* d, site of insertion of a flagellum in the surface of *Clostridium thermohydrosulfuricum;* e, insertion of a flagellum in the surface of *C. thermosaccharolyticum.* – Arrows = regular array of subunits. Bars: 0.1 μm. (a–c, from SLEYTR & MESSNER 1983; d, e, from SLEYTR 1978)

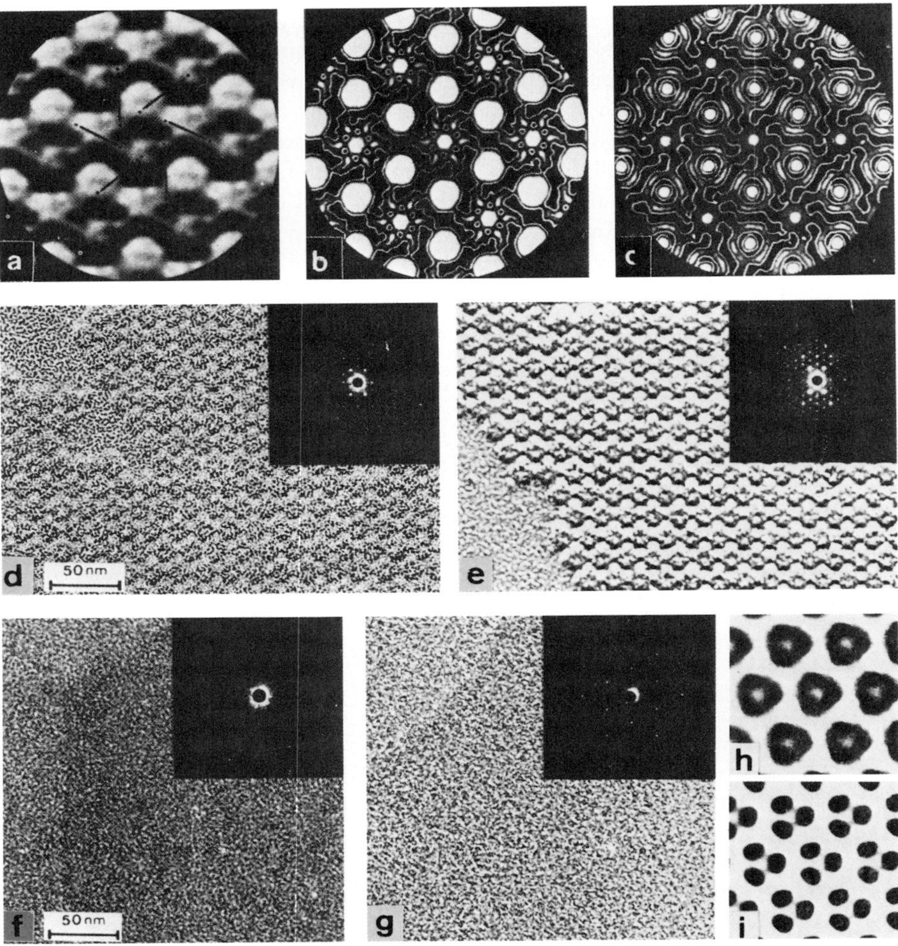

Fig. 46. Analysis of fine structural details on the HPI-layer (a–e) and the purple membrane (f–i). a, unidirectional; b, rotational; c, perpendicular shadowing; d, unidirectional shadowing, temperature – 80°C; e, as d, but at – 260°C; f–h, extraplasmatic side of the purple membrane after C-shadowing; f, perpendicular; g, rotational shadowing; h, digital filtered version of g; i, filtered image after perpendicular shadowing with Ta/W. (From GROSS et al. 1984)

ticum (SLEYTR & GLAUERT 1976). These heterologous reattachment experiments provided evidence that the reattachment process and the lattice orientation on intact cells are not influenced by any order within the binding sites of the peptidoglycan layer. Thus, information for morphogenesis of the S-layer resides in the molecular structure of its protomers (SLEYTR & GLAUERT 1975). For the maintenance of an ordered S-layer on a growing cell surface, the bonds holding the subunits together have to differ from those binding the subunits to the underlying peptidoglycan. The requirement for maintaining ordered arrays of subunits without gaps on a growing cell surface is a continuous synthesis of a surplus of subunits and their transfer to the sites of lattice growth. It was

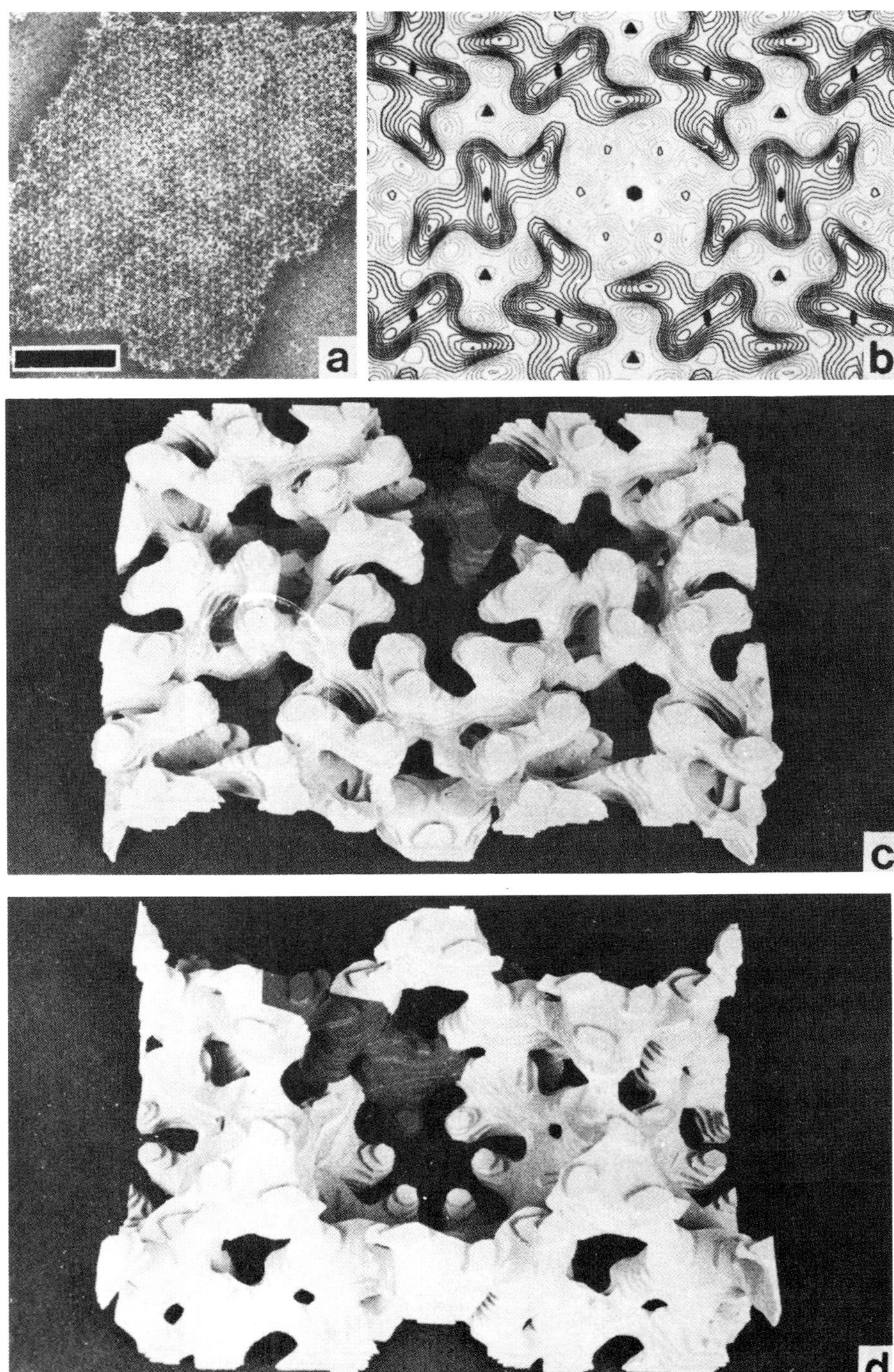

Fig. 47. S-layer of *Sulfolobus acidocaldarius* composed of a single glycoprotein species. a, S-layer fragment, negatively stained; b, in-plane projection; dark contours represent protein; light contours represent negative stain; c, d, model of three-dimensional structure exhibiting a remarkable porosity; external surface (c) fairly smooth; cellular side (d) very rough. Bar: 0.2 μm. (From SLEYTR & MESSNER 1983, after TAYLOR et al. 1982)

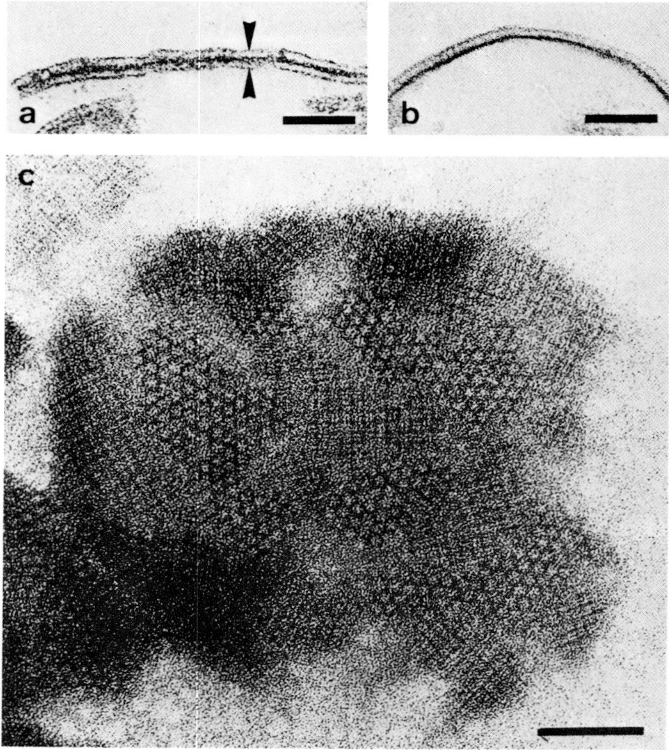

Fig. 48. a, b, thin sections of murein layers of *Clostridium thermosaccharolyticum* with reattached subunits of the S-layer. In a, reattachment on both sides (arrowheads); c, negatively stained preparation of a fragment of the cell wall from *C. thermosaccharolyticum* showing heterologous reattachment of S-layer subunits isolated from the same organism and from *C. thermohydrosulfuricum*. Bars: 0.1 µm. (From SLEYTR 1978)

assumed that, as with other secreted proteins, S-layer subunits are synthesized on plasma membrane bound ribosomes, and that they cross the membrane as growing chains rather than after the completion of their synthesis and subsequent tertiary folding. A considerable lateral diffusion may be required for the S-layer subunits in order to reach their final loci. Charge phenomena on the underlying layer may be responsible for a proper alignment of the subunits. Only a small excess of S-layer material has been found in the culture medium of *Bacillus sphaericus*, indicating strict control of subunit synthesis. An analysis of interrelationships between the regulation of synthesis of S-layer subunits and cell growth has been the subject of several investigations.

5.2.7.2 Chemical composition of S-layer subunits

Isolated S-layer subunits were found to have molecular weights between 40000 D and 200000 D. They are mainly composed of single homogeneous polypeptides with carbohydrates occasionally present as a minor component. A predominantly acidic amino acid composition has been determined for several purified S-layers (SLEYTR & MESSNER 1983).

5.2.7.3 Functional aspects

S-layers have probably evolved as a result of cell-environment interaction (SLEYTR 1978). They may act as a barrier or molecular sieve controlling the movement of external and internal factors such as toxic macromolecules. It was shown that strains of *Spirillum serpens* containing S-layers were resistant to attachment of *Bdellovibrio bacteriovorus*. S-layers may also protect the peptidoglycan from the action of lytic enzymes such as lysozyme.

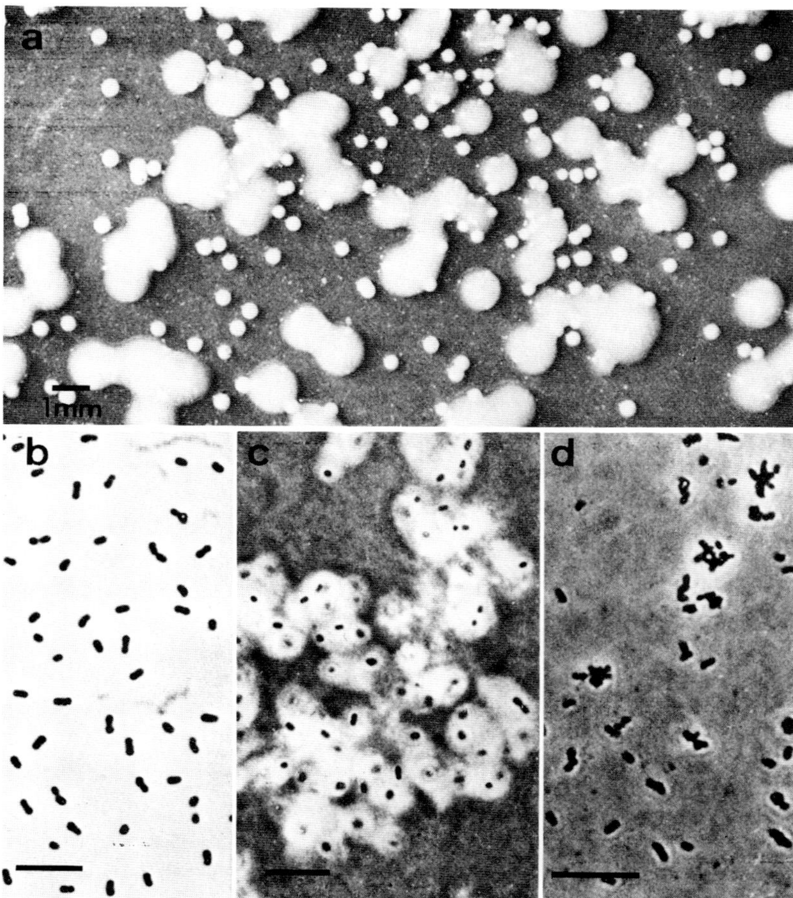

Fig. 49. Wildtype and slimeless mutant of the coryneform bacterium strain *7 C* grown on fructose agar. a, colonies of both cell types; b, unstained preparation of wildtype cells; c, indian-ink mount of the wildtype cells; d, indian-ink mount of mutant cells grown in fructose liquid medium. Phase contrast. Bars: 10 µm. (From ANDREESEN & SCHLEGEL 1974)

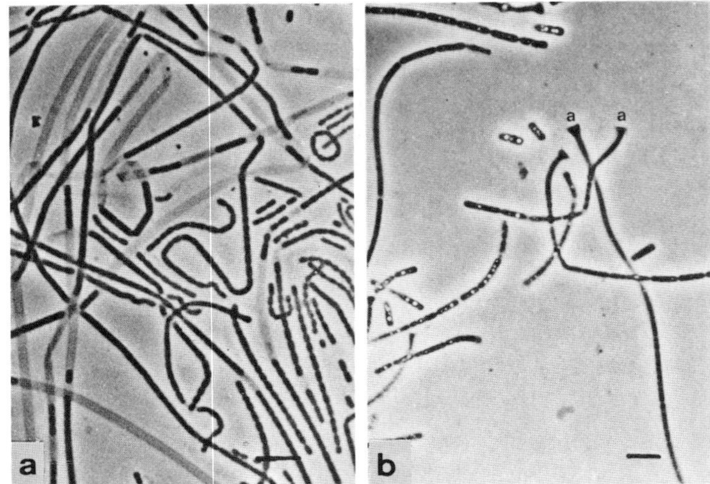

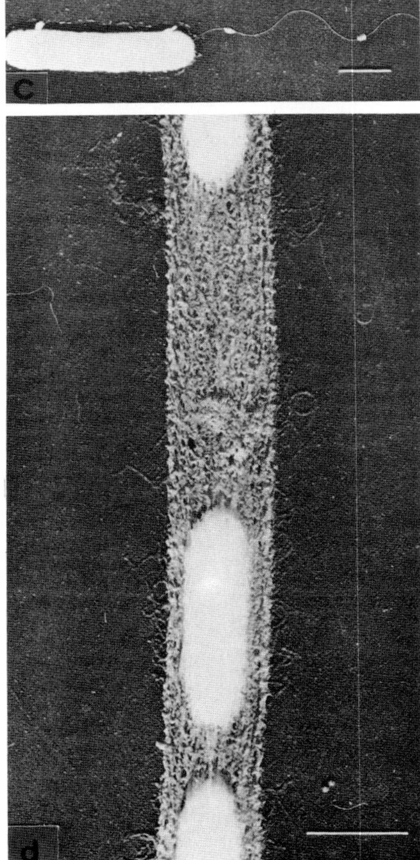

Fig. 50. Sheathed bacteria. a, b, cells (without and with sheath) of *Sphaerotilus natans;* holdfasts in b are marked by a. Bar: 10 µm. (From MULDER & DEINEMA 1981); c, *Leptothrix cholodnii;* single cell with one polar flagellum. Bar: 1 µm;
d, as c, but sheath encrusted with MnO_2. Bar: 10 µm. (From MULDER & DEINEMA 1981, after MULDER & VAN VEEN)

5.3 Capsules, sheaths, and slime

5.3.1 Ultrastructural aspects

The outermost surface layers of Gram-positive and Gram-negative bacteria are those cellular components which, besides pili and fimbriae (s. below) interact with the environment (DAHLBACH et al. 1981, DAZZO 1984, ESHDAT et al. 1978, FATTOM & SHILO 1984, FLETCHER & FLOODGATE 1976, ISMAIL & SOM 1982, OFEK et al. 1983, ROSENBERG et al. 1980, 1983, ROSENBERG & ROSENBERG 1981, SCHWARZMANN & BORING 1971, STAHL et al. 1983, SUTHERLAND 1972). Depending on the particular group and species of bacteria and their nutritional requirements, the surface of the cells can be formed by additional material surrounding the otherwise typical cell envelope (Figs. 49–54). When organized into a regular morphologically structured component, this is usually referred to as a capsule (Fig. 51) or glycocalyx (BRYANT et al. 1984, CAGLE et al. 1972, COSTERTON et al. 1981, GLOSSMANN & NEVILLE 1971) or, more seldom, as a sheath (Fig. 50) (MULDER & DEINEMA 1981). When loosely organized, or of variable consistency, it is designated as slime (Fig. 49). Some of these structures can be demonstrated under the light microscope by "negative-staining" with Indian ink. Difficulties in the examination of these structures by electron microscopy in the early days were due to the lack of adequate fixation and staining procedures. Application of ruthenium red (LUFT 1971, SPRINGER & ROTH 1973 a) has been shown to be useful in studying the presence of bacterial capsules. However, the three-dimensional structure of these outermost surface layers and their specific variations (BAYER & THUROW 1977, CAGLE 1975, COSTERTON 1984, MUDD et al. 1943, NOKHAL & MAYER 1979, PROGULSKE & HOLT 1980, REMSEN &

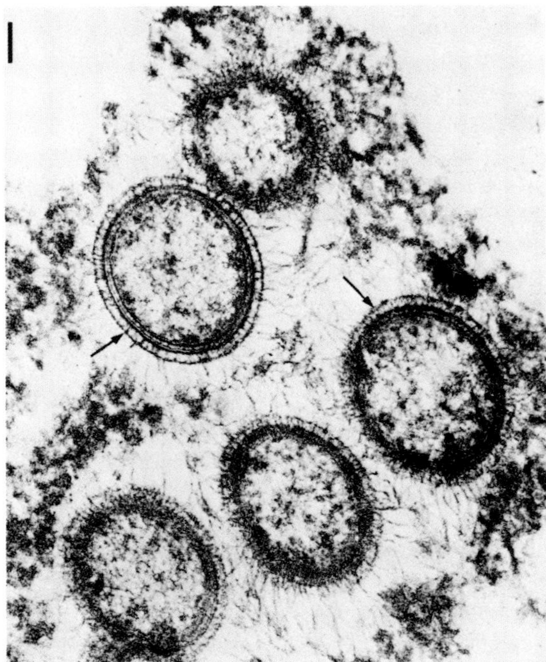

Fig. 51. Ruthenium red-stained microcolony from the bovine rumen showing thickened but extended capsular fibers emanating through a complex electron-dense layer (arrows) at the cell surface. Bar: 0.1 μm. (From COSTERTON et al. 1981)

Fig. 52

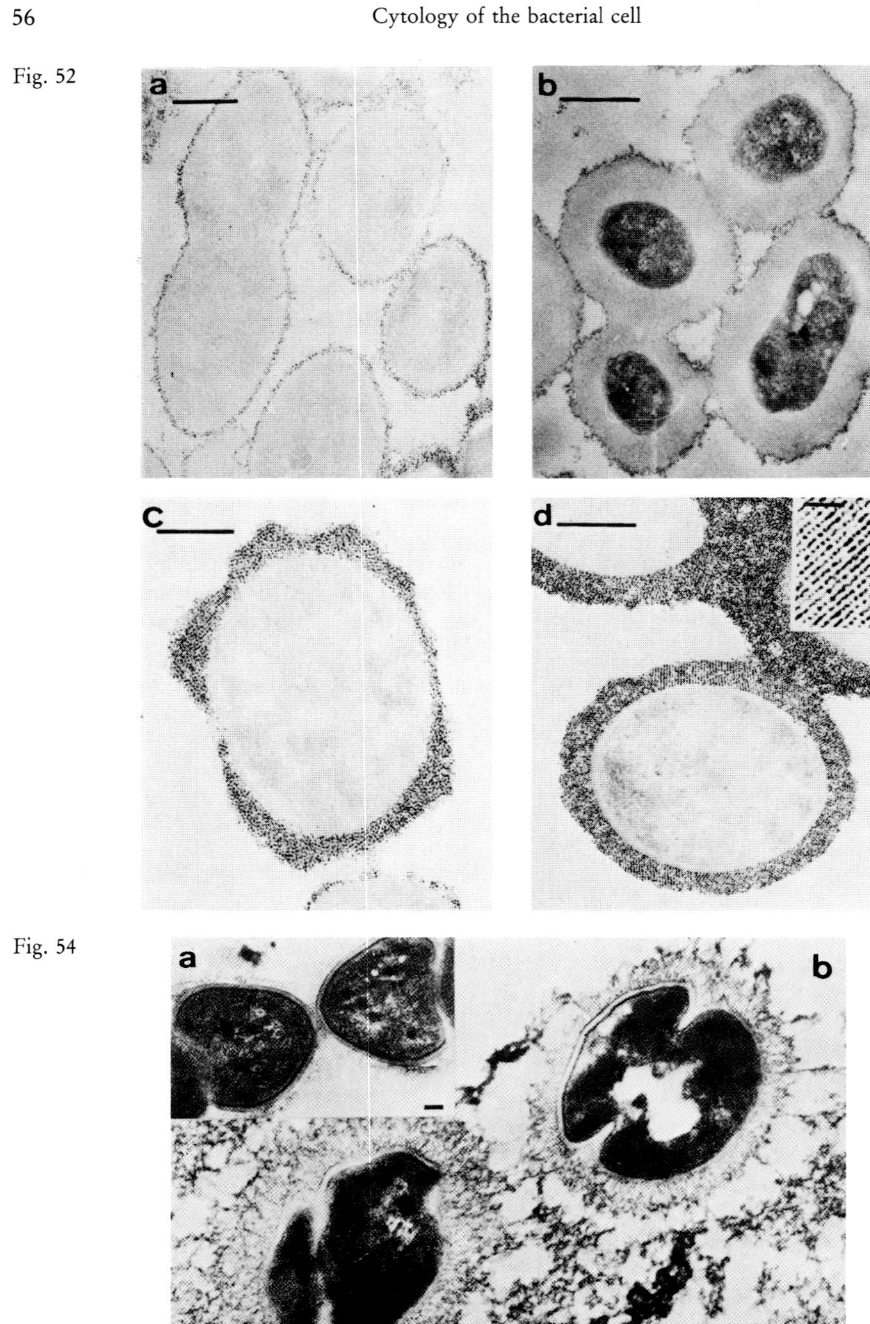

Fig. 54

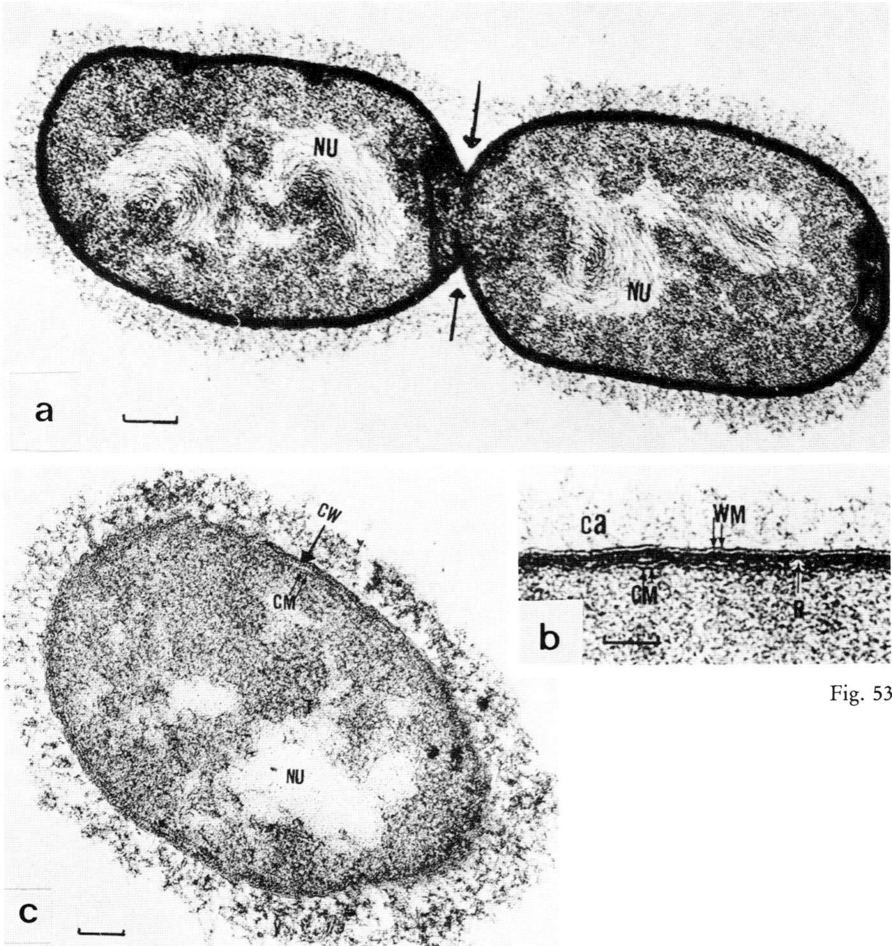

Fig. 53

Fig. 53. The capsular network of *Klebsiella pneumoniae* depicted after chemical fixation by using alcian blue lanthanum and tris-(1-aziridinyl)phosphine oxide-aldehyde-osmium procedures. a, c, low magnification also showing the nucleoplasm; b, higher magnification. – ca = capsule; CW = cell wall; CM = cytoplasmic membrane; NU = nucleoid; WM = wall membrane; R = peptidoglycan layer. Arrows in a, point to the constriction area of cell separation where scanty capsular material is seen. Bars: 0.1 μm in a and c, 50 nm in b. (From Cassone & Garaci 1977)

Fig. 52. Ultrastructural visualization of capsules by polycationic ferritin. a, Pneumococci; b, *Klebsiella* sp. after the capsular swelling reaction and subsequent prefixation with glutaraldehyde; c, uncontrasted cell of mucoid growing *E. coli* strain *(3551/78)*; d, *Klebsiella* sp. cells prefixed with glutaraldehyde. Inset: ferritin granules arranged in an ordered structure resembling fibrils. Bars: 0.2 μm in a and c, 0.25 μm in b, 0.5 μm in d, and 50 nm in the inset in d. (From Weiss et al. 1979)

Fig. 54. Outer cell envelope structures; a, group B *Streptococcus,* ruthenium re-stained; condensation of the capsular material; b, as a, but antibody-stabilized cells; extensive capsular material. Bars: 0.1 μm. (From Costerton et al. 1981)

LUNDGREEN 1966, WILKINSON 1958) is only revealed by applying a polycationized derivative of ferritin (Fig. 52) (SHANDS 1965, WEISS et al. 1979) which binds electrostatically to anionic sites on the surfaces of glutaraldehyde-fixed or unfixed cells, by fixation in a solution containing alcian-blue-lanthanum and tris-(1-asziridinyl) phosphine oxide (TAPO) (Fig. 53), or by incubation with specific anticapsular serum (Fig. 54) which has been prepared by intravenous immunization of rabbits. The advantages of the latter technique were demonstrated in a detailed study on the effect of increasing antibody concentrations on the capsular structure of *E. coli.* Dehydration for electron microscopy caused the capsules to collapse into thick bundles at acetone or alcohol concentrations above 50%. Pretreatment with capsule-specific immunoglobulin G prevented this collapse. A further example for the reliability of this technique is seen in work on the morphological stabilization of capsules of four prototypes of group *B* streptococci (MACKIE et al. 1979) where related but distinct antigenic types of capsular polysaccharides were stabilized and visualized by the application of specific antisera.

5.3.2 Chemical composition

Extracellular material making up capsules, sheaths and slimes usually appears as a highly hydrated, gel-like network of delicately meshed polysaccharide fibrils emanating from the cell wall (INGRAHAM et al. 1983, ROGERS et al. 1980). The polysaccharides are acidic in nature and can be isolated, to certain degrees, by washing at appropriate pH values and ionic strength. From *Rhodopseudomonas capsulata,* an acidic polysaccharide fraction has been obtained by fractionation of hot phenol-water extracts. The exopolysaccharides appear to be bound to the cell wall not by covalent bonds but by simple entanglement with the outer wall polymers. As the production of these exopolysaccharides is under genetic control, their occurrence has been found to be specific, in some cases, for certain growth phases, and their composition is strain-specific. This fact is widely used in diagnostics. Common sugars making up the polysaccharides are rhamnose, galactose, glucosamine, galacturonic acid or hyaluronic acid; their relative amounts may vary. In addition, fucose, 3-amino-3,6-dideoxygalactose, an unidentified amino sugar and an unidentified acidic component have been detected in *Rh. capsulata.* Similar findings have been reported for several strains of *Klebsiella pneumoniae* and *Klebsiella* spec. (CASSONE & GARACI 1977, GARACI & CASSONE 1974, SPRINGER & ROTH 1973b), for several species of *Brucella* (OBERTI et al. 1982), for chroococcacean cyanobacteria, for enteropathogenic strains of *E. coli* (MOOREHOUSE et al. 1977) and other bacteria.

The chemical structure of the *E. coli* polysaccharide has in part been elucidated by employment of capsule-degrading bacteriophages (FEHMEL et al. 1975, THUROW et al. 1974). It has turned out that the *E. coli K 29* polysaccharide consists of about 300 hexasaccharide-repeating units with a backbone of mannose, glucose, glucuronic acid, and galactose in a 1:1:1:1 ratio. A side-chain containing mannose, glucose, and pyruvate is attached at the glucuronic acid. Repeating subunits similar to these have been described for many of the linear polymers of capsules and slimes; molecular weights of 10^5 D and higher have been reported.

Capsules composed of polypeptides or cellulose have also been described (SCHLEGEL 1981). Examples are *Sarcina ventriculi* and *Acetobacter peroxydans* subsp. *xylinum* where cellulose is excreted into the medium as fine fibrils forming the "mycoderma aceti".

Some capsular components, such as xylans, are of industrial interest. Slime of certain bacteria, e. g. *Leuconostoc mesenteroides,* has been seen to interfere with scaled-up industrial processes such as sugar production (SCHLEGEL 1981). Within a short period of time, *Leuconostoc* converts sucrose into a gel-like material consisting of dextran. This occurs outside of the cell wall with the help of a hexosyltransferase. *Streptococcus mutans* and *S. salivarius*, occurring in "plaque", also convert sucrose into polyfructoses (laevans) by an extracellular hexosyltransferase. Thus, a matrix is formed within which acidic products of streptococcal fermentation processes, such as lactic acid, may be enriched, thus causing tooth decay (FULLER & LOVELOCK 1976, NEWMAN 1976).

5.3.3 Synthesis, interaction with bacteriophages

In a study on the polysaccharide capsule of *E. coli*, BAYER & THUROW (1977) analyzed the sites of synthesis of exopolysaccharides and bacteriophage adsorption to these compounds (CHOY et al. 1975, STIRM & FREUND-MÖLBERT 1971). From ultrathin sections and freeze-etching it was shown that, after an inducing downshift of temperature, the cells of a population did not start making capsular substances in synchrony. Instead, a considerable variation in amount was found shortly after 10 to 15 min downshift. Later (45 to 60 min), almost all cells showed at least some exopolysaccharide at their surfaces. Number and location of "antigen export sites" (i. e., exopolysaccharide synthesis and translocation sites) were studied using anticapsular IgG after a temperature downshift and subsequent growth for up to 20 min at 25°C. Between 10 and 40 polysaccharide export sites per cell were observed. After a prolonged period of growth, it appeared to be very difficult to decide whether the synthesized and transported material had been spreading laterally or had readsorbed and engulfed the cells. The first export products of polysaccharide synthesis appeared as filaments or delicate strands protruding from the outer membrane into the medium. Frequently, the emerging filaments could be seen at areas where the cell wall and cytoplasmic membrane adhered to one another, especially after slight plasmolysis (these sites have been shown before to be involved in a variety of envelope functions). Besides a filament-like appearance, an additional conformation of the newly synthesized polysaccharide was a cluster representing coiled filaments.

The serotype *K 29* polysaccharide of *E. coli* can be degraded by the *K*-specific bacteriophage *29* (FEHMEL et al. 1975, THUROW et al. 1974), a situation similar to that with *Klebsiella* where phages *11, 20* and *45* perform such a function. During phage infection of the cells, phage *29* has been observed to leave a tunnel-like path in the capsule, indicating its ability to degrade the polysaccharide strands. This finding has induced detailed studies on the chemical composition of polysaccharides by phage-mediated degradation (s. above, Chemical composition).

5.3.4 Functional aspects

The production of a capsule or slime layer by bacteria means expending energy. However, these structures confer multiple advantages on the cells. An obvious one is that they facilitate the attachment or adherence to biological and non-biological surfaces (Figs. 55 and 56). A role for the adhesion of bacteria has been postulated in the pathogenesis of both plants and animals. Enteropathogenic strains of *E. coli* have been investigated in detail. It has now become clear that the majority of strains responsible for different outbreaks of diarrhoea are nonfimbriate, suggesting the involvement of exopolysaccharides and lectins. It has been postulated that adherence, in general, primarily depends on hydrophobic interactions (DAHLBECK et al. 1981, FATTOM & SHILO 1984, OFEK et al. 1983, ROSENBERG et al. 1980), and that modifications in adherence might be caused by alterations in the degree of hydrophobicity. It was therefore suggested that further studies on adherence phenomena might be valuable for the development of vaccines. Capsules and slimes also function in virulence, render bacteria resistant against phagocytosis, prevent the reaction of

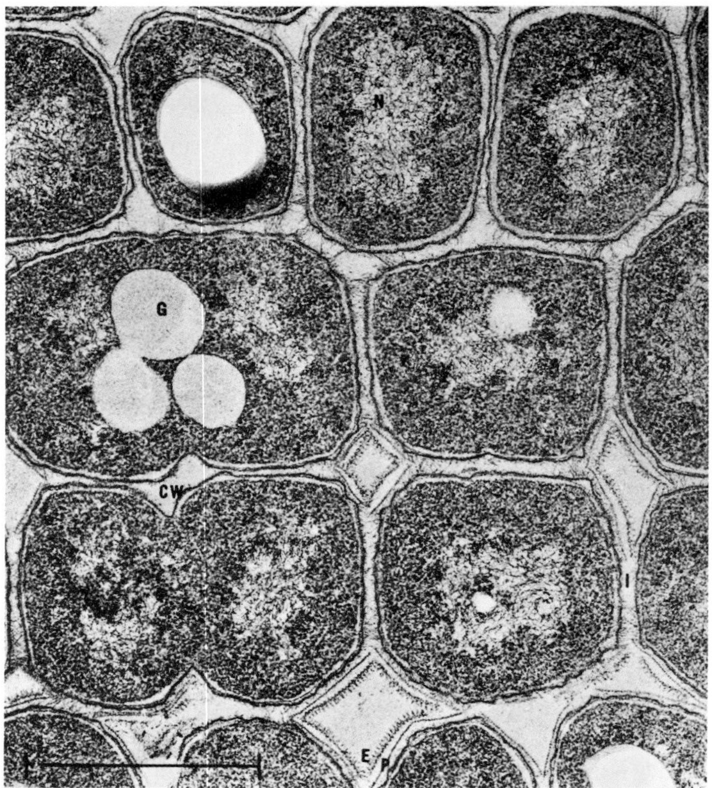

Fig. 55. Formation of bacterial cell aggregates; *Lampropedia hyalina*. – CW = cell wall; E = echinulate layer; G = poly-β-hydroxybutyrate granule; I = intercalated zone; N = nucleus; P = perforated layer. Bar: 1 μm (From STARR 1981, after PANGBORN & STARR)

complement and specific antibodies with antigens located beneath the capsule, and hinder the attack of bacteriophages that are specific for O-antigens and lipopolysaccharides in Gram-negative cells (ROGERS at al. 1980). On the other hand, capsular components have also been described as the receptor sites for some bacteriophages (*K 29*, s. above). Similarly, carbohydrates of bacterial surfaces were found to be specific receptor sites for lectins. An investigation of *Proteus mirabilis* has shown that copious quantities of extracellular slime surrounded swarm cells and that the slime appeared to provide a matrix through which the cells could migrate. Swarm cells were always found to be embedded in slime (STAHL et al. 1983). It is unclear whether extracellular slime production is also a prerequisite for migration (s. below). In *Galionella ferruginea*, polar excretion of slime might even provide a means for locomotion.

Sheath structures (MULDER & DEINEMA 1981) are defined as hollow cylinders formed from heteropolysaccharides surrounding filament-like groups of cells (Fig. 50). The physiological advantage here is an attachment of the cells to solid submerged surfaces. After cell division within the sheath, motile daughter cells can leave the colony. Examples are bacteria of the *Sphaerotilus-Leptothrix* group.

Slime has also been seen to cement single cells together as cell aggregates *(Zoogloea ramigera, Bacteriogloea)*. A similar effect has been observed for *Acetobacter peroxydans* subsp. *xylinum* and *A. aceti* and for regularly organized cell aggregates formed by *Sarcina ventriculi* and *Lampropedia hyalina;* in these cases, the connecting material was identified as cellulose (SCHLEGEL 1981; s. above). A further possible functional aspect of exopolysaccharides has arisen from investigations of the ultrastructure and physiological properties of budding bacteria such as *Pedomicrobium*-like organisms (Fig. 57). When these strains were grown in the presence of iron or manganese, the corresponding oxides accumulated on the cell surfaces (CULLIMORE & McCANN 1977, DUBINIA 1979, GEBERS & HIRSCH 1978, GHIORSE 1980, GHIORSE & HIRSCH 1978, 1982, JOHNSON & STOKES 1966, KHAK-MUN 1968, MACRAE & EDWARDS 1972, SCHWEISFURTH et al. 1978, TYLER 1970, TYLER & MARSHALL 1967a, b, c). Extraction of these oxides followed by ruthenium red staining of the samples revealed that polyanionic polymers, i. e. acidic polysaccharides, previously synthesized by and deposited on the cells were associated with the metals. The deposition of manganese oxide is probably dependent on specific oxidation factors produced by the viable cell. For iron oxide deposition, a mechanism is proposed in which positively charged, autoxidized iron hydroxides (e. g., $Fe(OH)^{2+}$, $Fe(OH)^{+}$), which are known to form during iron oxidation, accumulate on negatively charged extracellular polymers in a way similar to that which occurs during the staining of acidic polysaccharides with colloidal iron. Once deposited, the iron oxides themselves would carry negative charges and would attract positively charged iron hydroxide ions, a process which, in principle, could continue without any biological activity. Indeed, non-viable cells accumulated iron oxides.

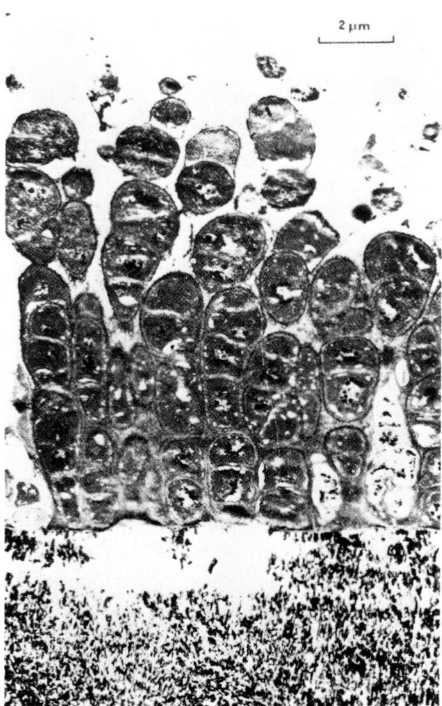

Fig. 56. Plaque formation; palisaded plaque. (From NEWMAN 1976)

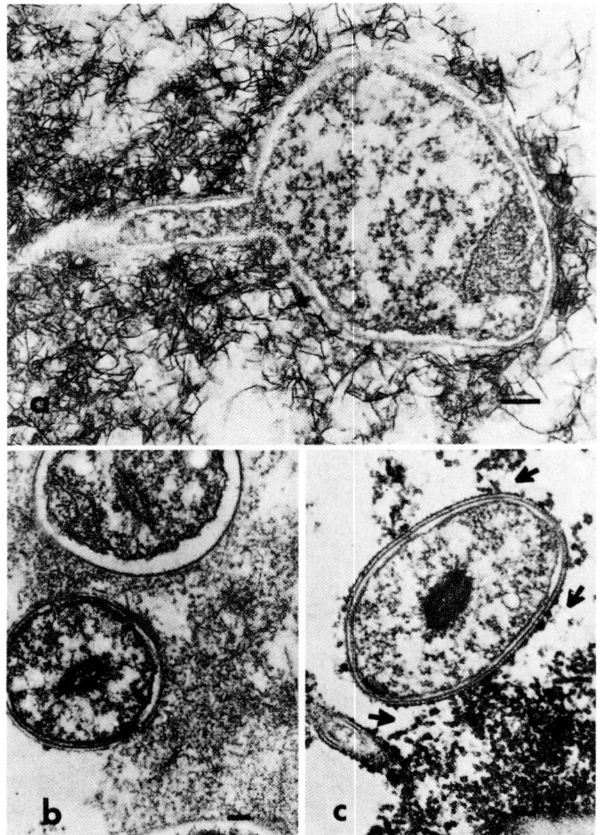

Fig. 57. a–c, *Pedomicrobium*-like budding bacteria; strain *868* grown with 16 mg Mn/l. Heavily encrusted cells in aggregates (a) were extracted with acidified leucoberbelin blue. An area of diffuse material (b) remained in place of the manganase oxides. This material became more electron-dense when stained with ruthenium red (c) indicating that it was composed of polymers. Arrows (c) indicate filaments inter-connecting spherical masses of polymers. Bars: 0.1 µm. (From GHIORSE & HIRSCH 1979)

5.4 The cytoplasmic membrane

5.4.1 Structure and components

After fixation with OsO_4 and staining with uranyl acetate and/or lead salts, the cytoplasmic membrane of bacteria resembles that of eukaryotes in ultrathin sections. Under these conventional preparation conditions it exhibits dark inner and outer layers, with an electron transparent central layer (INGRAHAM et al. 1983, ROGERS et al. 1980; further references therein). Usually the membrane measures around 6 to 8 nm in thickness. Depending on the type of bacterium and the fixation procedure, the cytoplasmic membrane has been found to be either asymmetric or symmetric (GIDDINGS & STAEHELIN 1978, RAZIN 1975). In electron microscopic examinations of frozen-hydrated *E. coli* and *Staphylococcus aureus* cells (DUBOCHET et al. 1983) at a resolution of 3 nm the cytoplasmic membrane showed an approximately uniform thickness. In *S. aureus* the density profile is characterized by an asymmetric double layer of dense material 5.5 ± 1 nm wide, bordering the cell wall at the cytoplasmic side, whereas an evaluation of *E. coli* envelopes did not exhibit any asymmetry of this kind. Occasionally, the wall and the cytoplasmic membrane of Gram-positive

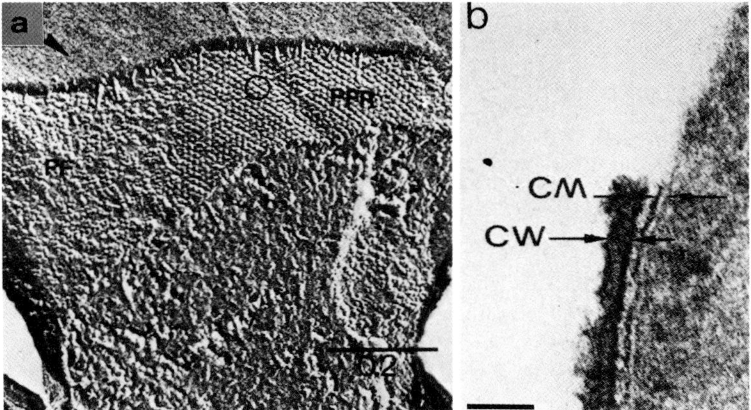

Fig. 58. a, broken cell of *Acetobacterium woodii* with surface striation (arrowhead), and cytoplasmic membrane showing particles with more or less random distribution (PF), and with strongly hexagonal arrangement (circle, PFR). (From MAYER et al. 1977); b, thin section of a freeze-fractured cell. Osmium tetroxide-fixed cells were frozen and fractured. After thawing cells were embedded and sections cut perpendicular to the direction of fracturing. It can be seen that cell wall (CW) and the outer leaflet of the cytoplasmic membrane (CM) have been removed simultaneously. This indicates that the fracture plane does not run between cell wall and cell membrane. Bar: 50 nm. (From NANNINGA et al. 1984)

bacteria appear to be physically anchored together. In Gram-negative bacteria, connecting strands or fibrils have been detected as areas of contact between envelope structures, especially at sites where bacteriophages were attached to the outside of the cell, and where lipopolysaccharides were found to be synthesized (s. above).

When freeze-fracture techniques were introduced for electron microscopic sample preparation, further insight into the structure of membranes (Fig. 58) was obtained (BRANTON 1966, TSIEN & HIGGINS 1974). This was due to the fact that under these conditions the lipid bilayer of the cytoplasmic membrane (and other membranes) was split apart, exposing the internal membrane structure. It became evident that proteins could be associated with the membrane in two different ways, i. e. as "peripheral" or "extrinsic" proteins, and "integral" or "intrinsic" proteins (CARLEMALM & KELLENBERGER 1982, ROHDE et al. 1984). Peripheral proteins associated with a membrane are attached principally by ionic bonds involving cations such as Mg^{2+}, and they can be removed from the membrane by altering the ionic strength of the suspending buffer. In contrast, integral proteins cannot easily be removed. On the average, proteins account for 60% to 70% of the dry weight of membrane preparations. The other major compounds are lipids (ROGERS et al. 1980, WEIBULL et al. 1983; further references therein). There are no covalent links between the membrane components. The fluid mosaic model proposed by SINGER & NICHOLSON (1972) meets all of the characteristic features so far observed from a wide variety of investigation techniques. From the cytological viewpoint, cytoplasmic membranes of different origin and functions appear very similar and do not reflect the existing complexities of these structures (exception s. below: cytoplasmic membrane functioning as photosynthetic membrane as well). Even the striking biochemical difference between eubacteria and archaebacteria regarding the type of linkage of residues in the lipid components of membranes is not seen at the level of the electron microscope. In some extreme halophiles the fatty acids are replaced by ether-substituted alkyl residues (WOESE

1981). Lipids containing glycerol ether-linked residues rather than unbranched molecules with ester linkages (common in typical eubacteria) have been detected in *Thermoplasma acidophilum*. A detailed study of a variety of methanogens has confirmed this difference: their lipids were also found to be branched glycerol ethers.

Other components of bacterial cytoplasmic membranes are glycolipids, glycophospholipids, hydrocarbons, quinones and isoprenic alcohols (ROGERS et al. 1980; further references therein).

5.4.2 Functional aspects

In many bacteria, the cytoplasmic membrane is the only type of membrane available for the location of a wide variety of membrane-bound enzymes. This fact clearly indicates the multipurpose function of this membrane. A detailed description of the function of the cytoplasmic membrane in bacteria is outside the scope of this book. However, some specific features should not be neglected here. One major aspect of the bacterial cytoplasmic membrane is that of its bioenergetic functions such as ATP hydrolysis and synthesis, oxidoreductions, cell motility and solute transport. In this respect a membrane-bound electron microscopically detectable structure is the $Ca^{2+}-Mg^{2+}$ activated ATPase (MUÑOZ 1982). Negative staining has revealed its presence as arrays of particles attached to the cytoplasmic surface of the cytoplasmic membrane (Fig. 59). Further insight into the composition of this complex enzyme has been obtained by labelling of selected polypeptide chains within the isolated aggregate (Fig. 59) (LÜNSDORF et al. 1984). This was done by application of monoclonal antibodies specific against polypeptides α or β. The complexes were visualized by negative staining. A second example is the enzyme CO dehydrogenase. This enzyme

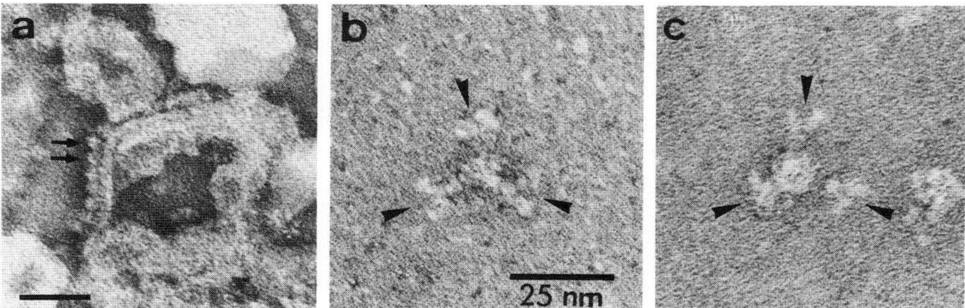

Fig. 59. a, bacterial ATPase (arrows) attached to the cytoplasmic membrane. Bar: 50 nm; b, c, localization of the α (b) and the β (c) subunits in *E. coli* ATPase by labelling of the respective subunits with monoclonal antibodies. Arrowheads point to antibodies. (a, original micrograph F. MAYER; b, c, from LÜNSDORF et al. 1984)

Fig. 60. a, the two faces of the purple membrane display different decoration behaviour. The membrane shown has folded over. It was decorated with 6 Å of silver and shadowed with 6 Å of Ta/W. Whereas the lattice periodicity is clearly revealed on one face (see also the corresponding optical diffraction pattern), the other face shows only a statistical distribution of silver particles. Bar: 50 nm. (From NEUGEBAUER & ZINGSHEIM 1978); b, the fine structure of purple membrane of *Halobacterium halobium*. Part of a three-dimensional map showing a region 5 nm thick spanning the membrane. The view shows protein molecules grouped around one of the threefold axes. The probable boundary of one of them is indicated by the broken line. The angle of view is such that the α helices furthest from the threefold axis partly overlap in the protein molecule outlined, but are seen almost end-on in the one on the right. Bar: 1 nm (From HENDERSON & UNWIN 1975) ▶

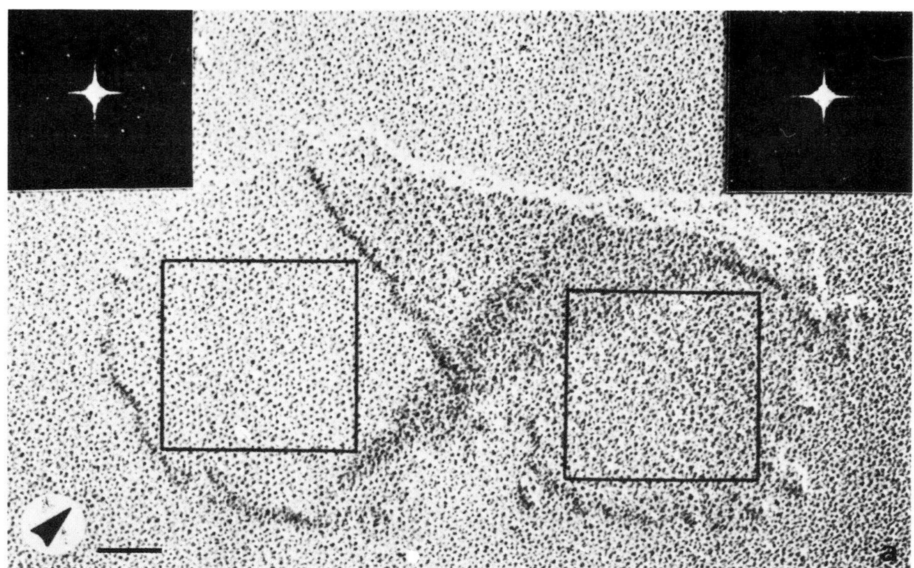

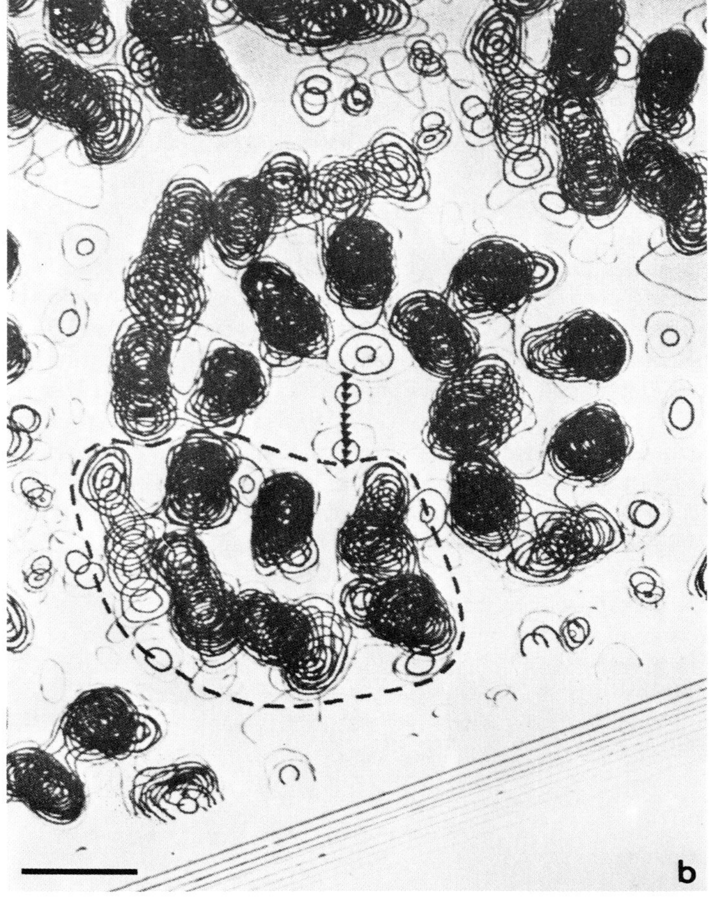

has been found to be loosely attached to the inner surface of the cytoplasmic membrane of *Pseudomonas carboxydovorans* cells in the exponential growth phase. In the stationary growth phase it is present both along the cytoplasmic membrane and in the cytoplasm. This enzyme is one of the best studied examples for dynamic interactions of enzymes with the cytoplasmic membrane (ROHDE et al. 1984).

In the atypical cyanobacterium *Gloeobacter violaceus*, the cytoplasmic membrane is the only membrane present in the organism (s. above), and it has the functions of a combined cytoplasmic and photosynthetic membrane (GUGLIELMI et al. 1981, RIPPKA et al. 1974).

Examples for the periodic arrangement of protein molecules within the cytoplasmic membrane are known. Usually these findings are only descriptive (Fig. 58) (MAYER et al. 1977) but this is not the case with the "purple membrane" (BOGOMOLNI et al. 1976, LANYI 1978, NIEDERMAN & GIBSON 1978, OESTERHELT & STOECKENIUS 1973, 1974, STOECKENIUS et al. 1979, USUKURA et al. 1980). The purple membrane is a specific patch-like differentiation of the cytoplasmic membrane found in *Halobacterium halobium* and several related strains. A structural model for the bacteriorhodopsin, a retinylidene protein contained in this membrane, has been proposed by HENDERSON & UNWIN (1975). The pigment, i. e. the retinal chromophore, enables the cell to utilize light energy for ATP synthesis, solute transport, and locomotion (BLAUROCK 1975, BOGOMOLNI et al. 1980, CASADIO & STOECKENIUS 1980, LOZIER 1982). In the first study of the structure of the purple membrane, BLAUROCK & STOECKENIUS (1971) examined freeze-fractured whole cells. They found that the cytoplasmic fracture face of the purple membrane patches displayed a textured appearance, with a strong indication of the hexagonal lattice which they also found to be present by X-ray diffraction of a suspension of isolated purple membranes. By contrast, the extracellular fracture face was smooth and seldom displayed the hexagonal lattice structure. More extensive X-ray analyses revealed that the purple membrane is a highly ordered crystal of space group P3, the thickness of the membrane corresponding to one unit cell and therefore to one molecule. The protein molecules were found to be oriented vectorially in the same direction across the membrane, and the two surfaces were crystallographically distinct (Fig. 60). The implication of this for the structure of the membrane is that the two surfaces must also be chemically distinct. Structural differences of the two faces of the purple membrane (BLAUROCK & KING 1977) were also found by metal decoration (NEUGEBAUER & ZINGSHEIM 1978), a view strongly supported by the results obtained by HENDERSON et al. (1978) who specifically labelled the protein and lipid on the extracellular surface of purple membrane. They applied a ferritin/avidin/biotin labelling procedure. Two biotin reagents of differing specificity were used, and the biotin was then visualized in the electron microscope using ferritin/avidin conjugate. For both of these reagents, the distribution of biotin on each side of the membrane was correlated with previous structural data obtained by electron diffraction and freeze-fracture electron microscopy. The following deductions were made: biotin labelling of the protein occurred only on the extracellular surface of the membrane; biotin labelling of lipid also occurred only on the extracellular surface of the membrane. The bottom of the model of purple membrane obtained by electron microscopy corresponds to the extracellular surface of the membrane. The extent of chemical asymmetry of both protein and lipid in the purple membrane is not a surprise, considering the vectorial nature of the proton pumping activity. Further studies have revealed that bacteriorhodopsin has a molecular weight of 26 000 D. Its complete amino acid sequence has been determined, and models for the three-dimensional arrangement of the amino acid chain forming the polypeptide have been proposed. STOECKENIUS & BOGOMOLNI (1982) used flash spectroscopy to investigate several bacterial isolates from salterns and natural brine pools. All of these were found to contain bacteriorhodopsin, together with a second type of very similar protein, halorhodopsin or

P_{588}, in much smaller amounts. A mutant strain lacking both pigments still showed the photosensory response of the wild type. It contained a third retinal pigment which undergoes a photoreaction cycle 1000-fold slower than the other pigments.

5.5 Intracytoplasmic membranes

Various bacteria produce morphologically and physiologically differentiated intracytoplasmic membranes. The most obvious examples are those of the photosynthetic bacteria (CLAYTON & SISTROM 1978; further references therein). Another type of conspicuous intracytoplasmic membrane observed in chemically fixed bacteria is the "mesosome" (NANNINGA 1968, 1975), originally called "peripheral body" or "chondrioid". Both types of membranes, together with other intracytoplasmic membranes of the "unit membrane" type, have in common their generation by a process of infolding from the cytoplasmic membrane.

5.5.1 The "mesosome"

A great variety of cellular functions which the bacterial cell must obviously be able to efficiently perform were attributed to "mesosomes". These structures have frequently been observed (GHOSH 1974, HIGHTON 1976, REMSEN 1968) in ultrathin sections of Gram-positive bacteria (Fig. 61), and occasionally in Gram-negative cells as well (Fig. 61). Mesosomes have been defined as "... vesicular, lamellar, or tubular packets of membrane enclosed by the invaginations of the plasma membrane", and "as seen in thin sections of whole or plasmolyzed cells, the mesosome or its derived vesicles have all of the features of unit membrane profiles". There is a wide variation in the overall appearance of mesosomes seen in thin sections of conventionally fixed bacteria from one species to another, and it has remained unknown what determines the "type" of mesosome seen in a particular bacterium. It has been discussed that "... a combination of factors including physiological growth conditions, the relative rates of wall and membrane synthesis, effects of temperature during growth, harvesting, and fixation of the specimens and perhaps the precise nature of the membrane protein and lipid components may govern the form of the mesosome in the cell". Several authors have, however, doubted the existence of mesosomes, proposing that these structures might instead be preparation artifacts (ACHTERRATH & LICKFELD 1972, BURDETT & ROGERS 1970, EBERSOLD et al. 1981, HIGGINS & DANEO-MOORE 1974, HIGGINS et al. 1976, LICKFELD & ACHTERRATH 1972, NANNINGA 1970, 1971). DUBOCHET et al. (1983) have studied the appearance of cellular structures of some typical Gram-positive and Gram-negative bacteria in frozen-hydrated sections prepared without any chemical fixation. They have confirmed that mesosomes do not exist in exponentially growing *Staphylococcus aureus* cells and that these structures are artifacts generated immediately after the onset of osmium fixation. Prior to these studies, GHOSH & NANNINGA (1976), by carefully investigating, by freeze-etching, the influence of chemical fixation, had already suggested that disorganization of protein-lipid associations by osmium tetroxide, in specific regions of the cytoplasmic membrane, induces the formation of mesosomal bodies. Membrane fractions obtained from cells in which mesosomal structures have been induced might, therefore, contain those regions of the cytoplasmic membrane which are especially susceptible to osmium-tetroxide-induced structural changes. This view assumes the existence of local chemical heterogeneities in the cytoplasmic membrane, and the question then arises which functional implications might be connected with them. In a report on bacterial anatomy in retrospect and prospect, NANNINGA et al. (1984) have discussed the various hypotheses concerning the origin of mesosomal structures. They suggested physical or chemical impairment as reasons for these artifacts (Fig. 62). A physical impairment of certain areas in the cytoplasmic membrane is thought to happen when the association of cytoplasmic membrane and cell wall is disrupted at a specific plasma membrane-cell wall site. This might occur during plasmolysis. NANNINGA (1975) had previously demonstrated that storage overnight of unfixed bacteria at about 6° to 9°C causes plasmolysis. Chemical impairment by OsO_4 fixation might lead to mesosomes in a situation where areas of the cytoplasmic membrane meet less physical resistance from the nucleoplasm (as compared to the cytoplasm), and parts of the cytoplasmic membrane might "stream" into these "weak" sites giving rise to mesosomal structures.

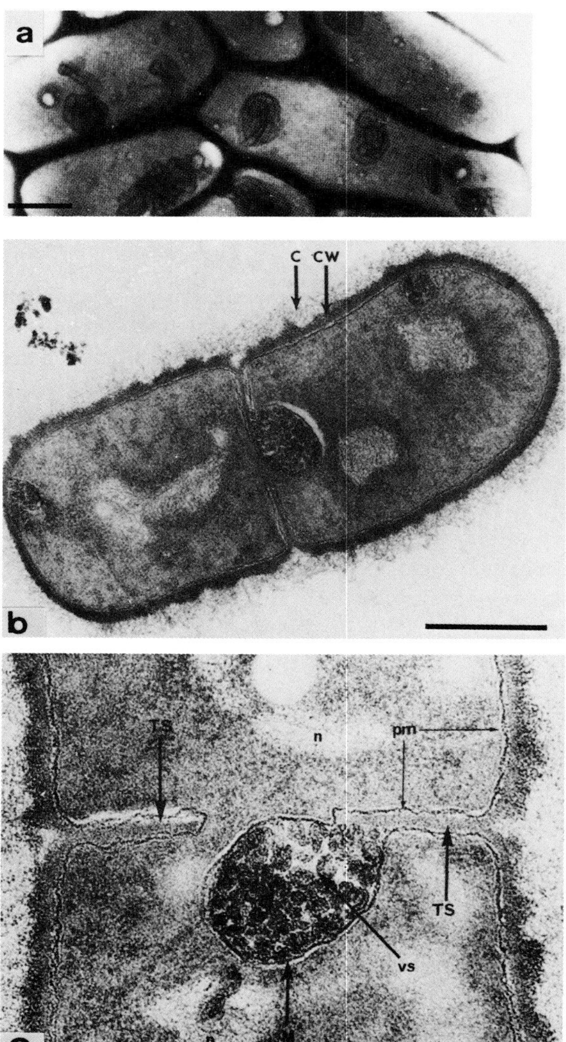

Fig. 61. Mesosome-like structures in bacteria. a, after negative staining; b, c, *Bacillus megaterium*. C = capsular material; CW = cell wall; M = mesosomes; n = nuclear material; pm = cytoplasmic membrane; TS = transverse septum; vs = mesosomal vesicles. Bars: 0.2 μm. (a, from MAYER 1978; b, c, from GHUYSEN & SHOCKMAN 1973)

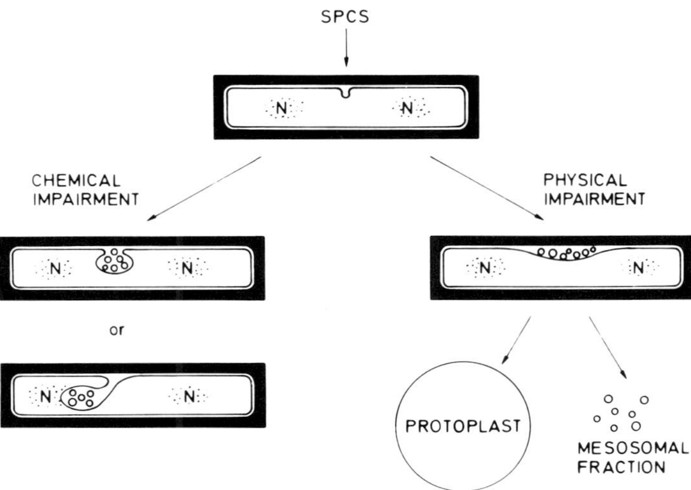

Fig. 62. Hypothetical scheme for the origin of mesosomal structures by chemical or physical impairment. – SPCS = specific plasma membrane – cell wall site; N = nucleoplasm. (From NANNINGA et al. 1984)

5.5.2 Intracytoplasmic membrane systems in photosynthetic bacteria

An evaluation of the literature available on the structure of the "chromatophores" in photosynthetic bacteria (Fig. 63) shows that these structures are membranous in all members of the Rhodospirillaceae, Chromatiaceae and Cyanophyceae (COHEN-BAZIRE & SISTROM 1966, DREWS 1978, KAPLAN & ARNTZEN 1982, MILLER 1979, 1982, NEUSHUL 1971, PFENNIG 1967, STAEHELIN et al. 1978a, STANIER et al. 1981, TRUEPER & PFENNIG 1981). The "chromatophores" of the green sulfur bacteria or Chlorobiaceae, however, are unique in that they are non-membranous; they are named chlorosomes. REMSEN (1978) has emphasized that, since the term "chromatophore" has been defined in different ways by different authors, it is no longer adequate to apply the term "photosynthetic membranes" for the various photosynthetic bacteria. Instead, the term "intracytoplasmic membrane systems" should be used for all photosynthetically active membrane structures. However, this view should be modified since VARGA & STAEHELIN (1983) have demonstrated that areas of the cytoplasmic membrane in *Rhodopseudomonas palustris* can be differentiated for photosynthesis as well.

Different groupings of photosynthetic bacteria have been proposed using the type and distribution of intracytoplasmic membrane systems (Fig. 63) and the mode of cell division, binary fission or budding, as discriminating factors (CLAYTON & SISTROM 1978). Accordingly, the major groups of photosynthetic bacteria are the Rhodospirillaceae or purple, non-sulfur bacteria, the Chromatiaceae or purple sulfur bacteria, and the Chlorobiaceae or green sulfur bacteria. These three groups, carrying out an anoxygenic photosynthesis using only one photosystem, are clearly separated from the cyanobacteria exhibiting oxygenic photosynthesis with two photosystems. However, that an organism possessing two photosystems may, under suitable conditions, grow and carry out photosynthetic CO_2 assimilation with only one photosystem has been demonstrated for *Oscillatoria limnetica*. In this bacterium, the water-cleaving photosystem II is inhibited in the

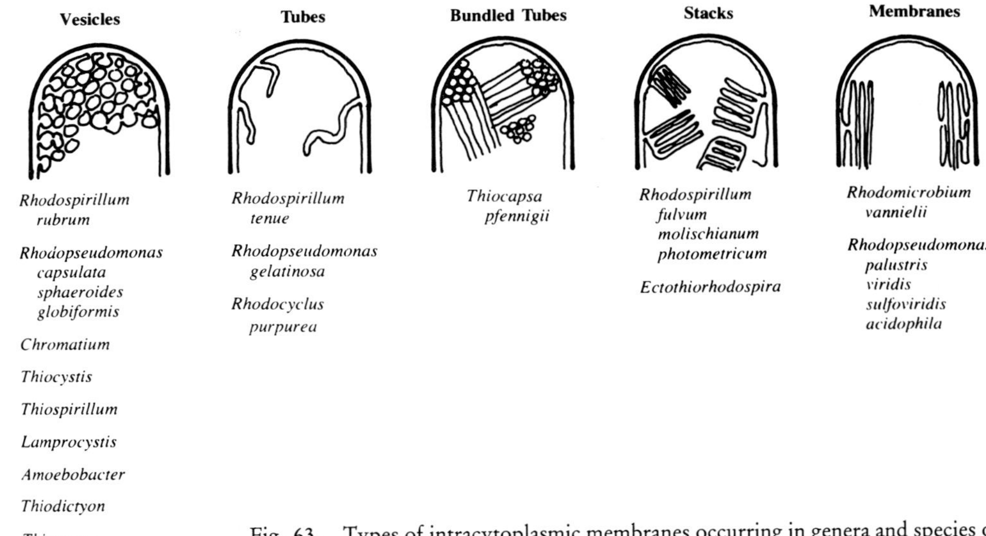

Fig. 63. Types of intracytoplasmic membranes occurring in genera and species of the Rhodospirillaceae and Chromatiaceae. (From TRUEPER & PFENNIG 1981)

presence of sulfide, which at the same time serves as electron donor for photosynthesis with photosystem I alone.

The photosynthetic membranes of the bacteriochlorophyll b-containing bacteria *Ectothiorhodospira halochloris* (REMSEN et al. 1968), *E. abdelmalekii, Rhodopseudomonas viridis, Rh. sulfoviridis* and *Thiocapsa pfennigii* have been investigated by ENGELHARDT et al. (1983) using electron microscopy and digital image analysis (Figs. 64 and 65). All species analyzed were found to have the photosynthetic complexes hexagonally arrayed in the membrane with lattice spacings around 13 nm. The basic organization of the complexes, i.e. a ring, probably containing the light harvesting polypeptides, surrounding a core (the reaction centre) appeared almost identical for all species investigated. MILLER (1982) has also described the photosynthetic membrane complex of *Rhodopseudomonas viridis;* he has shown a clear division into six subunits. Further investigations on the supramolecular organization of bacterial photosynthetic membranes have been performed (ENGELHARDT et al. 1984a, GOLECKI et al. 1979, KYLE 1983, REED & RAVEED 1968, SHIOZAWA et al. 1981, 1982, STAEHELIN 1976, STAEHELIN et al. 1977, VALKIRS & FEHER 1982), in addition to the development of these structures (GIESBRECHT & DREWS 1962, KAPLAN 1981, NIEDERMAN 1981, TAUSCHEL & DREWS 1967).

The structure of some phototrophic bacteria, i.e. the complexity and organization of their intracytoplasmic membrane systems, has been observed to vary, depending on growth conditions (FIRSOW & DREWS 1977, GOLECKI & OELZE 1980, GOLECKI et al. 1980, KAPLAN 1978, KAUFMANN et al. 1982, SCHUMACHER & DREWS 1979, Wakim et al. 1979). *Rhodopseudomonas capsulata* and other members of Rhodospirillaceae are facultative phototrophic bacteria, growing either aerobically in the dark, producing ATP by oxidative phosphorylation (chemotrophic) or anaerobically in the light, forming ATP by cyclic photophosphorylation (phototrophic). Lowering of oxygen partial pressure in the dark, or variation of incident light intensity under phototrophic conditions induce a differentiation of the membrane system: after transition from chemotrophic to photo-

Fig. 64. The structure of photosynthetic membranes of *Ectothiorhodospira halochloris.* a, cytoplasmic, b, periplasmic face of the photosynthetic membrane; metal shadowing; bar: 0.1 μm. c, d, averages of the two faces; lattice spacing close to 13 nm; e, f, Z-scaled three-dimensional representation of the six-fold symmetrized reconstructions. (From ENGELHARDT et al. 1984) ▶

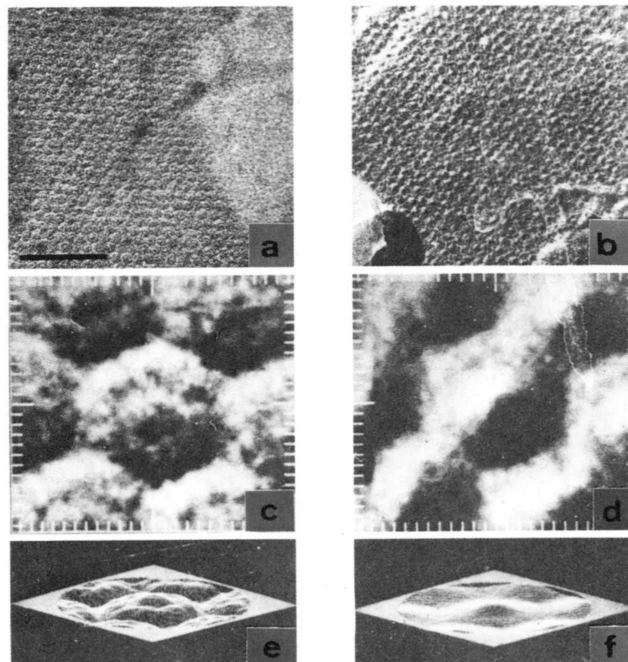

Fig. 65. a, negatively stained photosynthetic membrane of *Ectothiorhodospira halochloris;* bar: 0.2 μm; b, six-fold symmetrized photosynthetic unit as revealed after image processing by correlation averaging; lattice spacing 13 nm; c, dark-field STEM image of unstained membranes; bar: 0.1 μm; d, histogram of mass per area (kDa/nm²) of single and stacked membranes. The mass of the photosynthetic unit is calculated to be around 450 kDa. (From ENGELHARDT et al. 1984b) ▼

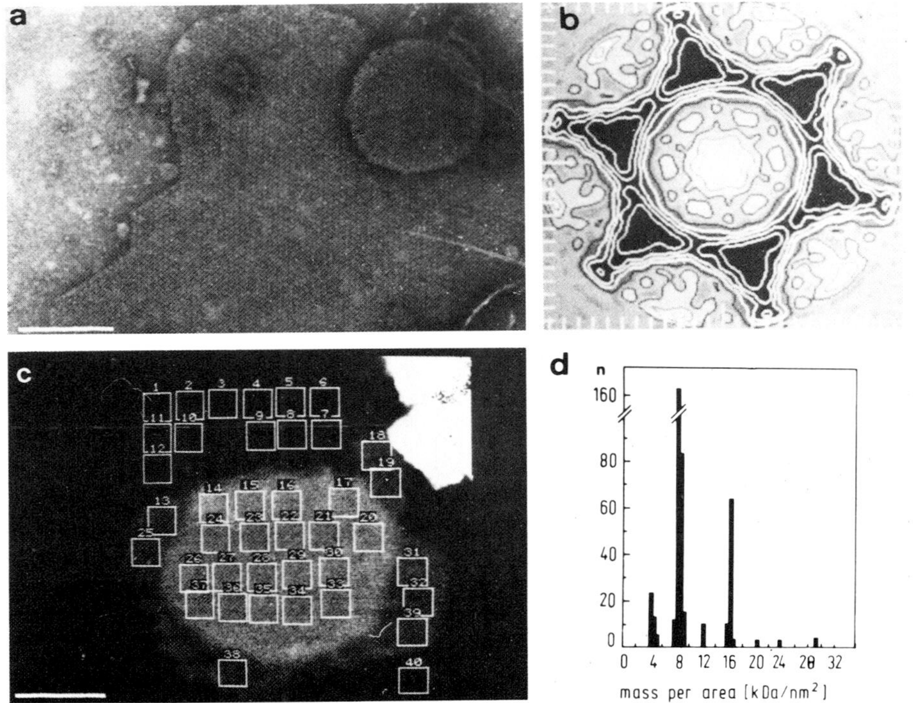

trophic growth conditions new photosynthetic units are synthesized, and intracytoplasmic membrane vesicles are formed by invagination from the cytoplasmic membrane. A decrease of light intensity in a phototrophic culture induces a rise in the amount of intracytoplasmic membranes. Clearcut dynamic changes in membrane composition have also been found at the molecular level. Although cytoplasmic and intracytoplasmic membranes form a continuous membrane system, the constituents of the photosynthetic apparatus and of the respiratory chain are inserted at specific sites and with different kinetics into the membrane. Transition from chemotrophic to phototrophic growth induces the synthesis of photochemically active reaction centers and light-harvesting bacteriochlorophyll-carotenoid-protein complex I. The light harvesting pigment complex II is synthesized afterwards with a different rate in cells of *Rhodopseudomonas capsulata, Rh. sphaeroides* and *Rh. palustris.* The stepwise formation of the pigment complexes proceeds as a simultaneous incorporation of pigments and of the respective proteins into the cytoplasmic membrane. Varga & Staehelin (1983) measured changes in intracytoplasmic membrane particle size distribution in *Rh. palustris,* in response to changes in light intensity during growth (Fig. 66). They found that, as light levels were decreased from 8500 to 100 lx, the average particle diameter in the protoplasmic face of intracytoplasmic as well as cytoplasmic membranes increased from 8.6 nm to 10.3 nm. They also observed a distinct periodicity in the sizes of the intramembrane particles found in the stacked regions with the larger size peaks becoming more pronounced as light intensity decreased. They suggested that, as light levels decrease, subunits of discrete size are added to a core particle. Fig. 66 presents a diagram of the proposed organization of the reaction center and the light harvesting systems I and II in *Rh. palustris.* As indicated, the size of the complex is assumed to vary with the amount of bound light harvesting system II.

In many cases it has been shown that photosynthetic membranes arise by invagination of the cytoplasmic membrane; they may be pinched off, forming detached vesicles, or they may remain connected to the cytoplasmic membrane, forming tubular invaginations or folded stacks of membranes (s. above) (Fig. 67–75). Remsen (1978) has reviewed in some detail all pertinent variations

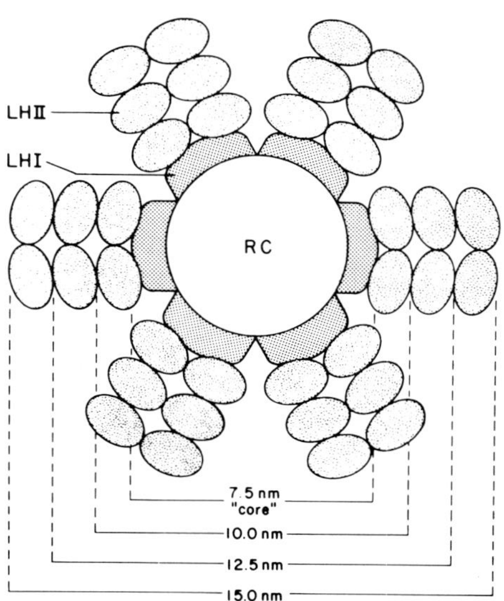

Fig. 66. Photosynthetic membranes in *Rhodopseudomonas palustris.* Diagram of proposed organization of reaction center (RC), light harvesting component I (LH I), and light harvesting component II (LH II). The size of the complex varies with the amount of bound LH II, giving rise to particles on freeze-fractured membranes that fall into distinct size classes as indicated. (From Varga & Staehelin 1983)

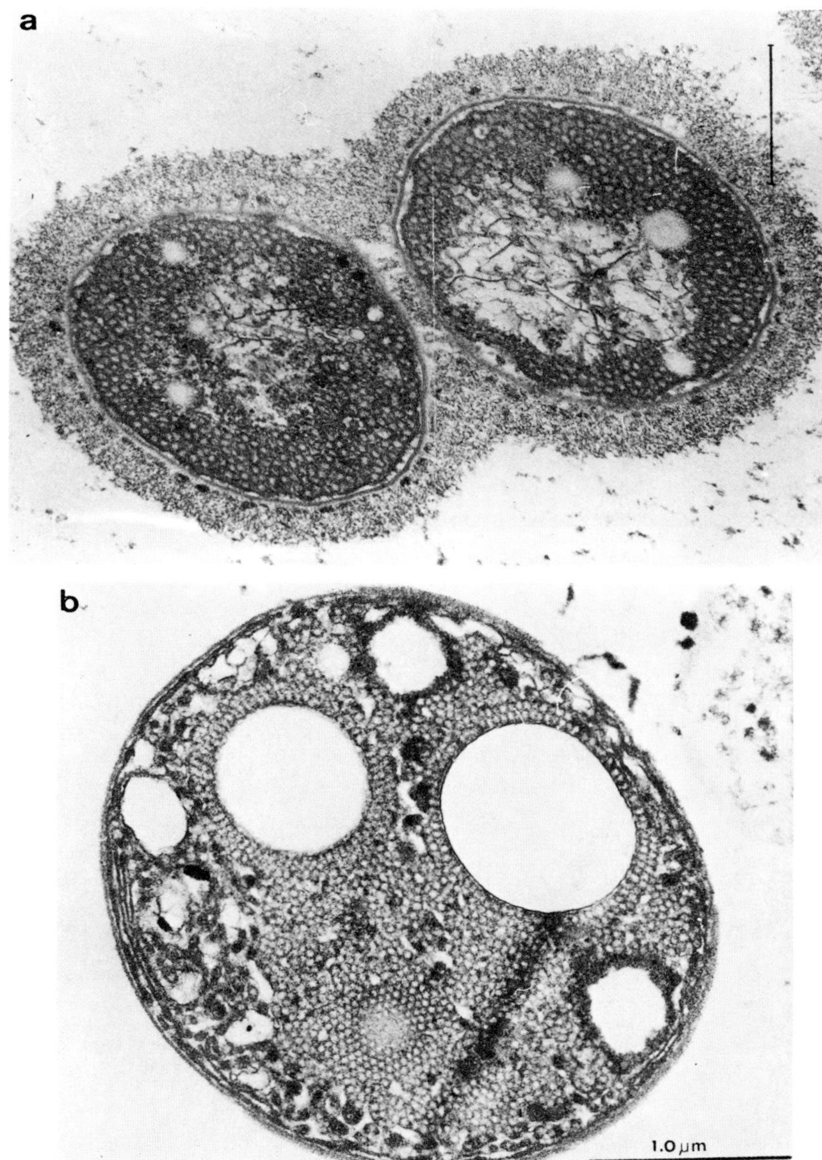

Fig. 67. a, *Thiocapsa rosea;* dense capsular material and dense packing of chromatophores. In the center of the cells are collapsed gas vacuole membranes. Bar: 1 µm. (From REMSEN 1978, after HIRSCH); b, *Thiocystis violacea* showing packed chromatophores, sulfur granules, and lamellar to tubular membranes characteristic of old cells of this species. (From REMSEN 1978)

and summarized the situation as follows: "Both vesicles and lamellar stacks of photosynthetic membranes occur in members of the Rhodospirillaceae. Those that have vesicles include *Rs.*

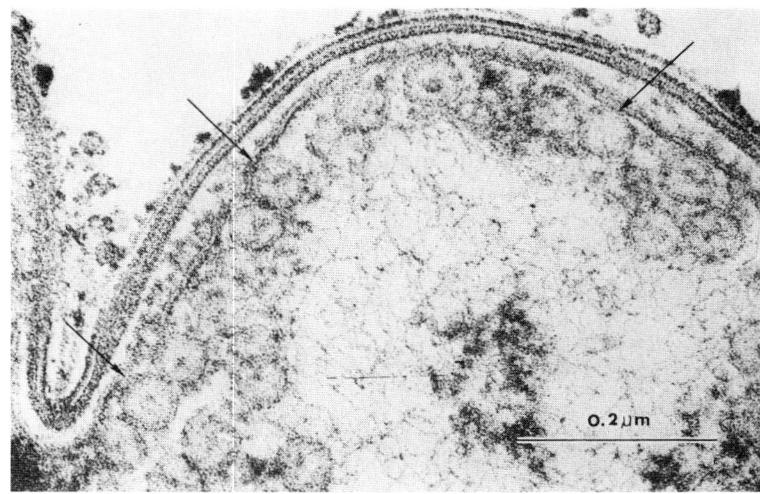

Fig. 68. *Thiocystis violacea* showing association of chromatophore vesicles with the cytoplasmic membrane (arrows). (From REMSEN 1978)

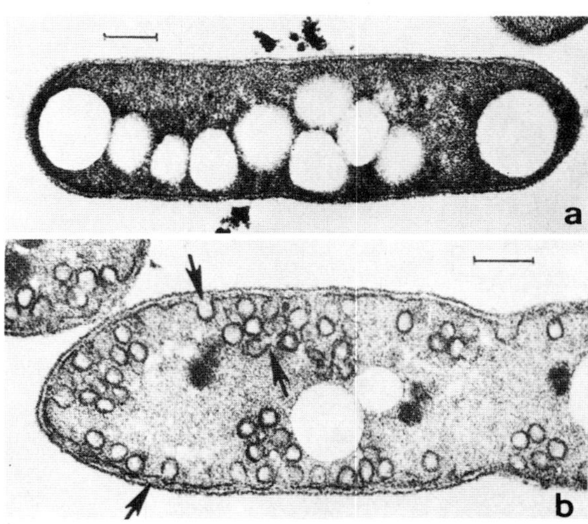

Fig. 69. a, section of *Rhodopseudomonas sphaeroides* after semi-aerobic culture in the dark. Cell has few distinguishable chromatophores, but is packed with poly-β-hydroxybutyrate granules; b as a, but cells grown anaerobically at moderate light intensity. Arrows point to membrane invaginations. Bars: 0.2 µm. (From REMSEN 1978, after PETERS & CELLARIUS)

rubrum and *Rhodopseudomonas sphaeroides;* most of the other species in this family have lamellar membranes. There are two modes of cell division in this group of bacteria: binary fission and budding. The brown *Rhodospirillum* species *Rs. fulvum, Rs. molischianum,* and *Rs. photometricum* reproduce by binary fission and have cytomembrane stacks randomly arranged near the periphery of the cell. In contrast, the organisms that reproduce by budding, such as *Rhodomicrobium vannielii, Rh. palustris,* and *Rh. viridis,* have membranes at only one end of the cell. During budding, the cytomembranes remain with the mother cell, and the daughter cell forms new ones by an invagination of the cytoplasmic membrane. *Thiospirillum, Thiocystis,* and *Thiocapsa* have vesicular membranes; only a single *Thiocapsa* species has tubular membranes. Both vesicular and

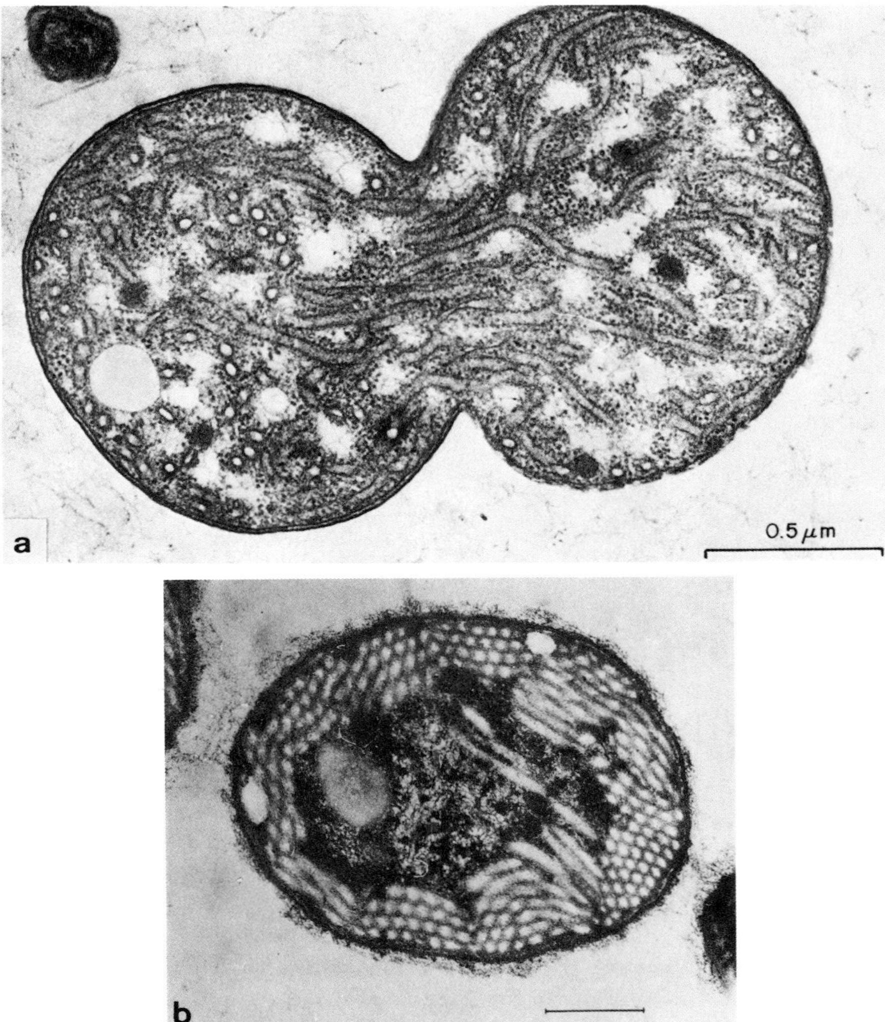

Fig. 70. Bacteria with intracytoplasmic membranes. a, *Nitrococcus mobilis* showing the arrangement of tubular membranes throughout the cytoplasm. (From WATSON et al. 1981, after WATSON & MANDEL); b, *Thiocapsa pfennigii* showing the stacking arrangement of tubular membranes around the periphery of the cell. Bar: 0.5 µm. (From REMSEN 1978, after EIMHJELLEN et al.)

lamellar membranes have been seen in some strains of *Thiocapsa* and *Chromatium*. The only Chromatiaceae that have stacks of membranes belong to the genus *Ectothiorhodospira*. All species in this genus deposit sulfur extracellularly rather than intracellularly, an activity shared by members of the Chlorobiaceae. The bacteria in this latter group, however, lack a true lamellar system, but instead have the unique "chlorosomes" located adjacent and parallel to the plasma membrane" (s. below).

CW
PM
R
N
LS

Fig. 71

CM
PFs
ICM
PFu
ICM
EFs
ICM
PFs
CM
PFu
intracytoplasmic
membranes
(ICM)
cytoplasmic
membrane
(CM)
CM
EF
cell wall
stacked membrane regions
unstacked ICM regions
unstacked CM regions

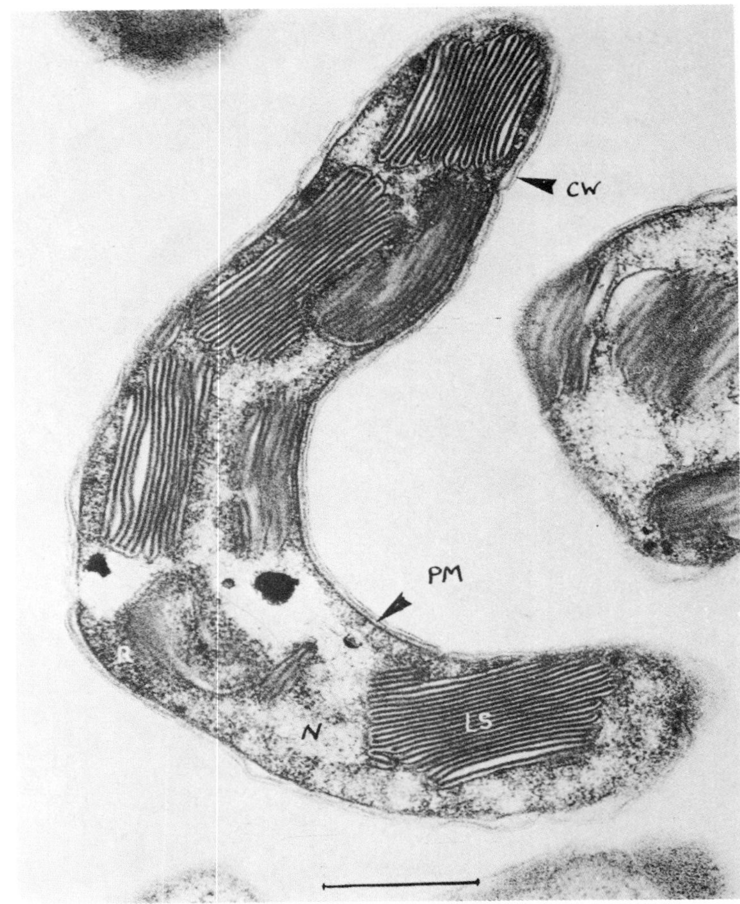

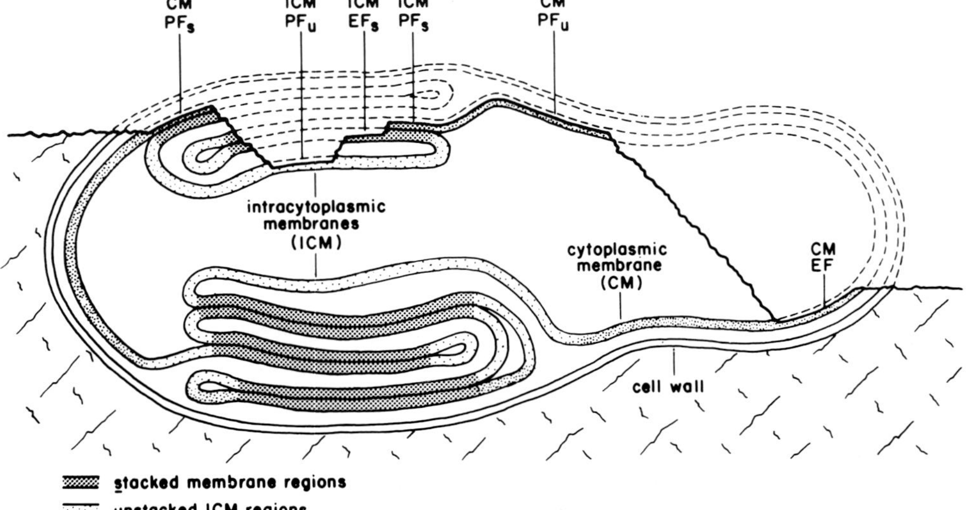

Fig. 73

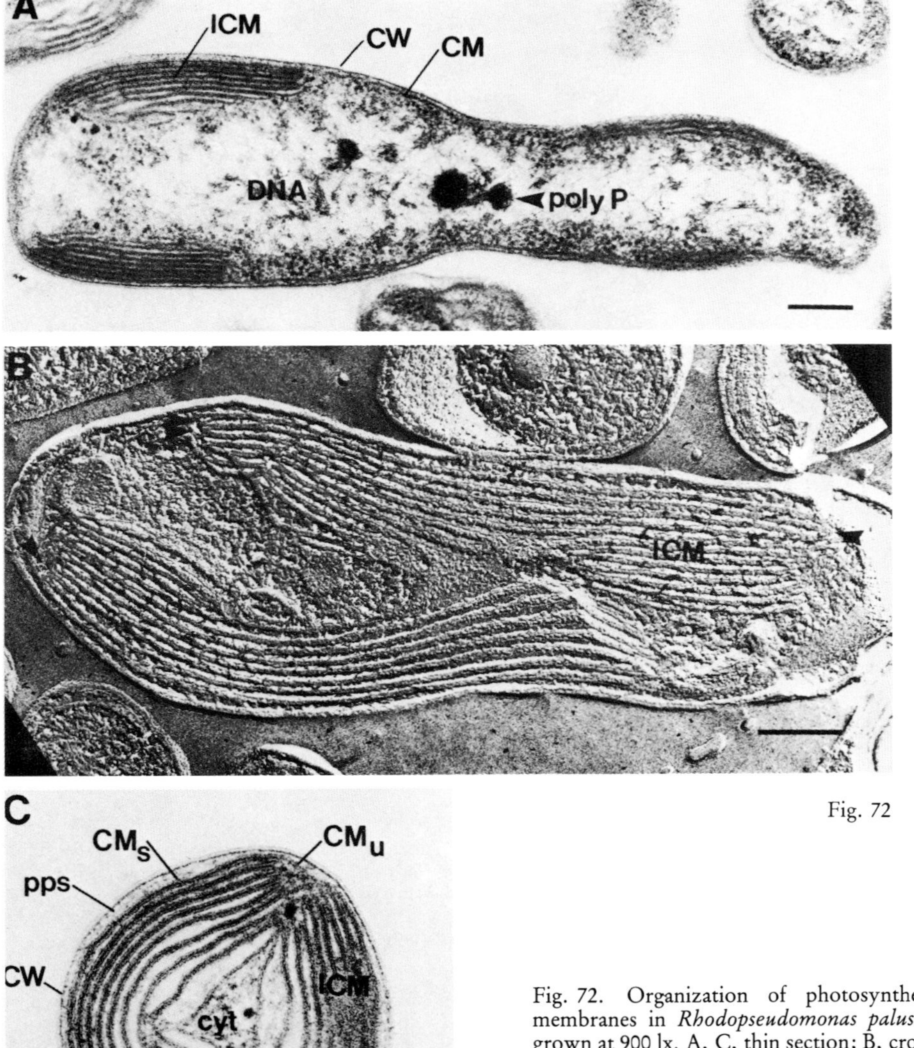

Fig. 72

Fig. 72. Organization of photosynthetic membranes in *Rhodopseudomonas palustris* grown at 900 lx. A, C, thin section; B, cross-fracture through a jet-frozen cell. Bars: 0.2 µm. CM = cytoplasmic membrane; CM_s = stacked membranes; CM_u = unstacked membranes; CW = cell wall; cyt = cytoplasm; ICM = intracytoplasmic membranes; poly P = polyphosphate granule; pps = periplasmic space. (From VARGA & STAEHELIN 1983)

Fig. 71. *Ectothiorhodospira mobilis* showing the general fine structure of the cell. – CW = cell wall; LS = lamellar stacks; N = nucleoplasm; PM = cytoplasmic membrane; R = ribosomes. Bar: 0.5 µm. (From REMSEN 1978, after REMSEN et al.)

Fig. 73. Photosynthetic membranes in *Rhodopseudomonas palustris*. Diagram illustrating the nomenclature for the freeze-fractured membranes of *R. palustris*. – CM = cytoplasmic membrane; ICM = intracytoplasmic membrane; EF = E-face; PF = P-face; s = stacked; u = unstacked. (From VARGA & STAEHELIN 1983)

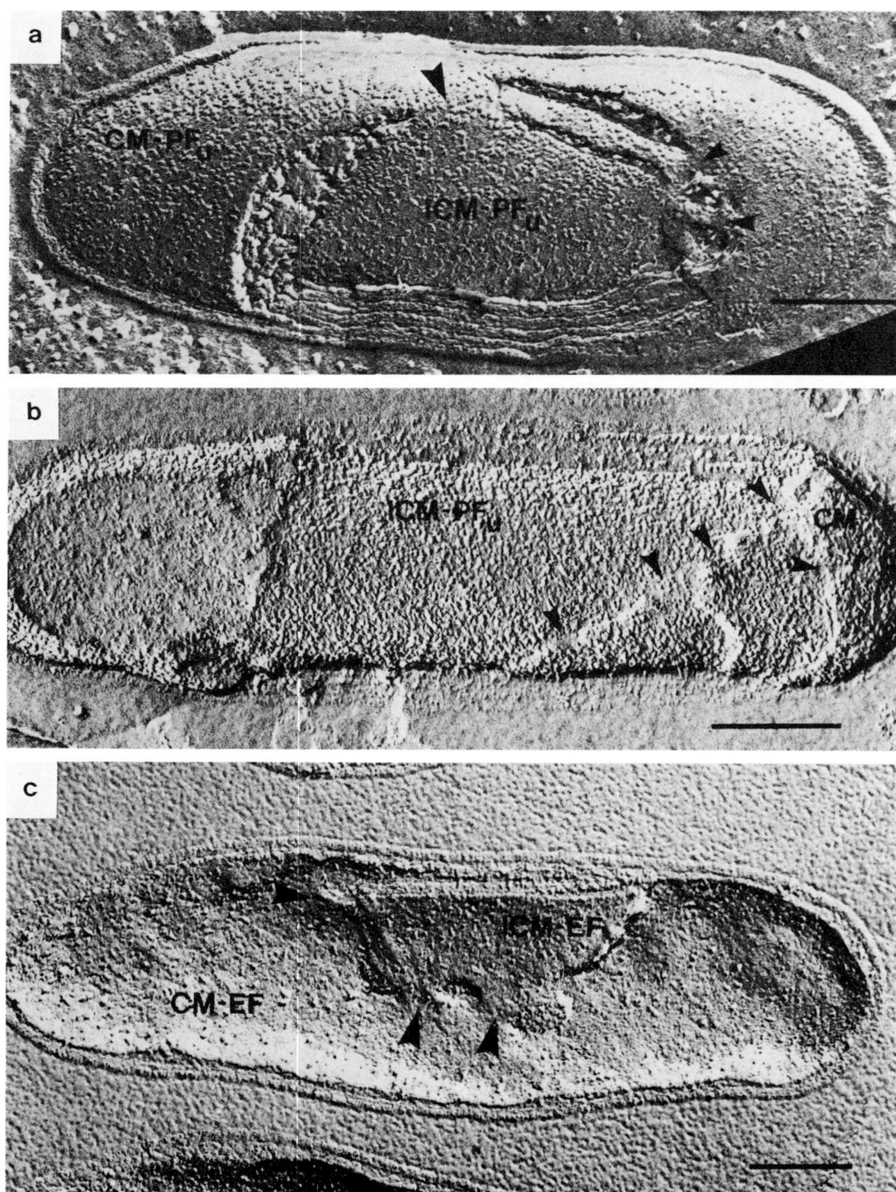

Fig. 74. Photosynthetic membranes in *Rhodopseudomonas palustris.* a, view illustrating the physical continuity (arrowheads) of cytoplasmic membrane and intracytoplasmic membranes; b, jagged appearance of the particles in comparison with those in a, due to preparation artifacts; membrane connections are shown by arrowheads; c, the multiple, flattened tubular connections between cytoplasmic membrane and the bottom intracytoplasmic membrane are clearly resolved (arrowheads). Abbreviations as in Fig. 73. Bars: 0.2 μm. (From VARGA & STAEHELIN 1983)

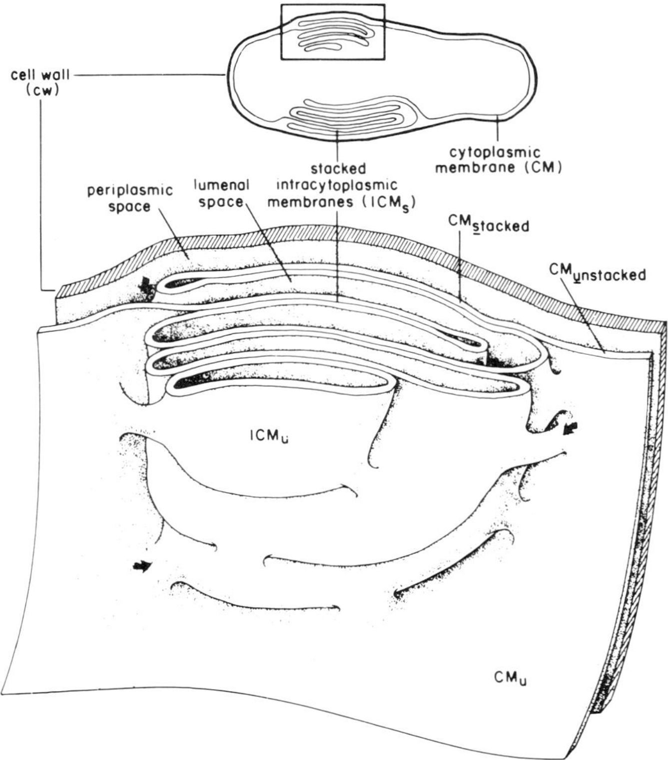

Fig. 75. Photosynthetic membranes in *Rhodopseudomonas palustris.* Drawing showing the spatial relationship between the cell membrane (CM), and intracytoplasmic membranes (ICM). Tubular connections between the CM and ICMs at different levels of the stack, als well as between adjacent ICMs, are shown. (From VARGA & STAEHELIN 1983)

Cyanobacteria are a group of photosynthetic bacteria (Fig. 76) with a degree of internal organization that is more complicated than that in most other bacteria (BUTLER & ALLSOP 1972, GANTT & CONTI 1969, GLAZER 1977, LANG 1968, LEFORT-TRAN et al. 1973, NIERZWICKI et al. 1982, RIPPKA et al. 1974, 1979, STANIER & COHEN-BAZIRE 1977, STANIER et al. 1971, WALSBY 1981, WOLK 1973). Their photosynthetic apparatus is contained in bodies known as thylakoids which consist of flattened membranous sacs and which are the sites of chlorophyll and carotenoids. The major photopigments, the phycobiliproteins, however, are contained in organelles termed phycobilisomes (s. below), and are attached in regular arrays on the surface of the thylakoids. The synthesis of the components of the photosynthetic apparatus, and their morphogenesis are not yet well unterstood, although studies have been undertaken to describe the major developmental patterns of several cyanobacteria in more general terms (GIDDINGS & STAEHELIN 1979, GOLECKI 1979, HOARE et al. 1971, JENSEN & SICKO 1973).

KUNKEL (1982) has reported previously undescribed "thylakoid centers" from which photosynthetic membranes (thylakoids) radiate in several cyanobacteria exhibiting an otherwise conventional structural organization with regard to the architecture of cytoplasmic and intracytoplasmic

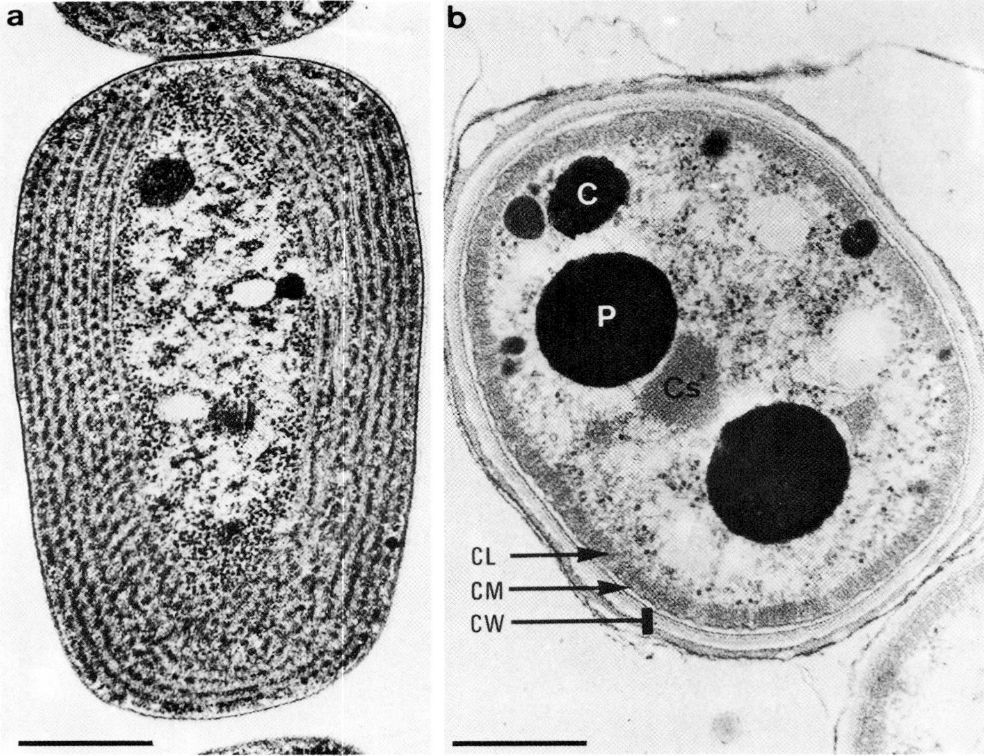

Fig. 76. a, filamentous cyanobacterium of the *Lyngbya-Plectonema-Phormidium* group, *PCC 7408*, with a cortical parallel array of thylakoids bearing phycobilisomes. Bar: 0.5 μm; b, a cyanobacterium without thylakoids, *Gloeobacter violaceus.* C = cyanophycin granules; CL= cortical layer of phycobiliproteins; CM = cytoplasmic membrane; Cs = carboxysome; CW = cell wall; P = polyphosphate granule. Bar: 0.5 μm. (From Stanier et al. 1981).

membranes. These structures were found to be cylinders 30 nm wide by 320 nm long, consisting of globular subunits oriented in non-parallel stacked arrays. Thylakoids were attached to the outer surface of the cylinder along its longitudinal axis. Kunkel discussed his observations with respect to the situation that the association between thylakoids and the cytoplasmic membrane is less frequently reported in cyanobacteria than in other bacterial cytoplasmic-intracytoplasmic membrane systems. He mentioned that observations of apparent thylakoid-cytoplasmic membrane continuity might be a result of inadequate preservation. Freeze-etching data have actually revealed that intramembrane particles of thylakoids and the cytoplasmic membrane vary in size and number, indicating that these two membranes are structurally distinct, in contrast to the situation observed in purple sulfur and non-sulfur bacteria. Kunkel (1982) has therefore concluded that thylakoid centers are functional entities due to their structure, location and thylakoid attachment, and he felt that his view is supported by some investigators who have suggested that the cyanobacterial photosynthetic membrane system is structurally and functionally distinct from the cytoplasmic membrane. Kunkel's interpretations should, however, be verified by studies on the development

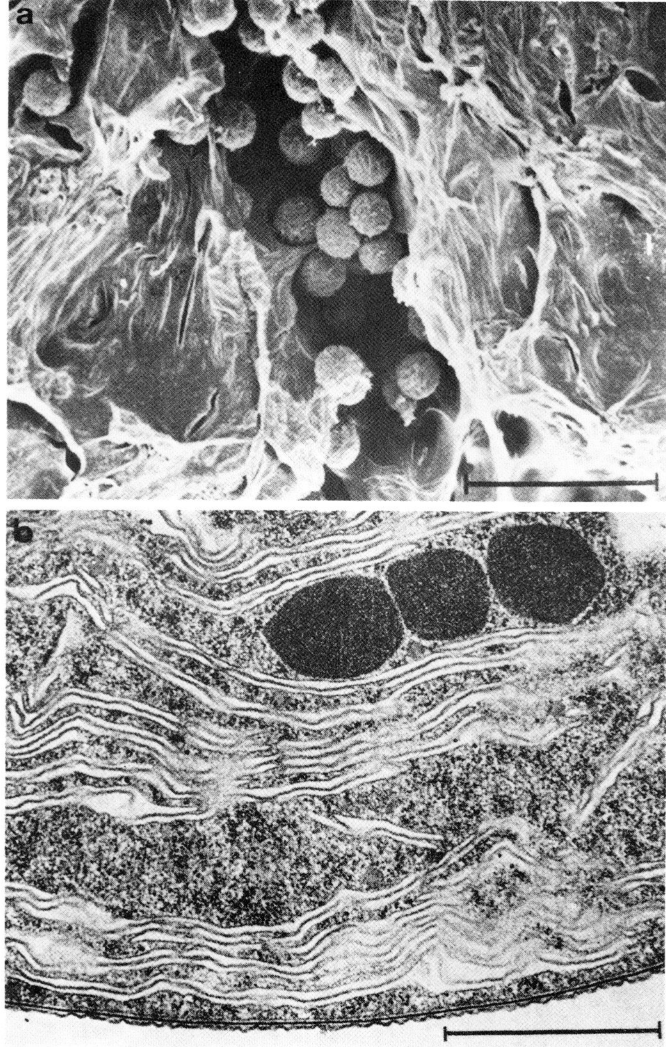

Fig. 77. a, *Prochloron* cells in oral groove on surface of *Didemnum carneolentum*. Bar: 50 μm. (From LEWIN 1981, after CHENG); b, portion of *Prochloron;* multilayered wall, paired thylakoids and polyhedral bodies. Bar: 1 μm. (From LEWIN 1981, after LEE)

of the thylakoid centers in order to provide further evidence on their possible functional significance.

A division established in 1977, the Prochlorophyta, with the genus *Prochloron* (Fig. 77), has in the meantime generated considerable interest (LEWIN 1981, STAEHELIN 1981). The cells of these bacteria contain chlorophylls *a* and *b*, and they lack accessory phycobilin pigments. A growing body of evidence indicates that there are close biochemical similarities, and presumably phylogenetic relationships, between *Prochloron* and the cyanobacteria, with respect to peptidoglycan, carotenoids, polar lipid composition, sterols, carboxysomes, and ribosomal RNAs. In addition to the fact that cyanobacteria do not contain chlorophyll *b*, other features distinguish *Prochloron* from cyanobacteria. Examples are the location of carboxysomes and DNA within the cell, and paired or stacked thylakoids. Phycobilisomes are missing in *Prochloron.*

The size distributions of photosynthetic particles on the inner thylakoid faces of *Prochloron* are much closer to those of chloroplasts of eukaryotic green algae or higher plants than to those of cyanobacteria. *Prochloron* does not contain gas vacuoles, nor cyanophycin or PHB. These similarities and differences have caused discussions on phylogenetic relationships between Prochlorophyta, cyanobacteria, green algae and plastids in higher plants as well.

Another exception from the usual physiological and cytological types of photosynthetic bacteria has been described by GEST & FAVINGER (1983). They named the organism *Heliobacterium chlorum* ("green organism"). It is a brownish-green anoxygenic photosynthetic bacterium, isolated from surface soil. It contains a specific form of bacteriochlorophyll ($BChlg_{Gg}$). The subcellular localization of the photosynthetic pigment is uncertain as neither chlorosomes nor extensively developed intracytoplasmic membranes of the kind produced by most photosynthetic bacteria were observed in ultrathin sections. Due to a unique constellation of properties shown by this organism, the authors suggest that this bacterium may represent a hitherto unknown genus.

5.5.3 Intracytoplasmic membrane systems in non-photosynthetic bacteria

Besides photosynthetic bacteria there are other groups which exhibit extensive intracytoplasmic membrane systems (Figs. 78–82) such as dinitrogen fixers, methane utilizing and methanogenic bacteria (s. below) (BALCH et al. 1979, MURRY & WATSON 1965, PANGBORN et al. 1962, VAN GOOL et al. 1969, WATSON et al. 1981, WATSON & WATERBURY 1971). As has been shown for intracytoplasmic membrane systems in photosynthetic bacteria, these membranes are also assumed to be formed by infolding of the cytoplasmic membrane (BOCK 1971, MARCUS & KANESHIRO 1972, OPPENHEIM & MARCUS 1970, PATE et al. 1973, POPE 1969). REMSEN et al. (1967) have studied *Nitrosococcus oceanus* (Fig. 81) by thin sectioning and freeze-etching and found well developed stacks of membranous sack-like structures. A similar situation was observed in *Nitrobacter* and *Nitrosomonas* (Fig. 78). The presence of intracytoplasmic membrane invaginations in cells of *Azotobacter* has also been reported by several authors. POST et al. (1982) have investigated morphological and ultrastructural variations in *Azotobacter vinelandii* growing in oxygen-controlled continuous culture. They have followed membrane development in response to growth at different oxygen tensions. The organisms were grown in a carbon-limited chemostat with either atmospheric nitrogen or ammonium as nitrogen sources. In both types of cultures not only the number of intracytoplasmic membrane vesicles per cell increased but the cell volume also enlarged with aeration. Whereas in nitrogen-fixing cells the proportion of intracytoplasmic membrane surface area per cytoplasmic membrane surface area increased from 1:2 to 3:1 the ratio stayed almost constant in ammonium-assimilating cells. The authors concluded that only under conditions of nitrogen fixation oxygen can mediate changes in the ratio of intracytoplasmic to cytoplasmic membrane surface areas. POST et al. (1983) have extended the analysis of the *Azotobacter* system. In spite of the fact that the ratio of intracytoplasmic to cytoplasmic membrane surface area is increased by a factor of six when the cells are grown in an oxygen-controlled chemostat from 1% to 100% air saturation, cells grown at 1%, 45%, and 80% air saturation exhibited only a single membrane fraction after equilibrium sucrose density gradient centrifugation. Total membrane

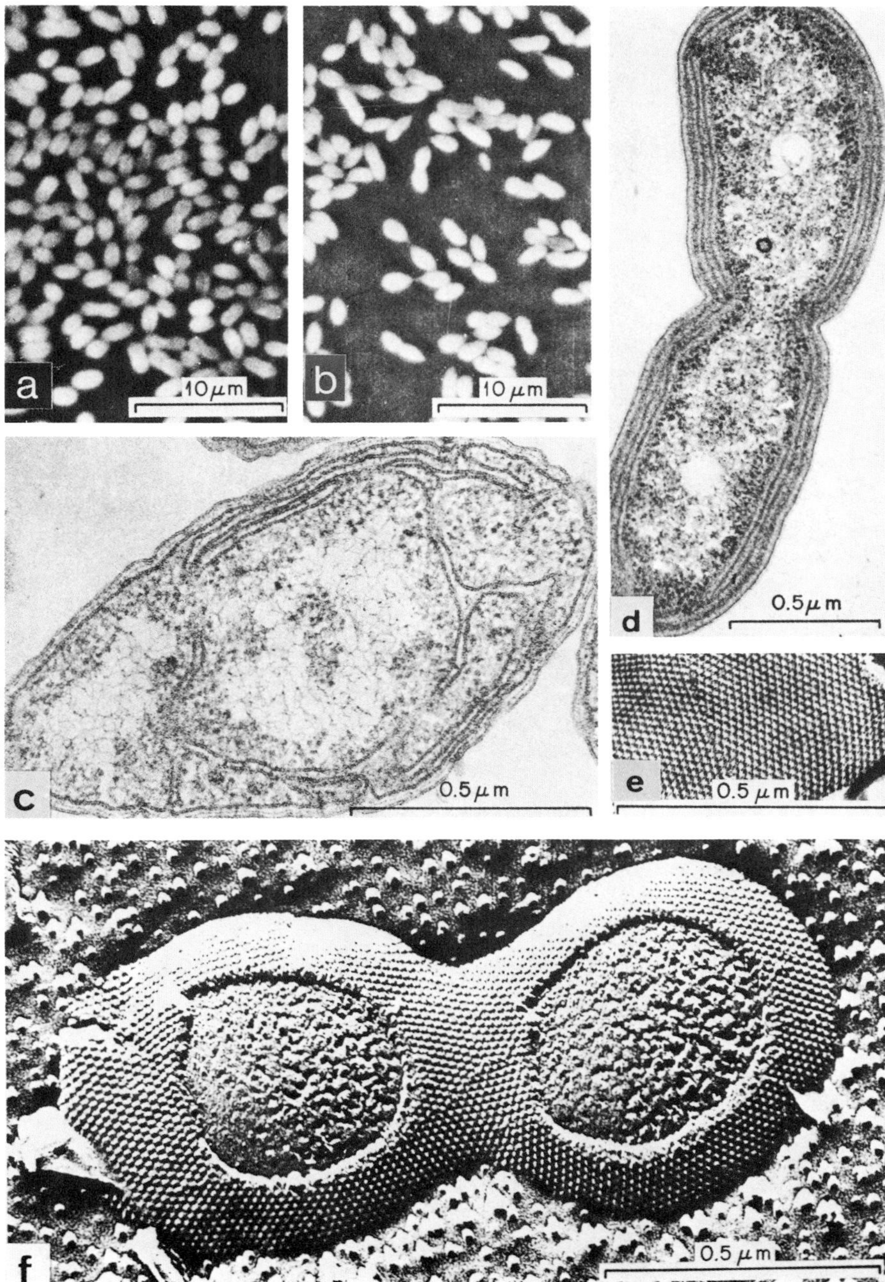

Fig. 78. Bacteria with intracytoplasmic membrane systems; cells of *Nitrosomonas*. a, c, *N. europaea;* terrestrial strain; b, *N. europaea;* sewage strain; d, e, marine *N. europaea* (after WATSON & MANDEL); f, marine *N. europaea* (after WATSON & REMSEN). (From WATSON et al. 1981)

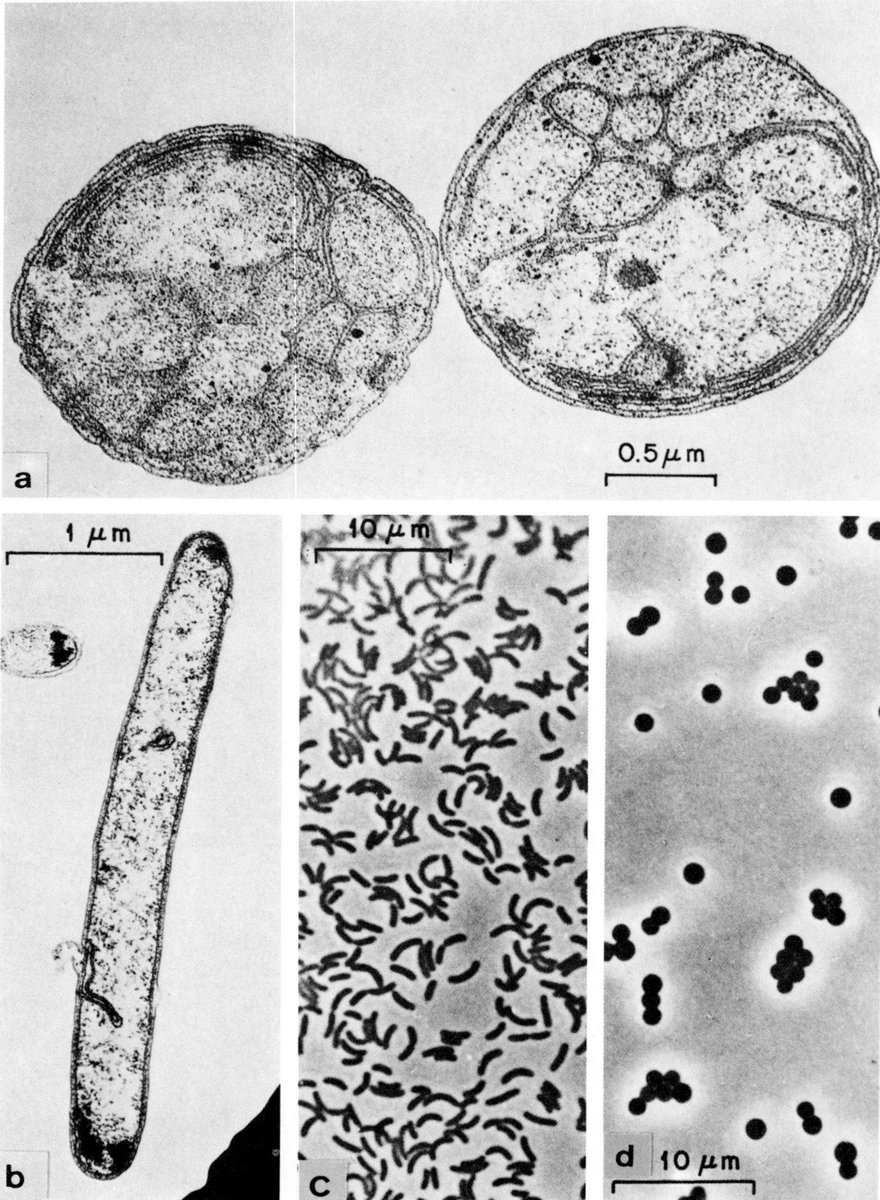

Fig. 79. Bacteria with intracytoplasmic membranes. a, *Nitrosococcus mobilis* (after Koops et al.); b, *Nitrosovibrio tenuis* (after Harms et al.); c, *N. tenuis* (after Harms et al.); d, *N. mobilis* (after Koops et al.). (From Watson et al. 1981)

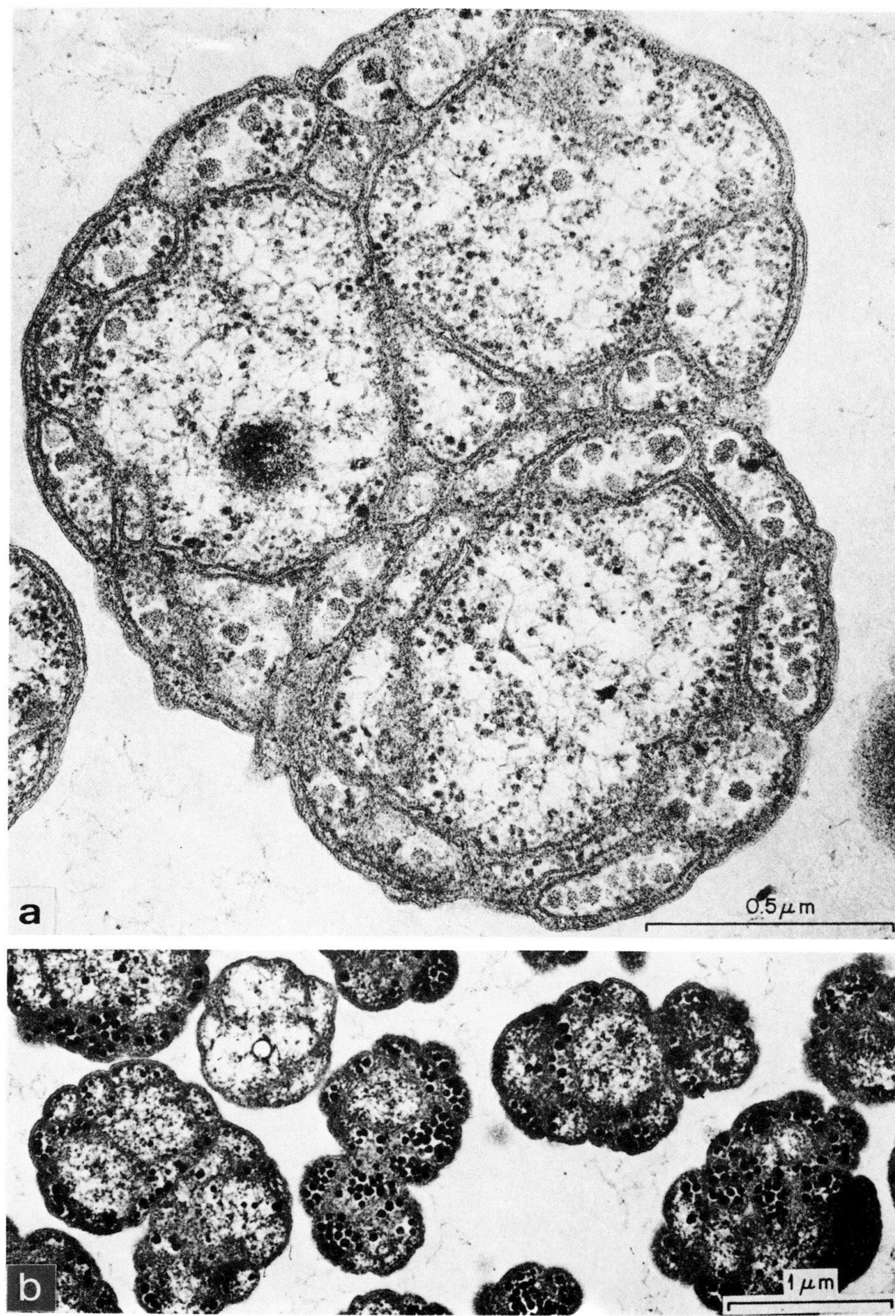

Fig. 80. Peripheral and central compartments in *Nitrosolobus multiformis*. a, cross section (after WATSON et al.); b, glycogen deposits (after WATSON et al.). (From WATSON et al. 1981)

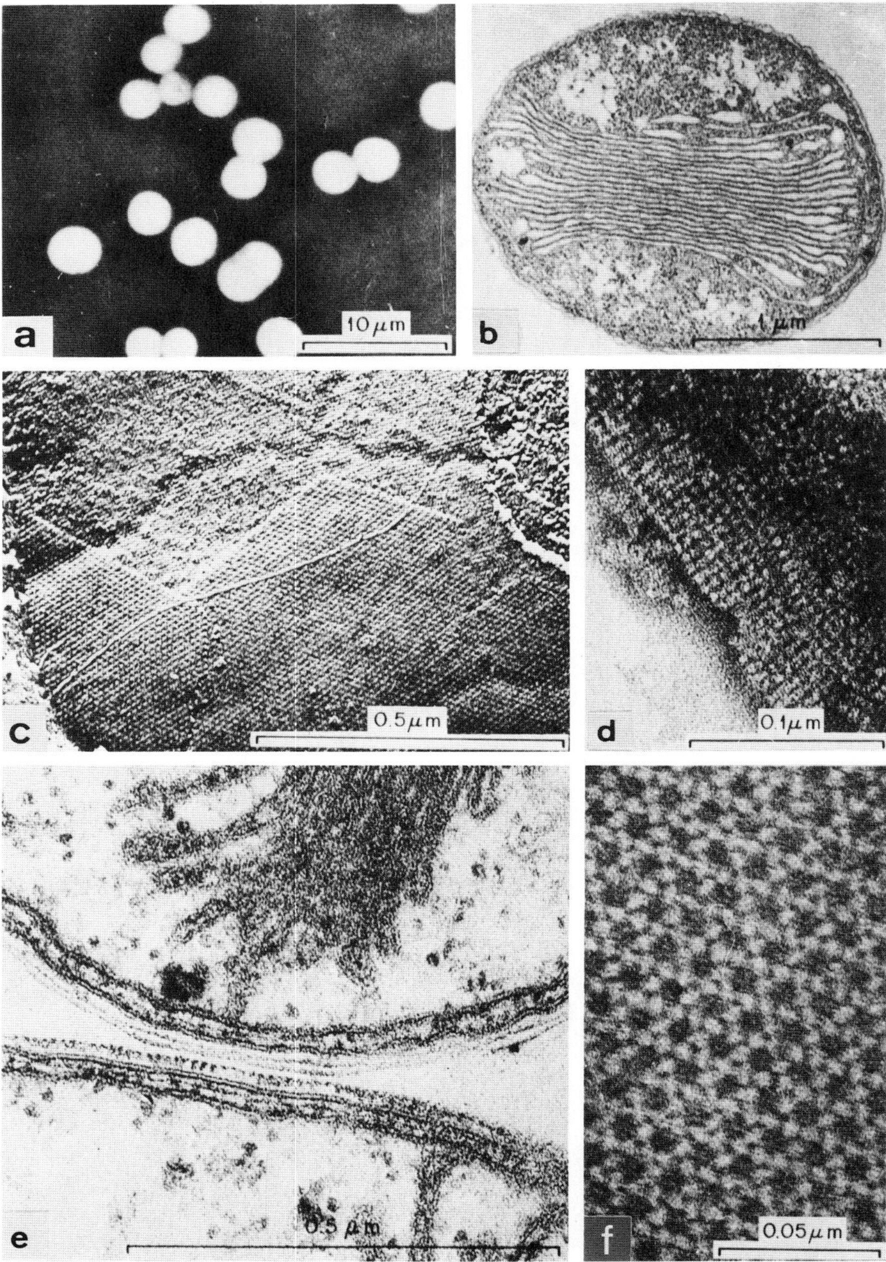

Fig. 81. Bacteria with intracytoplasmic membranes; *Nitrosococcus oceanus*. a, size and shape of cells; b, membrane system (after WATSON & REMSEN); c, outermost wall layers (after WATSON & REMSEN); d, inner wall layer; e, external wall layers (after WATSON & REMSEN); f, outermost wall layer (after WATSON & REMSEN). (From WATSON et al. 1981)

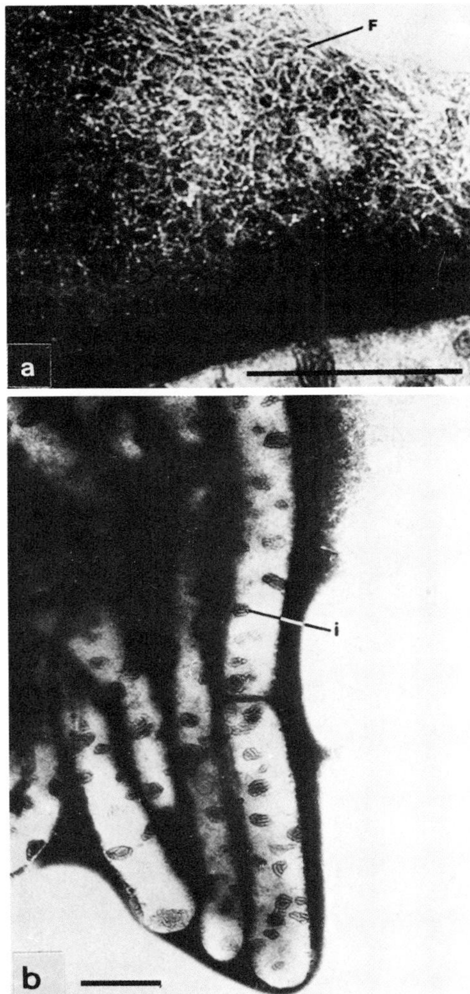

Fig. 82. Structural details of *Methanobacterium formicicum*. a, fibrillar meshwork (F) which binds clumps of the cells; b, clump of cells. Convoluted internal structures (i) like those found in *Methanobacterium* strain *M.o.H.* are found in all cells; Bars: 0.5 μm. (From LANGENBERG et al. 1968)

preparations were employed to study the proportion of respiratory reactions with cells grown at various oxygen concentrations. Maximum activities as well as substrate affinities of NADH, NADPH, and malate dehydrogenases increased from about 1% to 30% air saturation and stayed almost constant at higher oxygen concentrations. Activities of the three dehydrogenases changed largely in parallel. Essentially the same dependency on oxygen concentration was measured with the rate limiting step of the terminal respiratory pathway. Consequently, the ratio of the dehydrogenase activities to the activity of the rate limiting step of the terminal pathway stayed constant. The authors concluded that intracytoplasmic and cytoplasmic membranes do not differ with respect to respiratory activities, and they have proposed that both types of membranes may represent differently located parts of an otherwise identical membrane system.

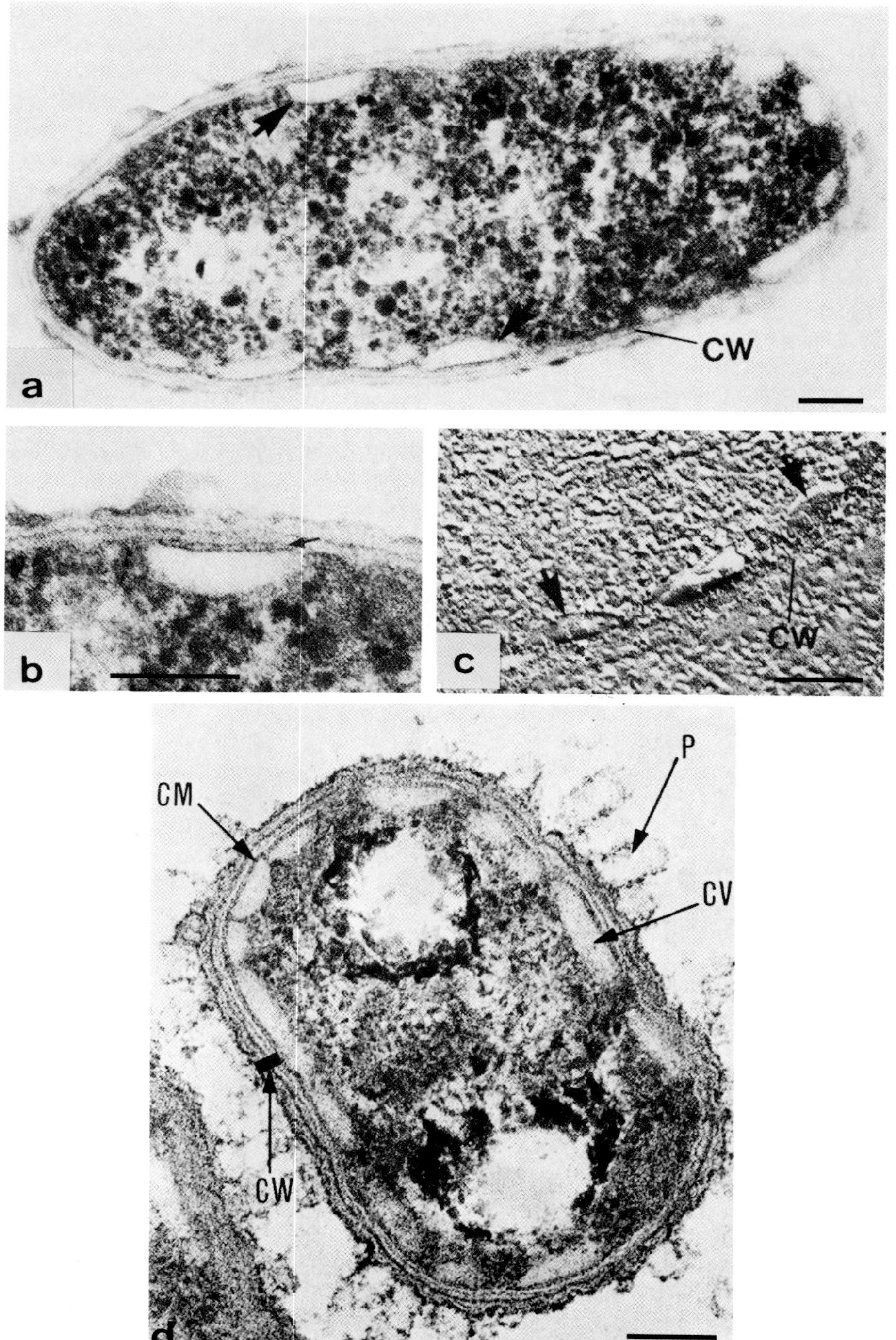
a
CW
b
c
CW
d
P
CM
CV
CW

5.6 Inclusion bodies

Various types of inclusion bodies have been described in bacteria, and different functions have been ascribed to them. According to SHIVELY (1974), bacterial inclusion bodies may be divided into two major groups based on the presence or absence of a surrounding membrane. Membrane-enclosed inclusion bodies are: chlorosomes, carboxysomes, gas vacuoles, magnetosomes, poly-β-hydroxybutyrate (PHB) granules, sulfur globules, and certain polyglucoside (glycogen) granules. Inclusion bodies lacking a membrane are: polyglucoside (glycogen) granules, polyphosphate (volutin, metachromatic) granules, crystals and paracrystalline arrays, and inclusions specific for cyanobacteria, i. e. cyanophycin granules and phycobilisomes. Intact and defective bacteriophages and R-bodies (s. below) should also be mentioned in this respect; however, they are not typical inclusion bodies.

5.6.1 Membrane-enclosed inclusion bodies

In ultrathin sections, the barrier surrounding all of these inclusion bodies appears to be of a "non-unit membrane" type. It is depicted as a dense layer 2 to 4 nm thick and, most probably, is composed entirely of protein; however, exceptions have been discussed (s. Chlorosomes).

5.6.1.1 Chlorosomes

Chlorosomes (CRUDEN & STANIER 1970) are found in the Chlorobiineae, i. e. in species of the Chlorobiaceae, the Chloroflexaceae, and in species of the genus *Chloronema.* Chlorobiaceae, i. e. green and brown sulfur bacteria, are non-motile, whereas the Chloroflexaceae, i. e. filamentous green bacteria, move by gliding. The species *Chloronema giganteum* and *C. spiroideum* making up the genus *Chloronema* were also observed to be gliding bacteria. It appears that in almost all genera of the green sulfur bacteria, green- and brown-colored species exist that morphologically, cytologically, and physiologically resemble each other. However, in all cases, the green and brown species were found to differ in terms of their photosynthetic pigments. The establishment of separate species for the green and brown counterparts in each genus, therefore, appears justified.

Chlorosomes have been found in close association with the inner surface of the cytoplasmic membrane (Figs. 83–85). They contain all of the antenna bacteriochlorophyll *c, d* or *e* within the cell. The reaction center bacteriochlorophyll *a* is located in the cytoplasmic membrane. Antenna bacteriochlorophyll *a* species is also located in the cytoplasmic membrane except in *Chloroflexus* where a minor bacteriochlorophyll *a* is associated with the chlorosome.

In both *Chlorobium* and *Chloroflexus,* the chlorosomes are randomly distributed along the protoplasmic side of the cytoplasmic membrane. Their dimensions vary somewhat, ranging from 40×70 nm to 100×260 nm.

The ultrastructure of chlorosomes ("chlorobium vesicles", "chlorobium type vesicles") in *Chlorobium limicola* (Fig. 86) and in *Chloroflexus aurantiacus* has been described by the freeze-fracture technique (SCHMIDT 1980, SPRAGUE et al. 1981 a, b, STAEHELIN et al. 1978 b, 1980). Here

Fig. 83. a–c, *Chlorobium limicola;* a, b, thin sections showing several chlorosomes appressed to the cytoplasmic membrane (arrows); c, portion of a freeze-fractured cell exhibiting similar features as seen in the thin sections. The chlorosomes (arrows) appear as oblong, amorphous bodies underlying the cytoplasmic membrane and the cell wall (CW). Bars: 0.1 μm. (From STAEHELIN et al. 1980); d, ultrathin section through a cell of *Chlorobium limicola.* – CM = cytoplasmic membrane; CW = cell wall; CV = chlorosomes; P = pili (fimbriae). Bar: 0.1 μm. (From STANIER et al. 1981)

the chlorosomes are seen to consist of an envelope layer (not a bilayer membrane, but perhaps a galactolipid monolayer about 2 nm thick) surrounding a varying number of rod-like structures that are oriented parallel to the major axis of the chlorosome. A baseplate between the chlorosome and the cytoplasmic membrane serves to anchor the chlorosome to a specific membrane attachment site. In *Chlorobium* the attachment site is marked by slightly larger than average intramembrane particles that are seen on the protoplasmic leaflet of the cytoplasmic membrane. Fig. 87 presents an interpretative model of a chlorosome.

A study on the development of chlorosomes in *Chloroflexus aurantiacus* has revealed that the number of chlorosomes per cell increases in response to decreased light intensity or lower oxygen tension. The same conditions led to a thickening of the chlorosomes but did not affect their length distribution. The thickening of the chlorosomes was accompanied by an increase in the bacteriochlorophyll *c*/bacteriochlorophyll *a* ratio. Parallel appearance of 5-nm particles (in a lattice config-

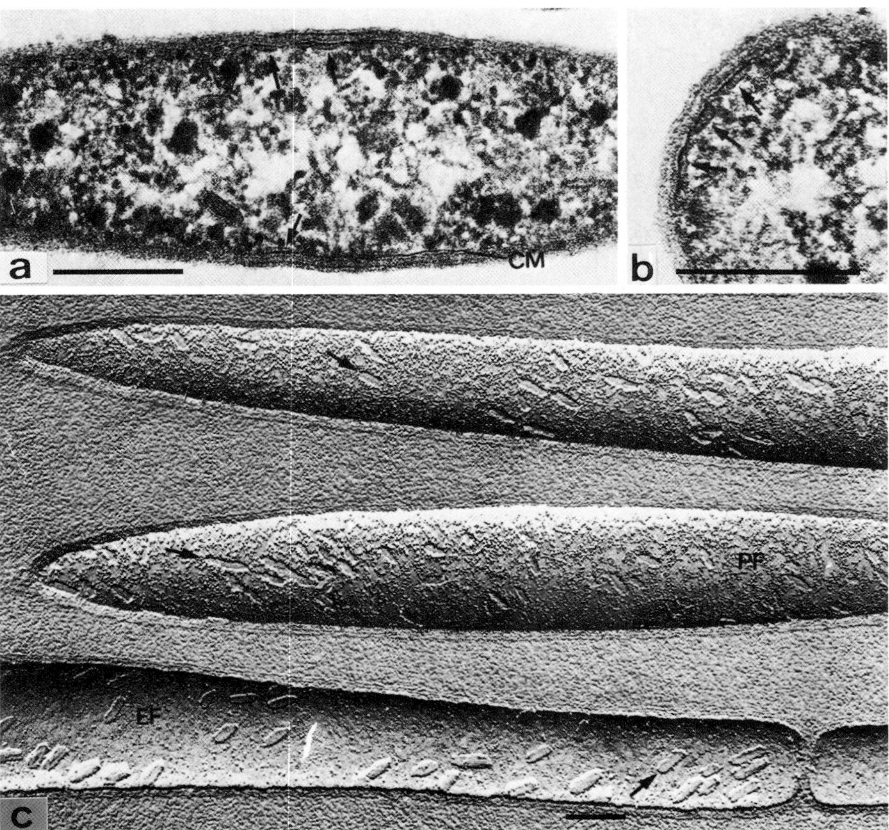

Fig. 84. Architecture of chlorosomes in *Chloroflexus aurantiacus*. a, chlorosomes appressed to the cytoplasmic membrane (CM) (arrows); b, chlorosomes (arrows) at higher magnification; c, where the elongated chlorosomes are attached to the cytoplasmic membrane (arrows) the fracture planes are deflected into the vesicles; nearly all chlorosomes in a cell have the same orientation. Bars: 0.2 µm. EF = E-face; PF = P-face in freeze-etching. (From Staehelin et al. 1978)

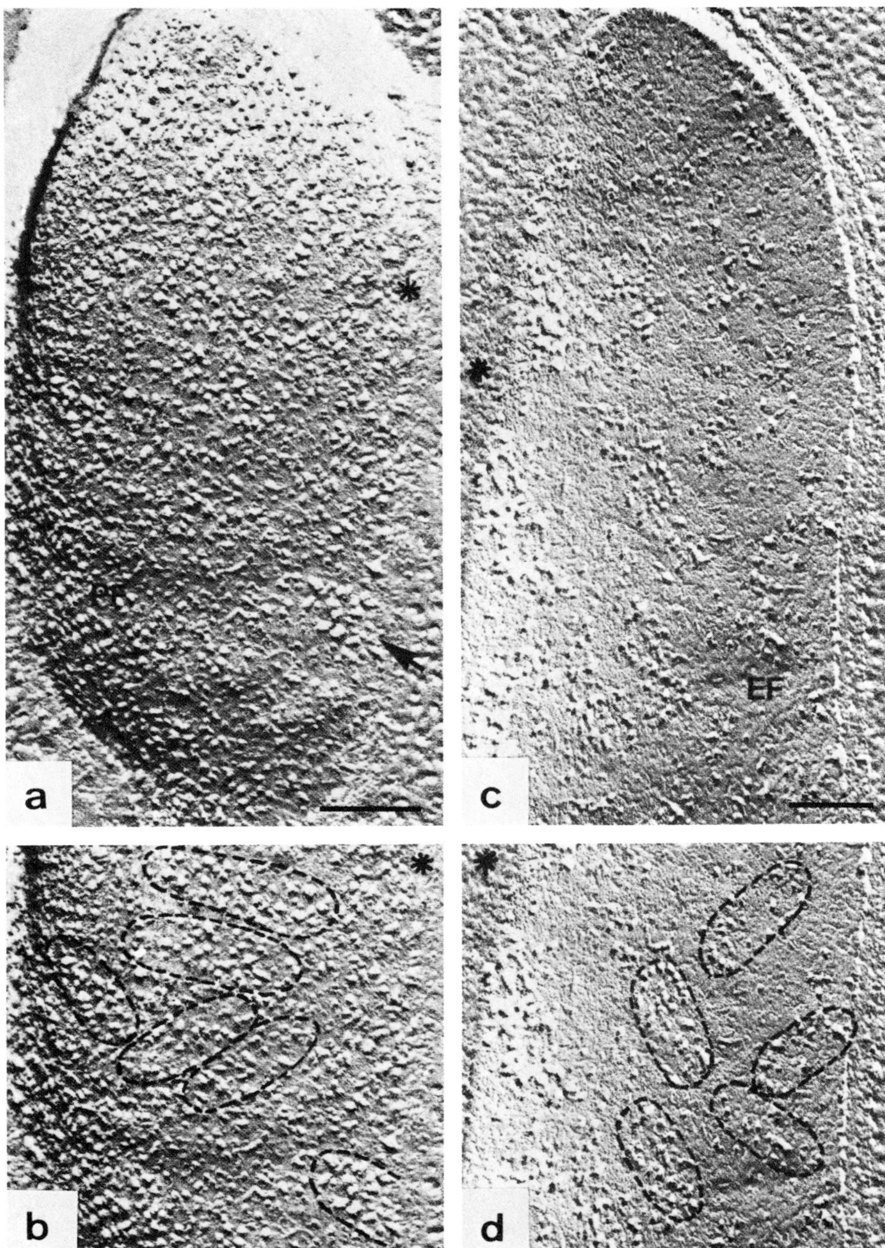

Fig. 85. Aspects of the cytoplasmic membrane of glutaraldehyde-fixed and glycerinated *Chlorobium* cells after freeze-fracturing. a, b, same sample; asterisks mark the same point of reference; in b, patches have been outlined, corresponding to attachment sites (arrow in a) of underlying chlorosomes; c, d, other aspect of the cytoplasmic membrane; same sample in c and d; asterisks mark the same point of reference; in d, patches of particles have been outlined; they can easily be recognized against the relatively smooth background. – EF = E-face. Bars: 0.1 μm. (From STAEHELIN et al. 1980)

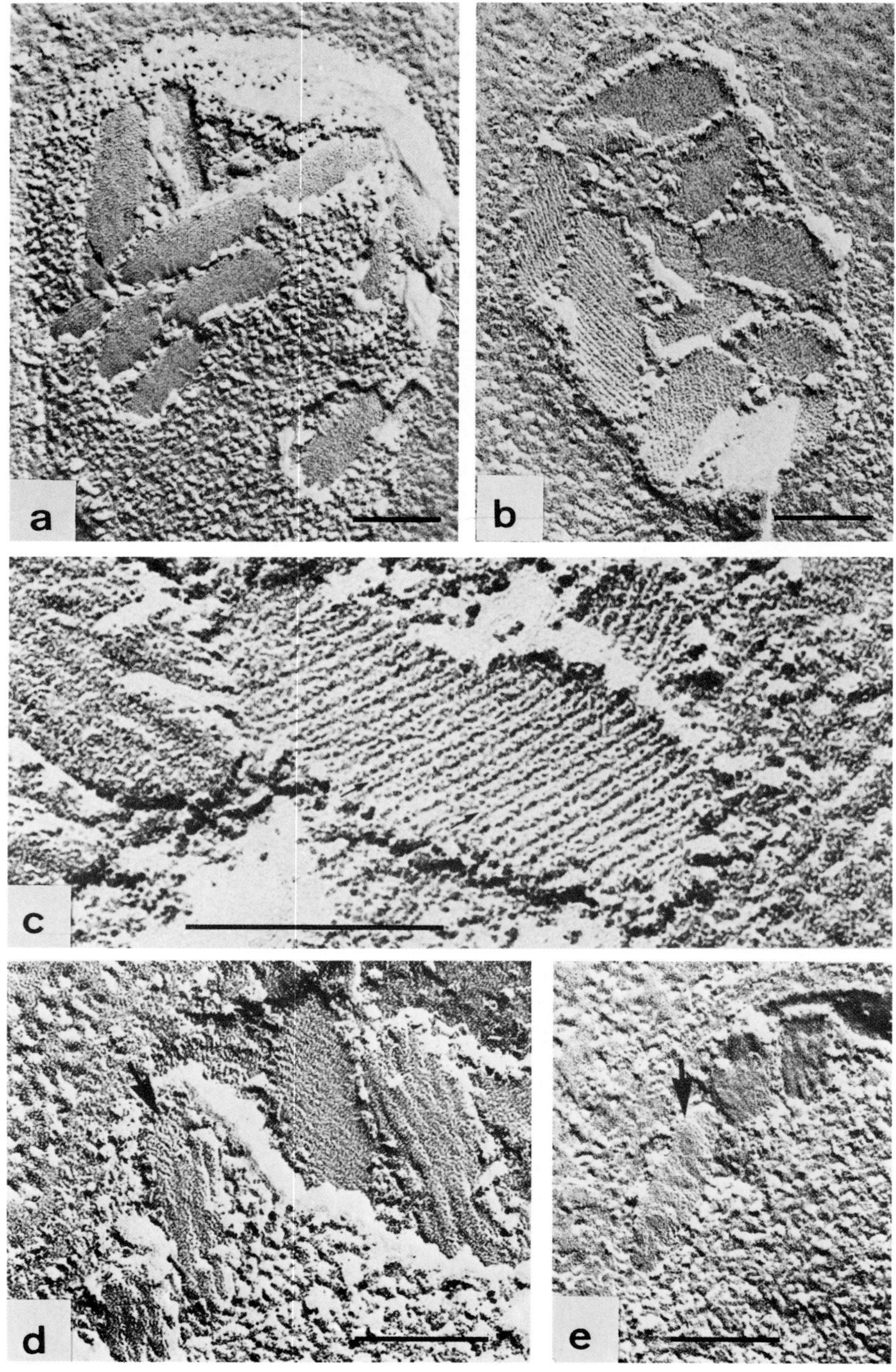
a
b
c
d
e

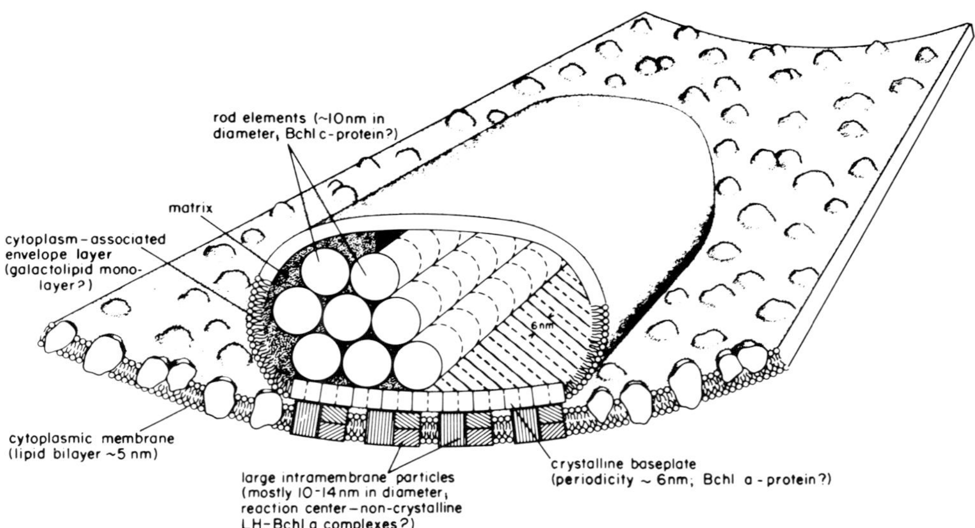

Fig. 87. Model of a chlorosome and its associated cytoplasmic membrane of *Chlorobium limicola* based on freeze-fracture observations, and on biochemical and biophysical studies. – Bchl = bacteriochlorophyll; LH = light harvesting. (From STAEHELIN et al. 1980)

uration) in the membrane attachment site, of the crystalline baseplate material (with a periodicity of 6 nm) adjacent to the membrane attachment site, as well as of the chlorosome envelope layer preceded the addition of longitudinally oriented rod-like elements (diameter 6 nm) to the chlorosome core. It was estimated that each chlorosome can funnel energy into approximately 100 reaction centers located in the cytoplasmic membrane.

5.6.1.2 Carboxysomes

Carboxysomes are inclusion bodies (Fig. 88) in many species of chemolithotrophic bacteria and in cyanobacteria (MURPHY et al. 1974, SHIVELY et al. 1970, 1973 a, b). They are polyhedral particles and contain the enzyme ribulose-1,5-bisphosphate carboxylase-oxygenase (RubisCO). Preliminary electron microscopic examination of carboxysomes indicated that they might be icosahedral particles about 100 nm in diameter containing the typical "doughnut"-shaped 10–11 nm RubisCO molecules bound by a 3 to 4 nm thick monolayer membrane or shell. Cryo electron microscopy recently revealed that carboxysomes are not icosahedrons but pentagonal dodecahedrons (VAN BREEMEN, personal commun.). The composition of the shell is not well known although the stability of its shape after disruption of the carboxysome and its rather high density as well as a lack of chloroform extractable material in the carboxysome indicate that it is composed of protein and/

◀ Fig. 86. Supramolecular organization of chlorosomes and their attachment sites in *Chlorobium limicola*. Freeze-fractured samples. a, view of the cytoplasmic membrane and of underlying chlorosomes; b, chlorosome baseplates; low magnification; c, baseplate at high magnification, with crystalline substructure; d, chlorosomes illustrating the closely packed, parallel rod elements; e, chlorosome (arrow) fractured along its cytoplasm-associated envelope layer. Bars: 0.1 μm. (From STAEHELIN et al. 1980)

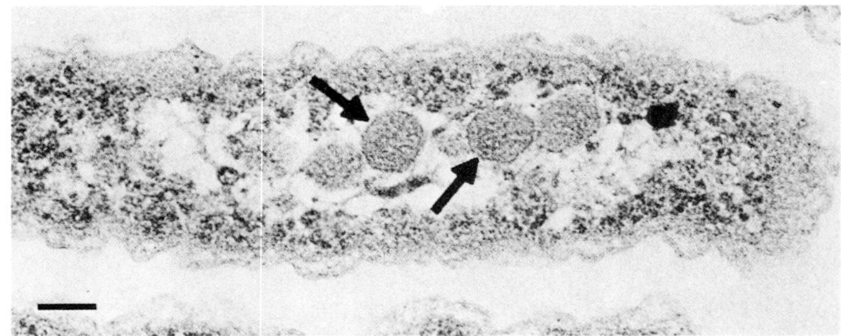

Fig. 88. Polyhedral bodies (carboxysomes; arrows) in *Thiobacillus neapolitanus*. Bar: 0.1 µm. (From SHIVELY 1974)

or glycoprotein. WESTPHAL et al. (1979) have found molecules of circular double-stranded DNA in carboxysome preparations of *Nitrobacter agilis* and *N. winogradskyi*. The structure and function of this DNA have not been elucidated. Most probably, this DNA is not an integral component of the carboxysome. The RubisCO molecules have occasionally been observed in a paracrystalline array within the carboxysome. Presumably in most cases, an initially ordered arrangement is disturbed during cell rupture and purification. However, an ordered arrangement of enzyme molecules might also be created as a drying artifact during preparation of negatively stained samples. In order to determine the relative distribution of RubisCO in the bacterial cell, i. e. in carboxysomes or in the cytoplasm, several attempts have been made to discriminate between the two possibilities. In cells of *Thiobacillus neapolitanus*, BEUDEKER et al. (1980) reported changes in the ratio of carboxysomal to cytoplasmic RubisCO. In addition, the same authors showed that RubisCO from carboxysomes has a higher specific activity than cytoplasmic RubisCO during CO_2-limited growth (BEUDEKER et al. 1981). These observations might suggest that carboxysomes are in some way involved in the regulation of carbon metabolism in this bacterium. BEUDEKER & KUENEN (1981) have reported that isolated carboxysomes of *T. neapolitanus* contain all of the enzymes of the Calvin cycle as well as malate dehydrogenase, aspartate transferase and adenylate kinase. The arrangement of enzyme activities within the carboxysome proposed by these authors would allow the carboxysome to serve as a specialized CO_2-fixing organelle ("calvinosome"). However, CANNON & SHIVELY (1983) investigated a homogeneous preparation of carboxysomes isolated from *T. neapolitanus* (Fig. 89) by means of density gradient centrifugation and agarose preparative electrophoresis. Analysis of carboxysomes by denaturing SDS polyacrylamide gel electrophoresis revealed the presence of 12 to 15 bands interpreted as representing 12 to 15 polypeptide types. Approximately 62% of the total protein was found to consist of the large and small subunits of RubisCO. Two polypeptides were found to be components of the carboxysome shell. Purified carboxysomal and cytoplasmic RubisCO were shown to be similar if not identical by several criteria including specific activity, carboxylase/oxygenase activity, and electrophoretic mobility. Ribose-5-phosphate isomerase, ribulose-5-phosphate kinase, and fructose-1,6-bisphosphatase could not be detected in the carboxysomes. Though the presence of other enzymes in the carboxysome could not be excluded, CANNON & SHIVELY (1983) have rejected the proposed "calvinosome" model due to the lack of three key enzyme activities (s. above), the absence of RubisCO stimulation by malate, the presence of only up to 15 polypeptide types and the apparent simplicity of the shell. Meanwhile, a similar view is also shared by KUENEN (personal commun.).

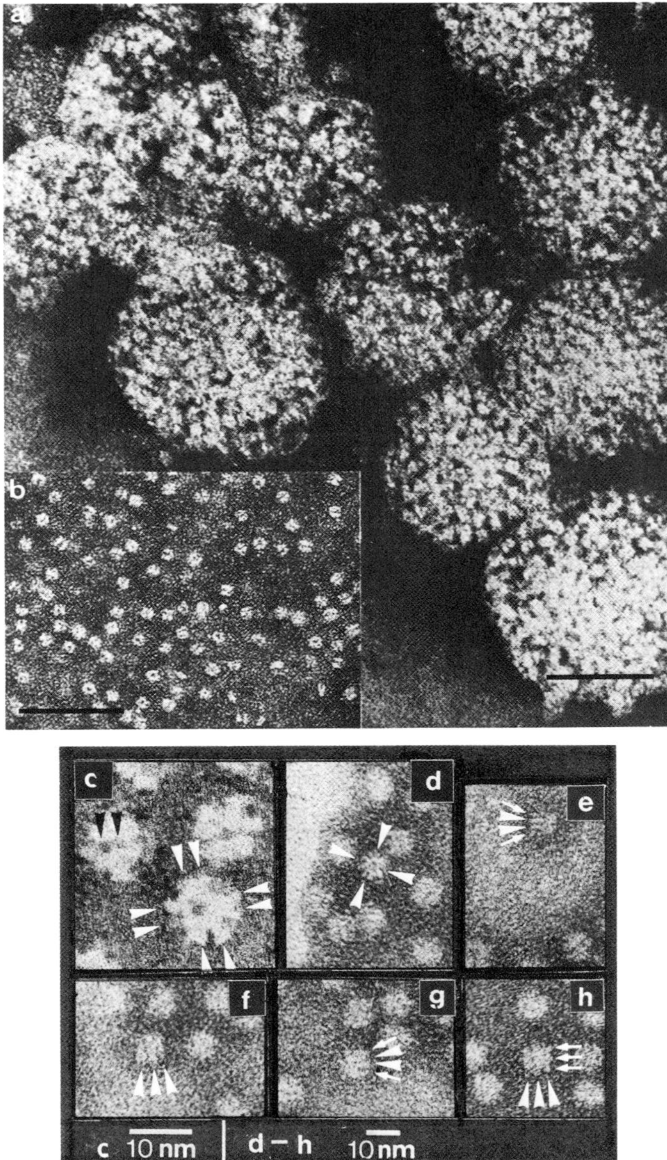

Fig. 89. Isolated carboxysomes from *Thiobacillus neapolitanus,* and isolated ribulose 1,5-bisphosphate carboxylase enzyme molecules. a, isolated carboxysomes with indication for particles inside; b, enzyme particles isolated from a sample of these carboxysomes; bars: 0.1 μm. (From SHIVELY et al. 1973); c–h, isolated ribulose 1, 5-bisphosphate carboxylase from *Alcaligenes eutrophus;* c, d, face views; e–h, side views. The white arrowheads and white arrows indicate positions of subunits, the rotational symmetry, or the layered structure of the particles; the black arrowheads mark central depressions with trapped negative stain. d–h, identical (low) magnification, c, higher magnification, as indicated. (From BOWIEN et al. 1976)

RubisCO enzyme molecules have been isolated from bacteria not containing carboxysomes, e. g. *Alcaligenes eutrophus.* The enzyme has been characterized by various techniques including electron microscopy (Fig. 89) and X-ray crystallography (BOWIEN et al. 1976, BOWIEN & MAYER 1978).

5.6.1.3 Gas vacuoles

Another non-unit membrane-bound type of inclusion body is the gas vacuole or gas vesicle (Fig. 90). This inclusion has been detected in many phototrophic and chemolithotrophic aquatic bacteria (VAN ERT & STALEY 1971, PARKES & WALSBY 1981, SHIVELY 1974, STOECKENIUS & KUNAU 1968,

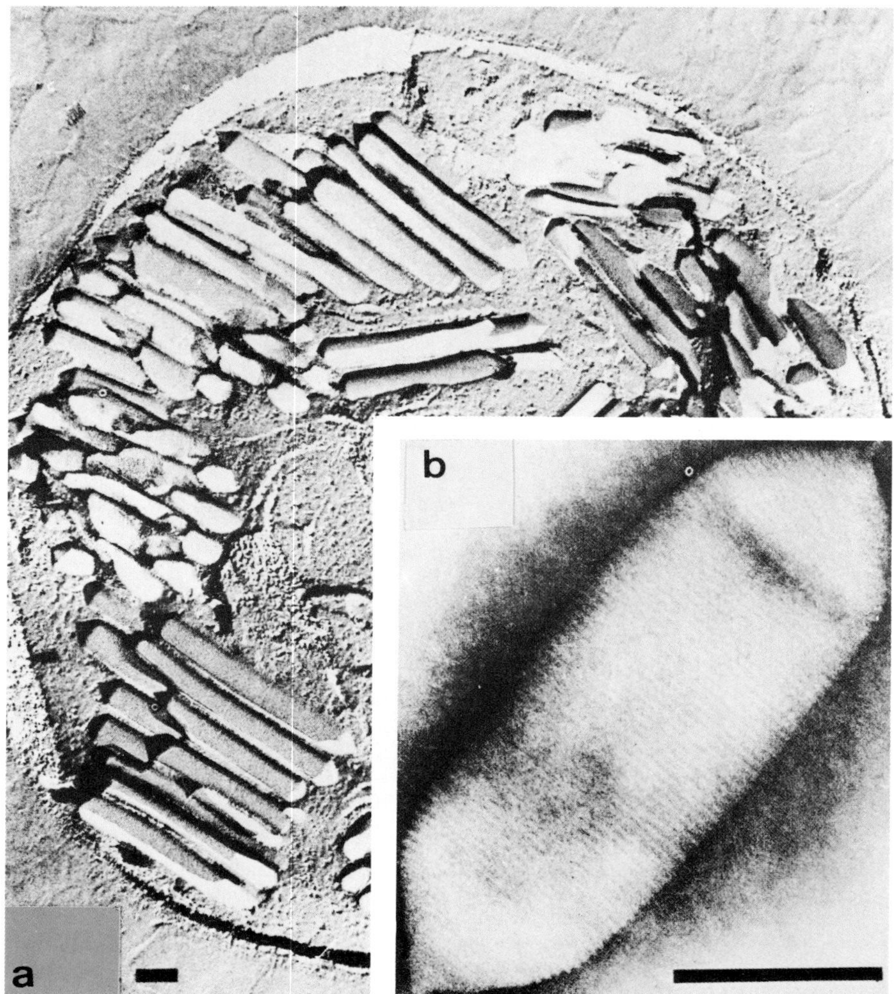

Fig. 90. Inclusion bodies in bacteria. a, freeze-etching of *Nostoc muscorum* showing gas vacuoles; b, isolated gas vesicle from *Anabaena flos-aquae*. Bars: 0.1 µm. (From SHIVELY 1974)

WALSBY 1972). In *Methanosarcina*, ZHILINA (1971) found, as a characteristic feature, densely packed gas vesicles which, depending on the direction of sectioning, were either shaped elliptically in longitudinal sections or like a honeycomb. Each vesicle, when filled with gas, was shown to represent a tubular structure with cone-shaped tips 70 to 80 nm in width and 200 to 400 nm in length. The bounding membrane of gas vesicles, according to measurements in ultrathin sections and from freeze-etching preparations, is about 2.5 nm thick, and it is organized into ribs 4 to 5 nm wide. Depending on the bacterial strain, the gas vesicles occur as more or less densely packed ordered aggregates. In *Methanosarcina* they were fund to lie in a single row, touching each other with their elongated sides. ZHILINA (1971) gained the impression that the gas vesicles slide off each other after division, which could explain their strictly oriented, single-row dense packing. The membranes of gas vesicles were found to be made up of protein subunits of a molecular weight around 14000 D, and it might be assumed that the vesicle shape is probably determined by properties intrinsic in the protein subunits, i. e. that the gas vesicles are self-assembly products. It has been proposed that gas vesicles have the important function, particularly in photosynthesizing bacteria, of maintaining the cells at a suitable bouyant depth in the water.

5.6.1.4 Magnetosomes

One of the most unusual observations in microbiology has been the discovery of magnetotactic bacteria (BLAKEMORE & FRANKEL 1981). A number of these organisms has been observed in nature, and their distribution appears ubiquitous. *Aquaspirillum magnetotacticum* has been grown in pure culture. The organism resembles other spirilla in general cell morphology. However, electron-dense particles within magnetotactic cells have been detected (Fig. 91), and these have been shown

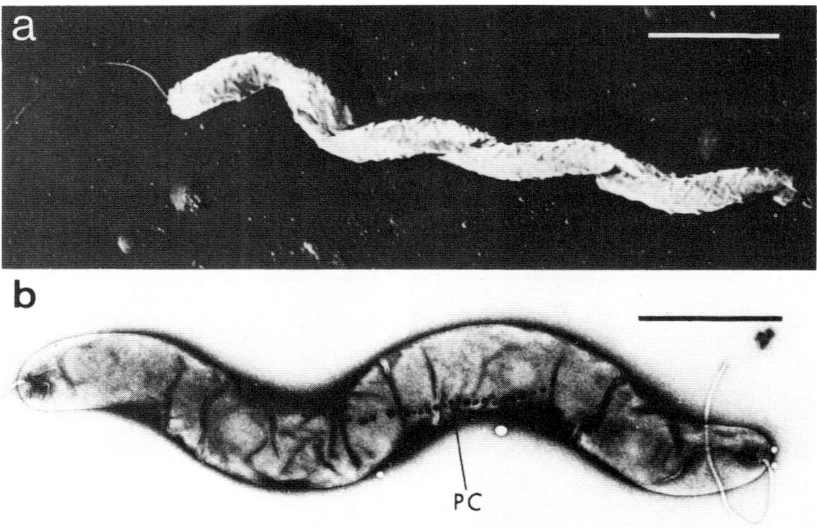

Fig. 91. Ultrastructure of a magnetotactic spirillum. a, critical point-dried; b, negative staining. – PC = electron-dense particles (magnetosomes). Bars: 1 µm. (From BALKWILL et al. 1980)

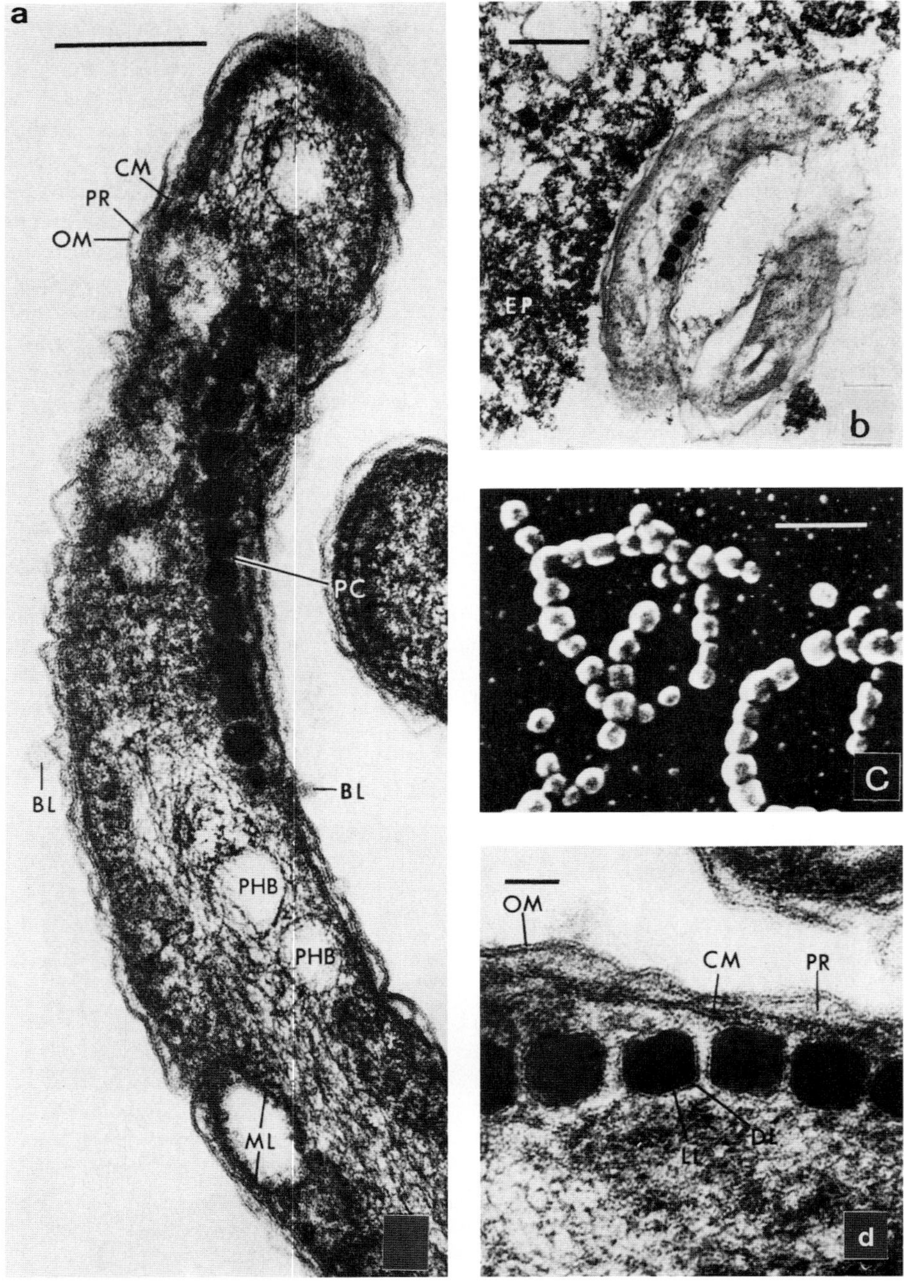

Fig. 92. Ultrastructure of a magnetotactic spirillum. a, general aspects of fine structure; b, ruthenium red-stained extracellular polysaccharide material; c, electron-dense particles released from magnetotactic cells; d, chain of electron-dense particles, – BL = membranous blebs; CM = cytoplasmic membrane; DL = electron-dense layer surrounding the dense particles; EP = extracellular polysaccharide; LL = electron-transparent layer; ML = membranous loop; OM = outer membrane; PC = chain of electron-dense particles; PHB = structures resembling polyhydroxybutyrate granules; PR = periplasmic region. Bars: 0.25 µm in a and b, 0.15 µm in c, and 50 nm in d. (From BALKWILL et al. 1980)

to contain iron. It was found that a chain of these particles which appeared associated with the inner surface of the cytoplasmic membrane traverse each magnetotactic cell in a specific arrangement. Each particle was surrounded by an electron-dense layer separated from the particle surface by an electron-transparent region. The term "magnetosome" was proposed for these cellular inclusion bodies by BALKWILL et al. (1980). The shape of these particles was more or less globular or pointed (Fig. 92). MANN et al. (1984) have analyzed the structure, morphology and crystal growth of the content of the magnetosomes which has been identified as magnetite (Fe_3O_4). They have demonstrated that the mature particles contain well-ordered single-domain crystals of magnetite, and have presented evidence for both crystalline and non-crystalline phases within individual magnetosomes. SPORMAN & WOLFE (1984) have shown that cells of a magnetotactic spirillum similar to *Aquaspirillum magnetotacticum* exhibit a chemotactic (aerotactic) response. Masses of cells formed bands in a slide preparation. Cells in a band reversed their direction of motility along the applied magnetic lines of force without turning; the tactile response of cells occurred in a similar manner. The authors proposed that, in nature, possession of an intracellular magnet allows the microaerophilic cell to persue the most efficient aerotactic behaviour; orientation along the geomagnetic lines of force eliminates the need for twiddling, long runs and short turns. Magnetotactic bacteria of various types from New Zealand and Tasmania were found to swim predominantly southward, i. e. in a direction opposite to that chosen by bacteria in the northern hemisphere. BLAKEMORE & FRANKEL (1981) stated that this behaviour represents a mechanism for migrating down to and remaining in sediments, indicating the survival value of magnetotaxis for microaerophilic cells in avoiding the more oxygenated (and hence toxic) surface water.

5.6.1.5 Poly-β-hydroxybutyrate (PHB) granules

The ability to accumulate PHB is common among a great variety of bacteria including the cyanobacteria (DUNLOP & ROBARDS 1973, ELLAR & LUNDGREN 1968, EMERUWA & HAWIRKO 1973, GRIEBEL & MERRICK 1971, GRIEBEL et al. 1968, JENSEN & SICKO 1971, SCHLEGEL et al. 1961, SHIVELY 1974, STOCKDALE et al. 1968, WANG & LUNDGREN 1969). In ultrathin sections, PHB granules (Fig. 93) appear transparent, probably due to the fact that their content is partially dissolved by application of organic solvents during sample preparation. In freeze-etched samples, PHB granules often appear as artificially stretched horn-like structures, a deformation caused by a certain residual viscosity of the content even at very low temperatures. Granule diameters between 100 and 800 nm have been reported. The surrounding membrane was found to be 2 to 4 nm thick, and it was speculated that it might represent (or arise from) one layer of the cytoplasmic membrane. The granules consist of 98% PHB, 2% protein and trace amounts of lipid and phosphorus. For the molecular weight of PHB values from 1000 D to 256 000 D have been reported, and indications have been obtained that PHB in the granules is in a crystalline configuration, with the molecules organized as right-handed helices stabilized by carbonyl-methyl interactions. The enzyme for polymerization of D-(–)-β-hydroxybutyrylCoA, PHB synthetase, and part or all of the depolymerizing complex (depending on the type of organism) are associated with the granules, probably as part of the surrounding membrane. The PHB polymer accumulates when carbon and energy sources are in excess. PHB is considered to be a cellular reserve of energy, or of carbon and energy. The reutilization of PHB, under conditions of energy and/or carbon starvation, occurs with the help of a complex system composed of an inhibiting factor, an activator, and the depolymerase. It has been shown that the initial step in the reutilization of PHB in *Bacillus cereus* is the rupture of the granule membrane. The products resulting from the action of the reutilization system are dimers of β-hydroxybutyrate which are further hydrolyzed by a dimer hydrolase.

5.6.1.6 Sulfur globules

Sulfur globules (Fig. 93) are generally observed in ultrathin sections of Thiorhodaceae and certain other apochloric sulfur bacteria as non-unit membrane-enclosed holes (FAURÉ-FREMIET & ROUILLER 1958, HAGEAGE et al. 1973, NICOLSON & SCHMIDT 1971, SCHMIDT et al. 1971, SHIVELY 1974, STUTZER & HARTLEB 1898), the sulfur having been removed during dehydration. The globules vary from 100 nm to over 1 μm in diameter and are commonly deposited in invaginated pockets of the cytoplasmic membrane, i. e. they are inside the cell wall, but outside the cytoplasmic membrane. The sulfur globules of *Thiovulum majus* and, in some instances, those of *Beggiatoa* and *Chromatium* have however been found in the cytoplasm proper. The globules have a monolayer

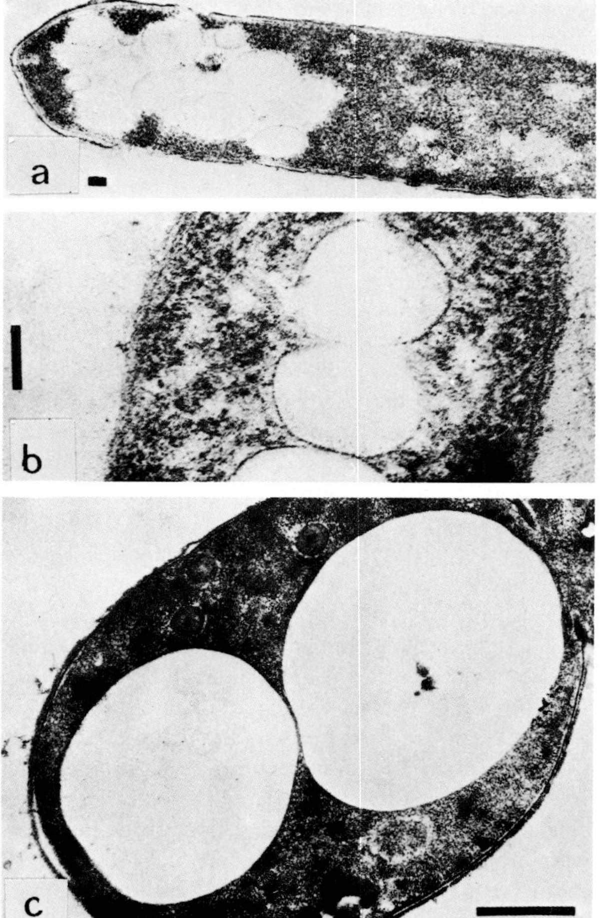

Fig. 93. Inclusion bodies in bacteria. a, polyglucoside granules in *Clostridium pasteurianum;* b, poly-β-hydroxybutyrate granules in *Thiobacillus (Ferrobacillus) ferrooxidans;* c, sulfur globules in *Chromatium weissii.* Bars: 0.1 μm. (From SHIVELY 1974)

membrane made up of globular subunits 2.5 nm in diameter. The subunits have been found to be protein molecules of molecular weigth 13 500 D. It has been speculated that the membrane provides binding sites for the enzymes responsible for sulfur metabolism. Analyses of globules in intact cells or isolated (wet state) have revealed that sulfur is in a liquid state. The sulfur is deposited as the cells grow in the presence of hydrogen sulfide. The sulfur is oxidized, i. e. the globules disappear, when hydrogen sulfide becomes limiting.

5.6.1.7 Membrane-enclosed polyglucoside (glycogen) granules

The occurrence of polyglucoside (glycogen) granules enclosed within a membrane is the exception for this type of material (Shively 1974). These granules have been observed in several species of *Clostridium*. The granules appear to be surrounded by a single-layered membrane and were seen first in cells of the early logarithmic growth phase and became more numerous as the culture aged. Their initial appearance correlated well with the derepression of ADP-glucose pyrophosphorylase. The polymorphic content of the granules is of the amylopectin type, and molecular weights of 180 000 D have been recorded. In *Clostridium pasteurianum*, the globules (Fig. 93) have a diameter between 160 and 300 nm, with polymer fibrils of diameter 12 to 16 nm in the granule matrix. The enzyme granulose synthetase associated with the granules has been suggested to be an integral part of the granule membrane.

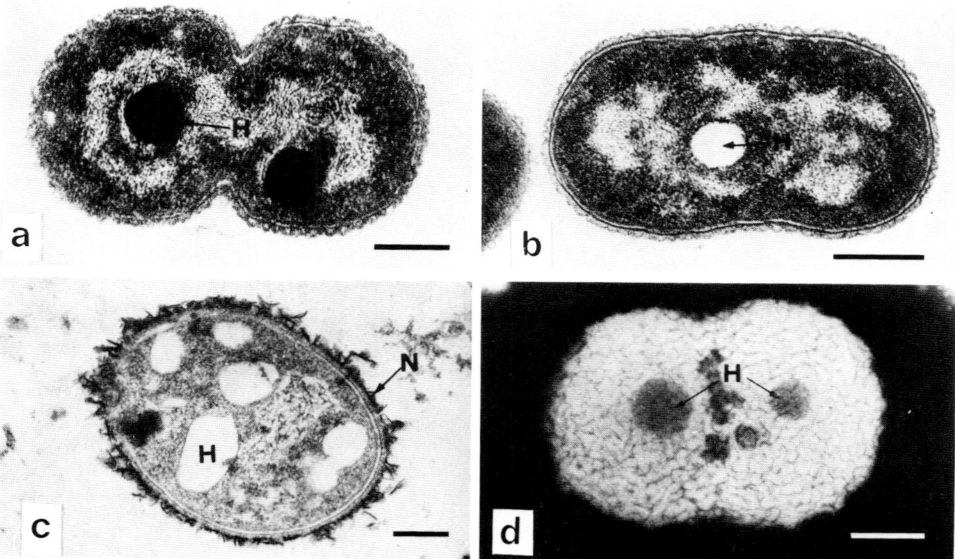

Fig. 94. *Acinetobacter* sp., a hydrocarbon oxidizing bacterium. a–c, ultrathin sections, d, negative staining. Cells were grown on hexadec-1-ene and fixed, dehydrated and embedded in different ways resulting in somewhat different appearance of the hexadec-1-ene inclusions (H). – N = Nickel naphthenate deposits. Bars: 0.2 μm. (From Kennedy et al. 1975)

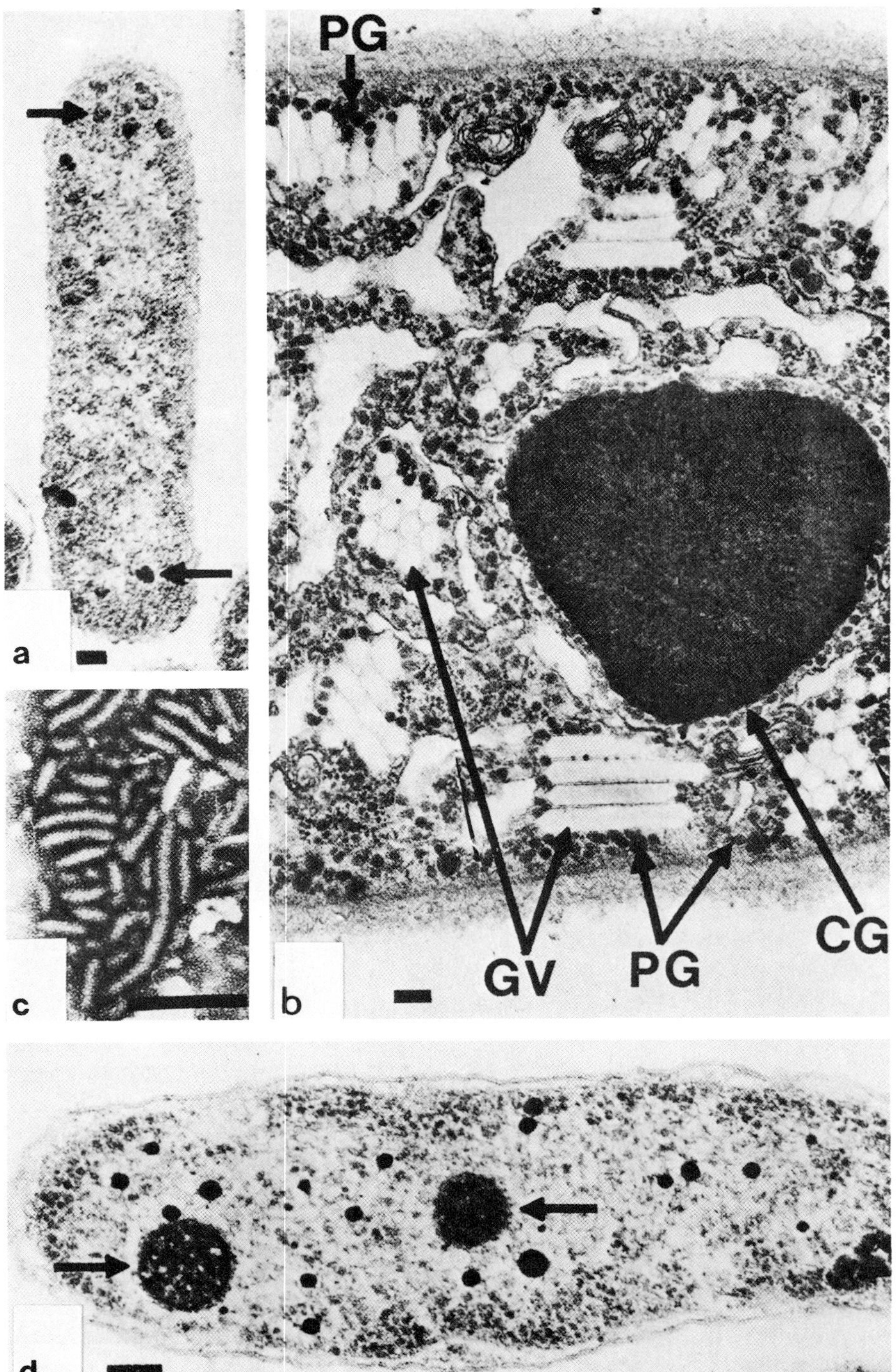

Fig. 95. Bacterial inclusion bodies. a, polyglucoside granules (arrows) in *E. coli;* b, polyglucoside granules (PG), cyanophycin granules (CG), and gas vacuoles (GV) of *Aphanizomenon flos-aquae;* c, isolated polyglucoside granules of *Oscillatoria rubescens;* d, polyphosphate granules (arrows) in *Thiobacillus novellus.* Bars: 0.1 μm. (From SHIVELY 1974)

5.6.2 Inclusion bodies without a membrane

5.6.2.1 Polyglucoside (glycogen) granules

Many bacteria store polymers of glucose (Fig. 95) (CHAO & BOWEN 1971, ELBEIN & MITCHEL 1973, LAISHLEY et al. 1973, ORPIN 1973, ROBERTSON et al. 1975, SHIVELY 1974) or other hydrocarbons (Fig. 94) (KENNEDY et al. 1975). With few exceptions (s. above), the polyglucoside is found dispersed throughout the cytoplasm as granules 20 to 100 nm in diameter. They commonly have an uneven appearance and are very electron-dense in thin sections stained with lead citrate. The granules of cyanobacteria may be spheres, rods or crystals and are generally located between the thylakoid membranes. The granules consist of highly branched, high molecular weight polymers which resemble glycogen or amylopectin. They are produced in response to excess carbon when nitrogen, sulfur, or phosphorus are limiting, or when the pH value is low. Granule formation in cyanobacteria is correlated with photosynthesis. The polymer is assumed to be a storage form of energy and/or carbon. Synthesis occurs by the following reactions:

$$\text{ATP} + \alpha\text{-glucose-1-P} \leftrightarrow \text{ADP-glucose} + \text{PP}_i$$
$$\text{ADP-glucose} + \alpha\text{-1,4-glucan} \leftrightarrow \text{ADP} + \alpha\text{-1,4-glucosyl-glucan}$$

5.6.2.2 Polyphosphate (volutin, metachromatic) granules

Polyphosphates are widely distributed among bacteria (CARR & SANDHU 1966, EIGENER & BOCK 1971, FRIEDBERG & AVIGAD 1968, HAROLD 1963, 1966, JENSEN 1968, 1970, NIEMEYER & RICHTER 1969, ROMANENKO 1984, SHIVELY 1974, TERRY & HOOPER 1970, WALTHER-MAURUSCHAT et al. 1977). They are commonly deposited as spherical, electron-opaque granules (Fig. 96) with diameters ranging from less than 50 nm to more than 1 µm. The granules are generally located in the

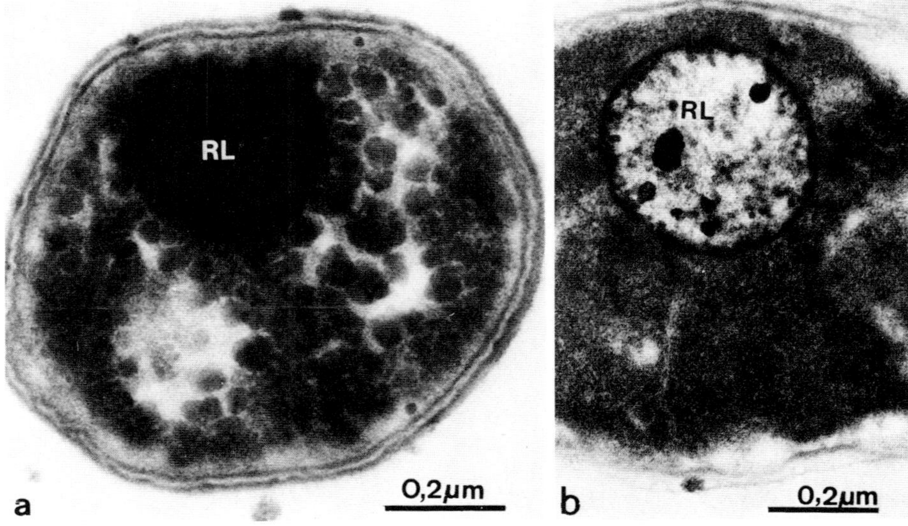

Fig. 96. Inclusion bodies (polyphosphate, RL) in a Gram-positive (a) and a Gram-negative (b) bacterium. In b, the inclusion body has a motteled appearance. (Original micrographs F. MAYER)

nucleoplasmic region of the cell, but have also been seen associated with other cell components. Due to evaporation caused by the electron beam, polyphosphate granules in ultrathin sections occasionally adopt a mottled appearance. In some cases, these granules have been described to have a surrounding membrane, but most probably this aspect is caused by initially loose, but later condensed material present around the granule. The bulk of the polyphosphate is made up of linear, high molecular weight molecules; more than 500 residues per molecule have been measured. The polyphosphate in the granules may constitute up to 50% of the total cell phosphorus. The granules occur at a fairly constant number during the exponential growth phase when phosphate is in excess, and decrease under phosphate limitation. Polyphosphate was shown to be synthesized by the enzyme polyphosphate kinase according to the following reaction:

$$ATP + (HPO_3) \leftrightarrow ADP + (HPO_3)_{n+1}.$$

It has been suggested that the polymer strands are precipitated onto cytoplasmic material and become tightly coiled into granules. The primary function of the polyphosphate appears to be phosphate storage, but energy storage or its role as a regulator of phosphate economy have also been discussed. Polyphosphate can be degraded by different enzymes such as polyphosphate kinase, polyphosphate-adenosine monophosphate-phosphotransferase, polyphosphate glucokinase, polyphosphate fructokinase, and various polyphosphatases.

5.6.2.3 Crystals and paracrystalline arrays

Crystals and paracrystalline arrays lacking a surrounding membrane have been reported to occur in a variety of bacteria (Duncan et al. 1973, Glatron et al. 1972, Herbert et al. 1971, Kim & Barksdale 1969, Lecadet & Dedonder 1971, Norris & Proctor 1969, Pope et al. 1968, Santo

Fig. 97. Freeze-etching of the parasporal crystal of *Bacillus thuringiensis* var. *alestii*. Bar: 0.1 μm. (From Shively 1974)

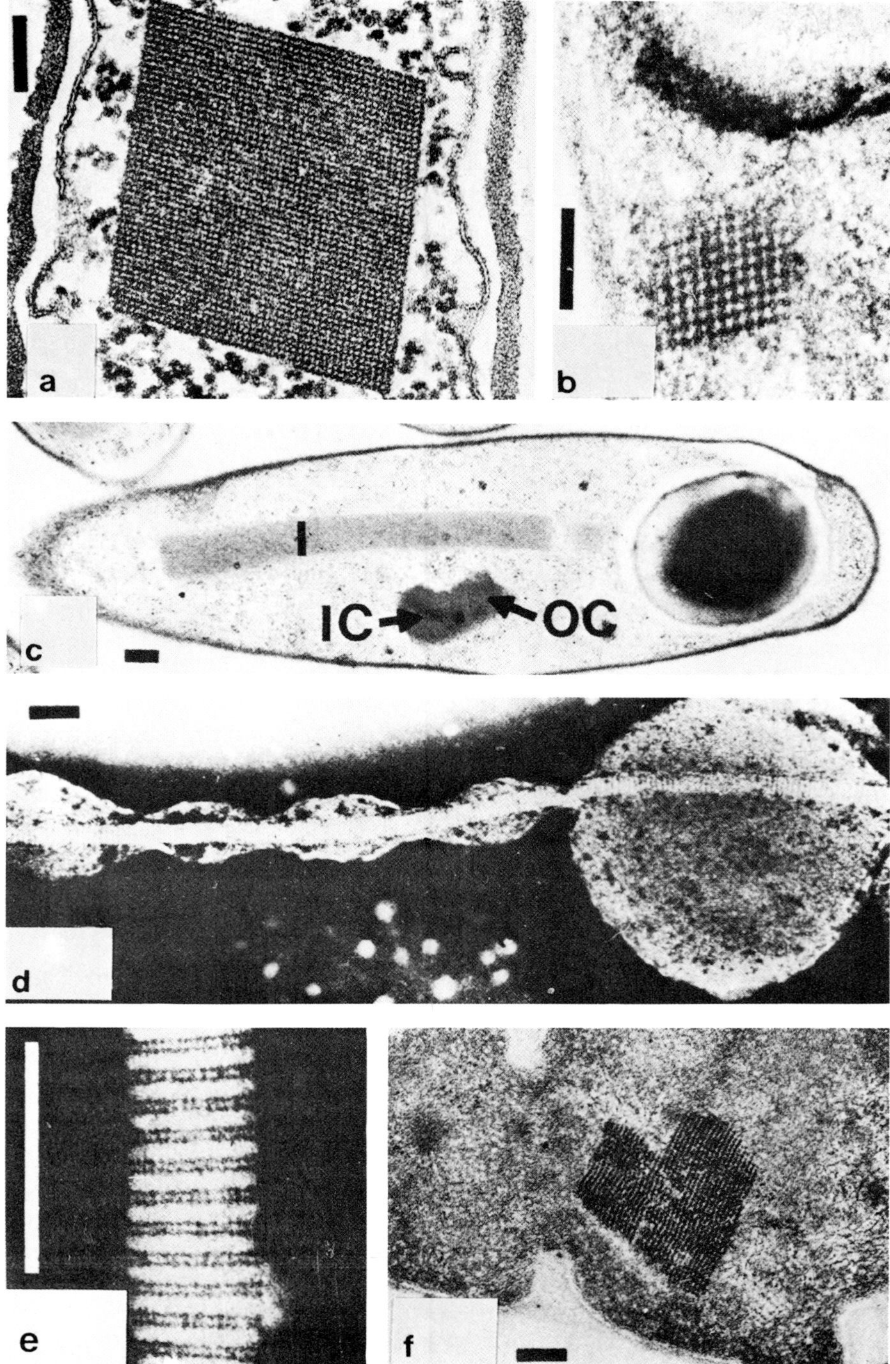

Fig. 98. Bacterial inclusions. a, crystal of RNA-polymerase mutant of *Bacillus subtilis;* b, parasporal crystal of *Clostridium cochlearium;* c, paracrystalline inclusion in mutant *8–4* of *Clostridium perfringens.* – I = inclusion; IC = inner spore-coat material; OC = outer spore-coat material; d, axial fiber in a variant *(rho)* form of *Mycoplasma;* e, segment of an isolated axial fiber of a variant *(rho)* form of *Mycoplasma;* f, crystal in bacterium *22M.* Bars: 0.1 μm. (From SHIVELY 1974)

& Doi 1973, Sayles et al. 1970, Shively 1974, Somerville 1971, Somerville et al. 1968) such as *Bacillus thuringiensis* (Fig. 97), in an RNA polymerase mutant of *Bacillus subtilis* (Fig. 98), in *Clostridium cochlearium* (Fig. 98), and *C. perfringens* (Fig. 98), in a variant form of *Mycoplasma* (Fig. 98), in *Thiobacillus intermedius,* in several cyanobacteria, in a root nodule endophyte, in the symbiotic bacteria of the leafhopper, and in several other bacteria (Shively 1974; further references therein). The origin and possible physiological functions of these inclusions are poorly understood. It is speculated that some of them might be of viral origin; some of them are interpreted to be enterotoxin. Different forms have been observed, ranging from defined bipyramidal octahedra *(B. thuringiensis)* to axial fibers *(Mycoplasma).* The best characterized crystalline inclusion body is one commonly found in *B. thuringiensis* because of its toxicity to *Lepidoptera* larvae. It is composed of rod-shaped subunits 4.7 nm by 11.8 nm with a molecular weight of 230000 D. The subunits break down to smaller polypeptide chains upon treatment with alkali. The crystal proteins have been found to be synthesized by RNA-mediated protein synthesis, and crystal formation may be initiated on the spore exosporium or on cytoplasmic membrane invaginations. Since the formation of the crystal is linked to sporulation, the crystal proteins may result from the overproduction of spore-coat proteins.

5.6.2.4 Cyanophycin (structured) granules

Cyanophycin granules have been observed in many cyanobacteria (Lang et al. 1972, Shively 1974, Simon 1971, 1973a, b). In ultrathin sections, they appear as tightly packed, undulating, flattened sacks. They lack a limiting membrane and are variable in size and shape. Common diameters are around 500 nm. The granules isolated from *Anabaena cylindrica* have been reported to consist of 25000 D to 100000 D molecular weight polypeptides containing aspartic acid and arginine in a 1:1 ratio. Experiments with addition of chloramphenicol indicated that these polypeptides are not synthesized by a mRNA-ribosomal system. The enzymes responsible for the synthesis must be present in the cells of exponentially growing cultures. The synthesis is an energy dependent process; (3,4-dichlorphenyl)-1,1-dimethyl urea (DCMU), a specific photophosphorylation inhibitor, blocks the synthesis of the polypeptide after the addition of chloramphenicol. The cyanophycin granules are assumed to be cellular nitrogen reserves. During normal growth, amino acids are incorporated into protein, but upon cessation of growth, nitrogen fixed from the atmosphere is stored in the polypeptides of cyanophycin granules.

5.6.2.5 Phycobilisomes

Phycobilisomes are supramolecular complexes (Figs. 99–101) in cyanobacteria (and red algae) (Cosner 1978, Gantt 1980, Gantt & Lipschultz 1977, Gantt et al. 1979, Glazer et al. 1979, Koller et al. 1977, Shively 1974, Tandeau de Marsac & Cohen-Bazire 1977, Williams et al. 1980, Wollman 1979, Yamanaka 1978) containing water soluble accessory pigments (Bogorad 1975, Bryant et al. 1976, 1981, Gantt 1977, Glazer & Bryant 1975, Glazer & Fang 1973, Grabowsky & Gantt 1978, Gray & Gantt 1975, Lichtlé & Thomas 1976, Muckle et al. 1978, O'Carra & O'H Eocha 1976). Regular arrays of these organelles are found on the surface of the thylakoid membranes of these organisms. Typical phycobilisomes isolated from cyanobacteria possess a triangular core assembled from three stacks of disc-shaped subunits. Each stack contains two discs which are 12 nm in diameter and 6 to 7 nm thick. Each of these discs is probably subdivided into halves 3 to 3.5 nm thick. Radiating from each of the two sides of the triangular core

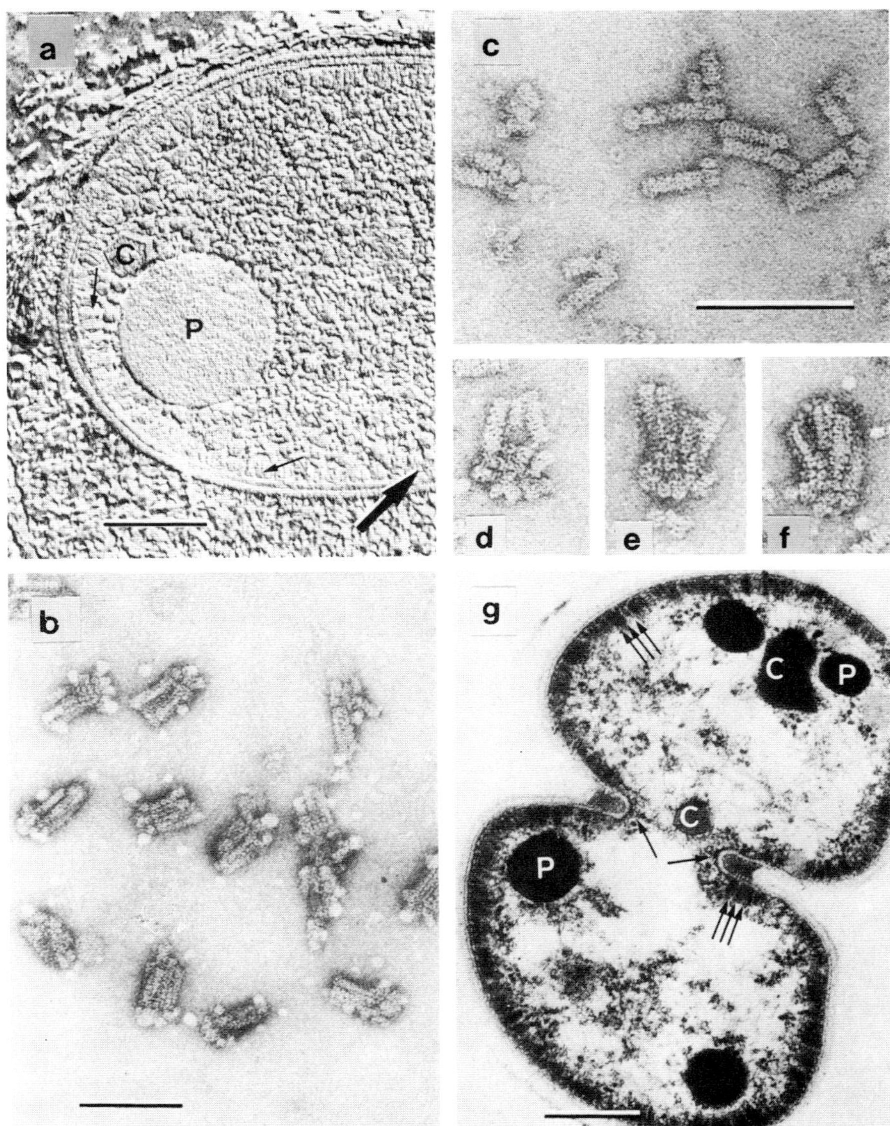

Fig. 99. a, cross-fractured cell of *Gloeobacter violaceus* exposing the rod-shaped elements in the cortical area adjacent to the inner surface of the plasma membrane; a 6 nm repeat across their long axis is visible in some areas indicated by small arrows. The cell contains a large polyphosphate granule (P) and a small carboxysome (C); b, phycobilisomes isolated from *G. violaceus,* negatively stained. c, rod-shaped elements contained in a 0.5 M sucrose fraction of a gradient; d, e, f, partially disrupted or flattened *G. violaceus* phycobilisomes occasionally present in sucrose gradients; the base of the particles appears lobed with no clear fine structure; g, thin section of a fixed *G. violaceus* cell in the process of division. The three grouped arrows show individualized phycobilisomes. P = polyphosphate granules; C = carboxysomes. Bar: 100 nm in a, b and c; 300 nm in g. d, e, f, same magnification as b. (From GUGLIELMI et al. 1981)

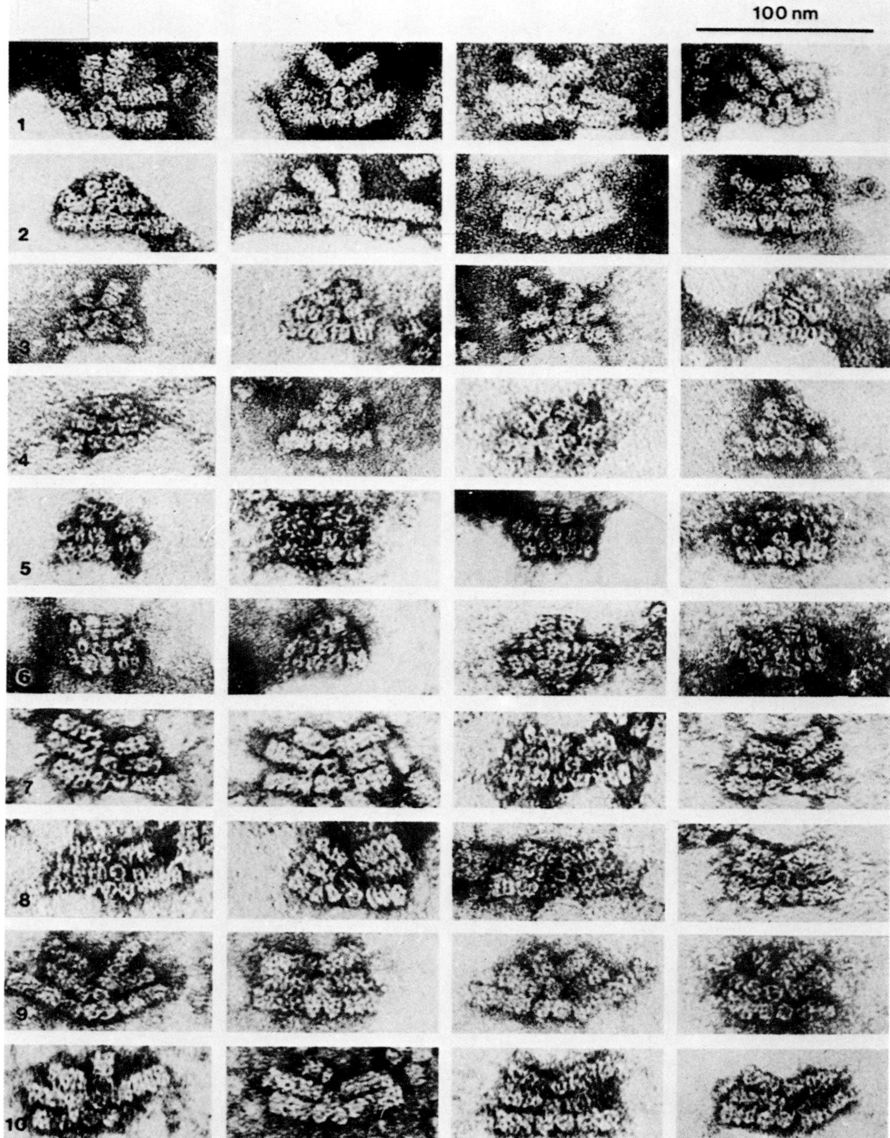

Fig. 100. Selected face views of phycobilisomes isolated from three species of cyanobacteria showing the effects of chromatic illumination on phycobilisome structure. Phycobilisomes from *Synechocystis 6701* grown in green light (rows 1 and 2) and red light (rows 3 and 4); from *Synechococcus 6312* grown in white light (rows 5 and 6); and from LPP-7409 grown in red light (rows 7 and 8) and in green light (rows 9 and 10). (From BRYANT et al. 1979)

are three rods 12 nm in diameter. Each rod consists of stacks of 2 to 6 disc-shaped subunits 6 nm thick. These discs are subdivided into halves 3 nm thick. The average number of discs of 6 nm thickness varies among strains. For certain chromatically adapting strains, the average rod length has been found to be dependent upon the wavelength of light to which the cells had been exposed during growth (Fig. 100). Analyses by different techniques suggest that the triangular core is composed of allophycocyanin (COHEN-BAZIRE 1977, GYSI & ZUBER 1976, LEMASSON et al. 1973, LEC et al. 1977) and that the peripheral rods contain phycocyanin (BERNS & EDWARDS 1965, DOBLER et al. 1972, EISERLING & GLAZER 1974, GLAZER 1976, GLAZER & HIXSON 1975, KAO et al. 1975, KESSEL et al. 1973, KOBAYASHI et al. 1972, TORJESEN & SLETTEN 1972) and phycoerythrin (when present) (ABAD-ZAPATERO et al. 1977, FUJITA & SHIMURA 1974, GANTT 1969, GLAZER & HIXSON 1977, MUCKLE & RÜDIGER 1977). Fig. 101 presents a scale model as proposed by BRYANT et al. (1979). These authors have suggested that a self-assembly process, in which the rod-binding sites are symmetrically determined by the core discs and limited by steric hindrance, can be envisaged for the phycobilisome. A somewhat different model has been proposed by MÖRSCHEL et al. (1977) and by KOLLER et al. (1978).

It has been suggested that the orderly array (s. above) of phycobilisomes observed on thylakoid surfaces may reflect an orderly arrangement of the reaction centers within these membranes. Freeze-fracture studies of cyanobacteria and some other organisms have revealed regularly aligned rows of particles on the E-face of the thylakoid membranes. A close correspondence between these arrays of particles, which may reflect the location of photosystem II reaction centers, and the rows of phycobilisomes on the stromal surfaces of the thylakoids is evident from freeze-fracture studies. Phycobilisome assembly, therefore, may be initiated at specific sites determined by regularities in the chlorophyll-proteins of the thylakoids. Phycobilisomes isolated from the atypical cyanobacterium *Gloeobacter violaceus* (GUGLIELMI et al. 1981) show a fine structure and composition

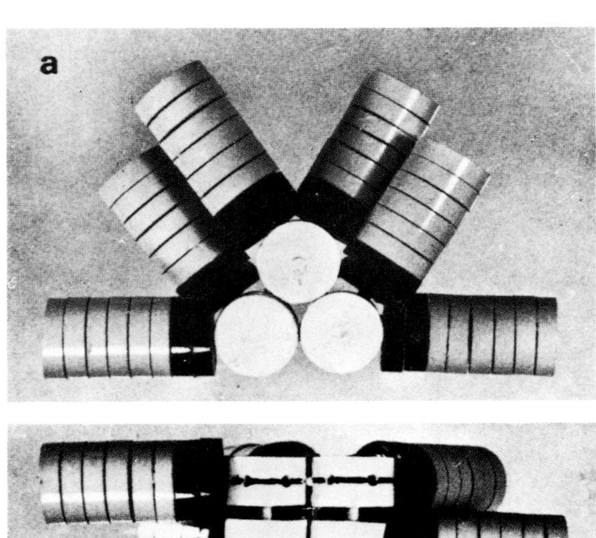

Fig. 101. Two views of a proposed scale model of the hemidiscoidal phycobilisome. White, black, and grey discs represent allophycocyanin, phycocyanin, and phycoerythrin, respectively. Single discs represent phycobiliprotein trimers $(\alpha\beta)_3$ with a diameter of 12 nm and a thickness of 3 nm.
a, face view; b, bottom view. (From BRYANT et al. 1977)

differing somewhat from typical phycobilisomes. Their location and arrangement within the cell are also different. It has been shown that this bacterium has a very specific architecture: the cytoplasmic membrane is the only membrane present in this photosynthetic organism, and this membrane has the dual function of the cytoplasmic and a photosynthetic membrane.

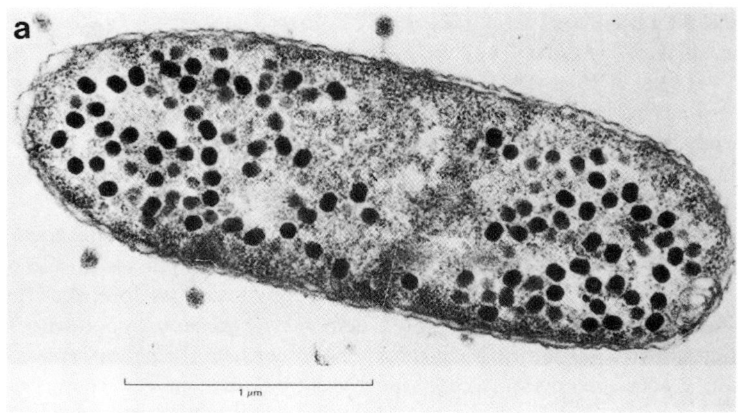

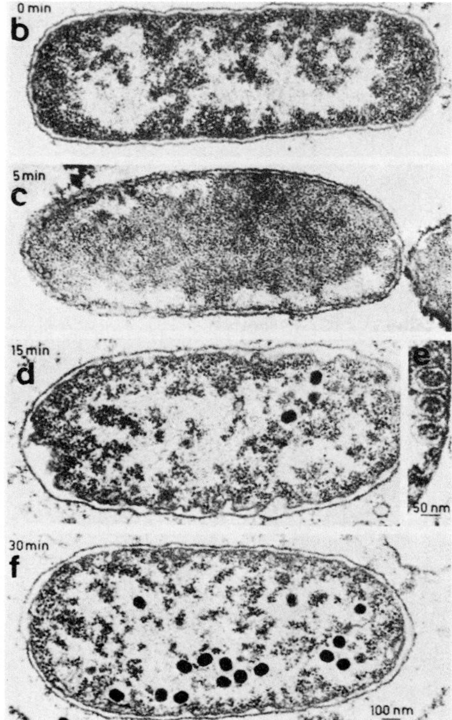

Fig. 102. a, a bacterial cell that has been infected with bacteriophages T4 for more than one hour. The cell is about to burst open; b–f, growth of virus inside a bacterial cell; e, the shells are first made separately. (From Alberts et al. 1983, after Kellenberger)

5.6.3 Intact and defective bacteriophages in bacteria

Different stages of bacteriophage synthesis and assembly within bacterial cells (Fig. 102) which harbor such viruses can be visualized in thin sections (BRADLEY & DEWAR 1967, CALENDAR 1970, KELLENBERGER et al. 1958, SIMON 1969, SIMON & ANDERSON 1967). Bacteria with a latent potential to produce bacteriophage (bacteria which are lysogenic) are well known. It has become apparent that not only intact but also defective lysogeny is a wide-spread phenomenon. A bacterium harboring a defective bacteriophage genome often has the ability to produce electron microscopically detectable non-viable phage-like particles or defective bacteriophages (Figs. 103 and 104). These structures have been defined to be particles which contain either all or some of the normal phage components, but which fail to form plaques on any known host. They have also been called high-molecular weight bacteriocins. The detection of "naturally" occurring defective bacteriophages relies heavily on electron microscopic observations which are necessary for the visualization and characterization of phage-like particles. GARRO & MARMUR (1970) divided the defective bacteriophages into three groups: particles which contain almost exclusively DNA of the bacterial host; particles containing phage-specific DNA; and particles which may not contain DNA and which resemble bacteriophage tails or tail components without a head. The last one of these groups is the one which usually is noticed in electron microscopic samples. It has been reviewed by several authors, together with other "bacteriocins" (CAMPBELL 1968, FREDERICQ 1963, IVÀNOVICS 1962, LOTZ 1976, further references therein, MAYR-HARTING et al. 1972, NOMURA 1967, OZEKI 1968, REEVES 1972, REICHLE & LEWIN 1968, YAMAMOTO 1967). These defective bacteriophages can be divided into three groups: particles which resemble sheathed phage tails, particles resembling "long

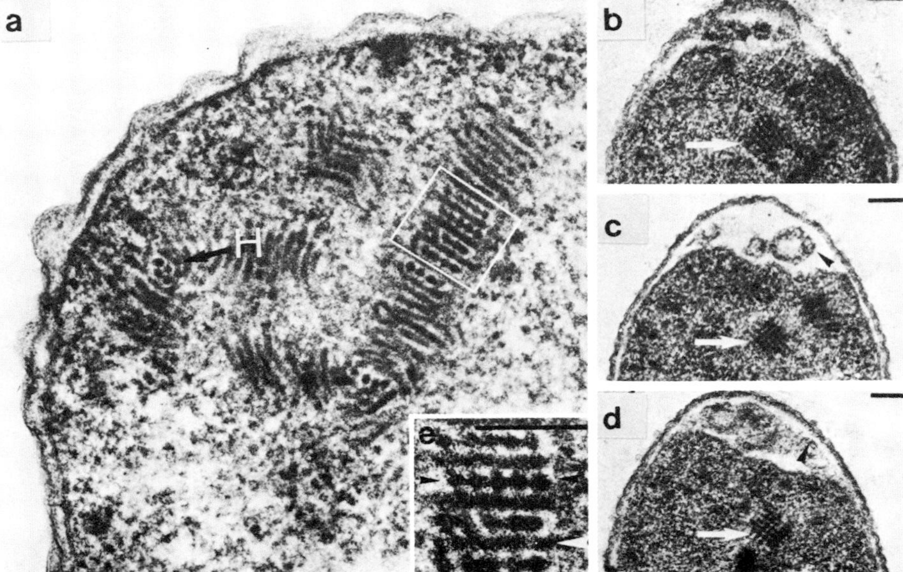

Fig. 103. *Serratia marcescens 16 b A^+* after induction with mitomycin C leading to intracellular bacteriocin production. a, hexagonal aggregations (H) of defective bacteriophage particles. Inset in a (= e): indications for fine structure of the particles resembling bacteriophage tails; b, c, d, serial sections through a cell pole with bacteriocins (white arrows) and membrane fragments (arrowheads). Bars: 100 nm. (From ACKER 1976)

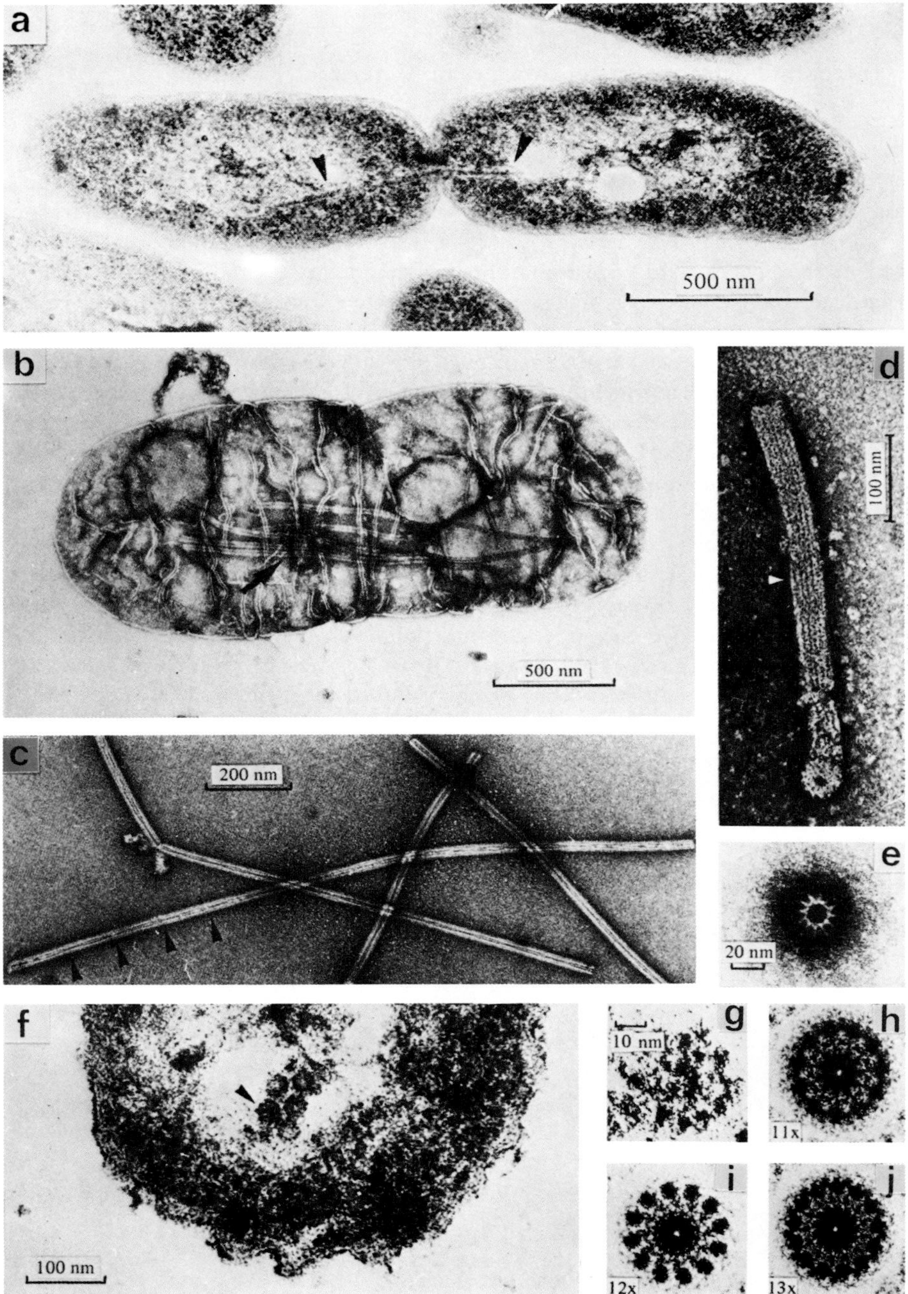

Fig. 104. Defective bacteriophages (high-molecular weight bacteriocins) in *Alcaligenes euthrophus H 16* cells. a, thin sectioned dividing cell with a polysheath (arrowheads) in the region of cell fission; b, bundle of polysheaths (arrow) in a cell treated with detergent prior to negative staining; c, isolated polysheaths; arrowheads indicate apparent "constrictions" of the inner channel (these constrictions are optical artifacts); d, isolated polysheath; a region is marked (white arrowhead) where six longitudinal rows can be discerned; e, short piece of polysheath standing upright, displaying 12-fold rotational symmetry; f, cross-sectioned bundle of polysheaths; g–j, Markham rotation technique confirms a 12-fold rotational symmetry. (From Walther-Mauruschat & Mayer 1978)

tubes", and particles resembling sheathless phage tails. Electron micrographs have been presented demonstrating the occurrence and the arrangement of such particles within bacterial cells (ACKER 1976, AULING et al. 1977, RUCINSKY & COTA-ROBLES 1973, SHINOMIYA 1972, WALTHER-MAURUSCHAT & MAYER 1978). Studies have been performed in order to obtain an insight into the mechanism by which these particles carry out their bactericidal activities (BOWMAN et al. 1973, COETZEE et al. 1968, DUCKWORTH 1970, GOVAN 1974, JAYAWARDENE & FARKAS-HIMSLEY 1970, KAGEYAMA et al. 1964, KAZIRO & TANAKA 1965a, KAZIRO et al. 1964, KROL & FARKAS-HIMSLEY 1972, LURIA 1964, NOMURA 1963), including the induction of synthesis, investigation of specific receptors on bacterial cell surfaces (IKEDA & EGAMI 1969, 1973, SABET & SCHNAITMAN 1973, SMIT et al. 1969), states of adsorption, different modes of action, i. e. killing of sensitive cells by injection or mere contact, and attempts to characterize chemical composition and bactericidal components by fractionation of isolated defective bacteriophages. The origin of these particles is assumed to have occurred by progressive reduction of phage-specific genetic information (BRADLEY 1967, GARRO & MARMUR 1970, HOMMA et al. 1967, HOMMA & SHIONOYA 1967, ITO & KAGEYAMA 1970, KAGEYAMA 1970a, b). The physiological role of defective bacteriophages might be to help to protect the bacterial cell against additional (potentially lethal) virus infection, either by conversion of the cell envelope, by prophage-encoded restriction endonucleases, or by prophage-encoded repressor proteins.

5.6.4 Intracellular refractile structures (R-bodies)

Inclusion bodies called "refractile bodies" (R-bodies) in killer stains of paramecia containing endosymbionts (e. g., *kappa*) have been described (Fig. 105) from light microscopic investigations (PREER & STARK 1953) and in early studies using ultrathin sectioning techniques (DIPPELL 1958, HAMILTON & GETTNER 1958). PREER (1981) and PREER et al. (1974) reviewed the *Paramecium-kappa*-R-body systems. Isolated R-bodies from *kappa* endosymbionts of *Paramecium aurelia* were studied by ANDERSON et al. (1964) using negative staining. This investigation indicated that R-bodies are hollow cylinders assembled or rolled from a ribbon-like sheet of material into a lamellar structure. When seen in profile, the R-body was approximately square with a side of about 0.5 µm. Unwound R-bodies were also observed. They revealed that the length of the ribbon-like material extended to about 15 µm and that it had a characteristic twist or folding pattern. From the occurrence in ultrathin sectioned endosymbionts of spherical dark-staining phage-like structures found to be present, together with R-bodies, it has been speculated that R-bodies might be related to bacteriophages. It has been discussed that some of the observed structures might represent incomplete or abortive assemblies of phage-like elements. Free-living bacteria with structures possessing the same features as R-bodies (Fig. 106) have been isolated (LALUCAT & MAYER 1978, LALUCAT et al. 1979, 1982, WELLS & HORNE 1983). Studies by LALUCAT et al. (1980) have shown that R-bodies from the free-living bacterium *Pseudomonas taeniospiralis* strain *2K1* have a toxic effect on susceptible paramecia. This bacterium has been reported to be infected with a defective bacteriophage, which appears to be similar to the occurrence of phage-like inclusion bodies associated with endosymbionts containing R-bodies (s. above). Experiments of QUACKENBUSH & BURBACH (1983) have provided direct evidence that the genetic determinants for R-body synthesis in *Caedibacter taeniospiralis 47 (47 kappa)* are plasmid-born. These investigations were done by the cloning and expression of DNA sequences associated with the killer trait of *Paramecium tetraurelia* stock *47*, which had been introduced into *E. coli 294* by transformation. Electron micrographs of ultrathin sections of these *E. coli* cells clearly revealed the presence of typical R-body structures.

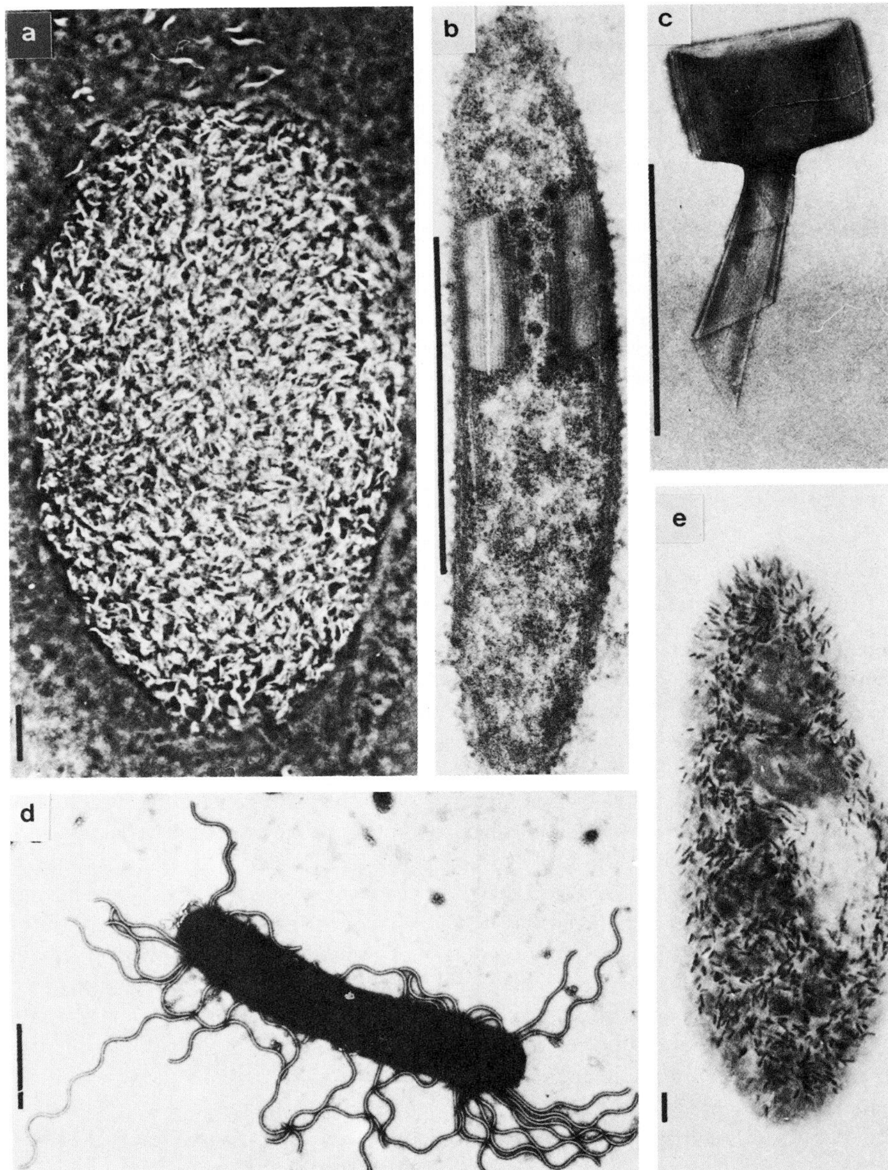

Fig. 105. Prokaryotic symbionts in *Paramecium*. a, vegetative macronucleus of stock *562 Paramecium biaurelia* containing *alpha* in spiral form; few *alpha* also in the cytoplasm (bar: 10 μm; after PREER); b, bright *kappa* of stock *7 Paramecium biaurelia;* dark-staining phages inside the coiled refractile (R-) body (bar: 1 μm; after PREER & JURAND); c, isolated R-body beginning to unroll from the inside (bar: 1 μm; after PREER et al.); d, *lambda* of stock *327 Paramecium octaurelia* (bar: 1 μm; after PREER et al.); e, whole mount of stock *239 Paramecium biaurelia* containing *lambda,* seen as dark-staining rods in the cytoplasm (after PREER et al.). (From PREER 1981)

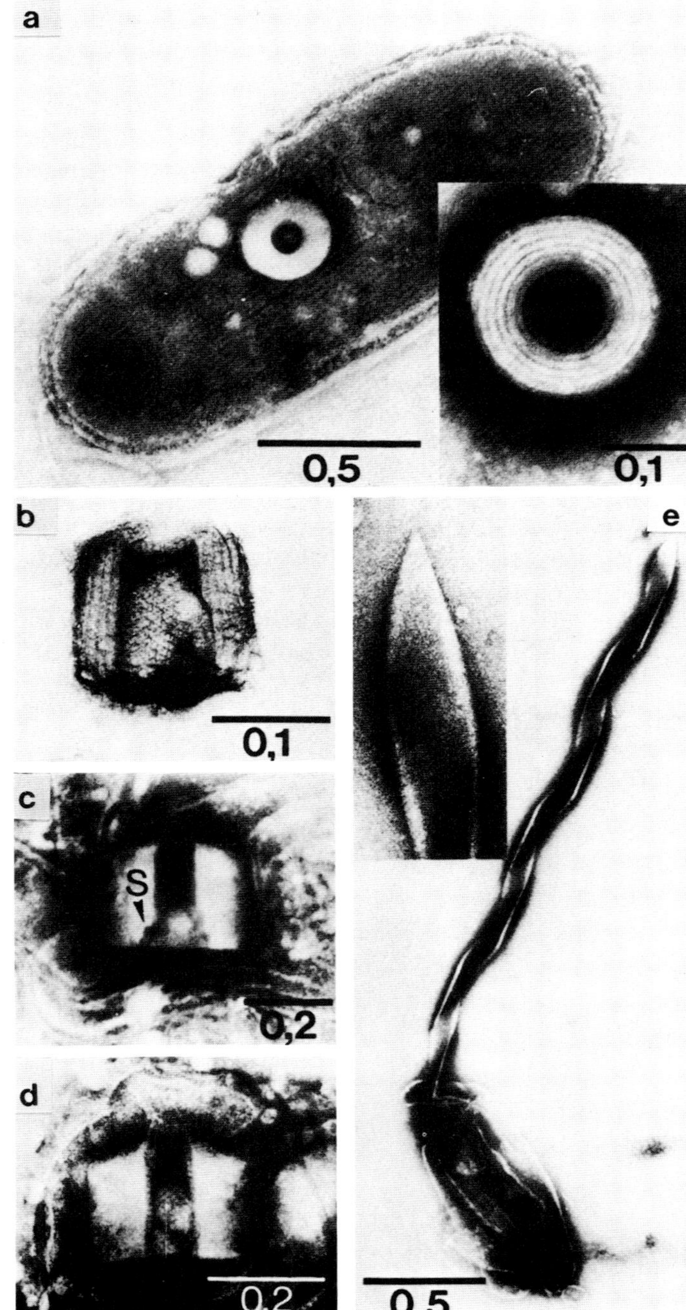

Fig. 106. R-bodies in the free-living hydrogen bacterium *Pseudomonas taeniospiralis* strain *2K1*. a and inset in a, autotrophic cell in the stationary growth phase with one R-body in face view. Inset: R-body in face view at high magnification; b, R-body in side view, surrounded by membrane remnants. Inner side with a regular striation; c, R-body in side view, with "steps" (S); d, R-body in side view; the inner side with a regular striation appearing like a furrow; e and inset in e, unrolled R-body sticking out from a lysed cell. Acute end (inset) not revealing any structural subunits.
All samples negatively stained. Dimensions are given in µm. (From LALUCAT et al. 1979)

5.7 The bacterial ribosome

Ultrathin sections of bacterial cells usually exhibit the nuclear region and the cytoplasm, the latter filled with ribosomes. The ribosomes show a diameter of around 20 nm. Often, they appear to be arranged as polysomes (Cancedda & Schlesinger 1974). Theories for the export of proteins from the interior of the cell to the outside require that some of the ribosomes should be in close association with the cytoplasmic membrane. However, attempts aimed at the identification of a specific population of ribosomes attached to the cytoplasmic membrane have not yet been successful. Among other factors, uncertainties in the proper control of the concentration of Mg^{2+} in the suspending fluid used for the isolation of membrane-ribosome complexes might have been one of the reasons why such associations found during these experiments have not been very meaningful. As demonstrated by Miller et al. (1970), polysomes can be isolated in a state where they are attached to the chromosomal DNA. However, neither the electron micrographs of ultrathin sections or freeze-etched samples allow any discrimination between possibly existing different classes of ribosomes.

A wealth of information is available on the three-dimensional structure and the composition of *E. coli* ribosomes (Boublik 1985, Nanninga 1973, Tietz & Hoppe 1984, Verschoor et al. 1984, Wabl et al. 1973), and their 30 S (Breitenreuter et al. 1984, Hochkeppel et al. 1977, Nanninga et al. 1972, Stöffler-Meilicke & Stöffler 1984, Traub & Nomura 1968, Wagenknecht et al. 1984) and 50 S subunits (Spiess 1979, Stöffler 1984, Wabl 1974), and models have been developed for bacterial ribosomes and their subunits (Figs. 107–113). A wide variety of techniques have been applied in order to obtain data on the numbers of different polypeptide chains in the subunits, their location, the types and locations of ribosomal RNA molecules and the interactions between these

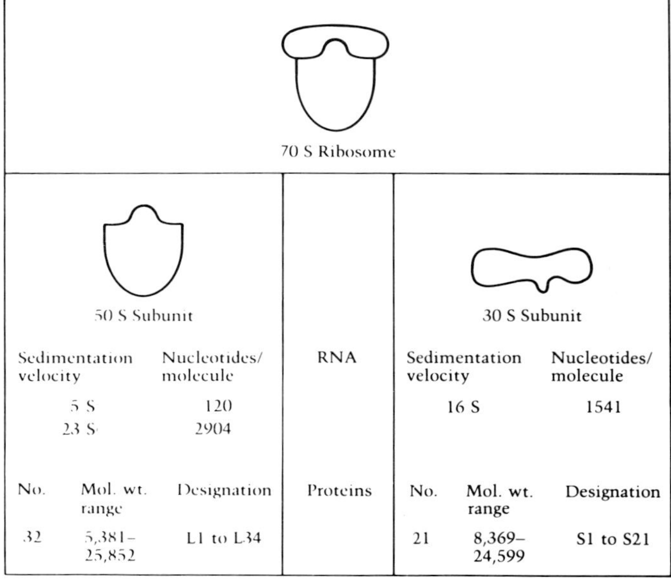

Fig. 107. The molecular structure of the *Escherichia coli* ribosome. (From Ingraham et al. 1983)

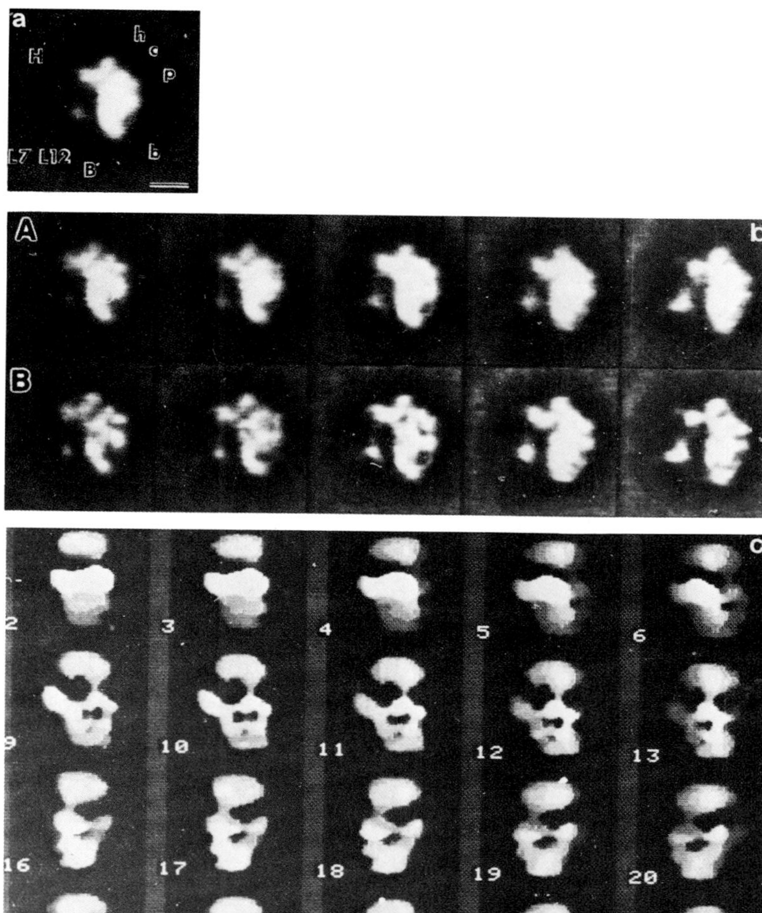

Fig. 108. Ribosome structure. a, total average of 95 aligned 70 S monosome images in the overlap view. 30 S subunit: h = head; c = cleft; p = platform; b = base; 50 S subunit: H = head; L = stalk; B = base; b, subset averages of 70 S monosomes; A, without filtration; B, with high-pass filtration in order to enhance the edges of the particles. (From VERSCHOOR et al. 1984); c, continuous stereographic representation of a 3D reconstruction of the 30 S subunit of *E. coli* ribosome. (From VAN HEEL 1984)

structural components. One of the major approaches for the localization of many of the ribosomal proteins exposed at the surfaces of the two types of subunits involves immunoelectron microscopy. Figs. 111 and 112 demonstrate some typical findings, indicating the positions of the polypeptide chains as deduced from exposed antigenic determinants localized by binding of specific IgG antibodies. This technique does not allow the identification and localization of those parts of the individual polypeptide chains which are "buried" within the particle. Therefore, several other approaches to protein placement have been applied. One is the production of neutron maps. For

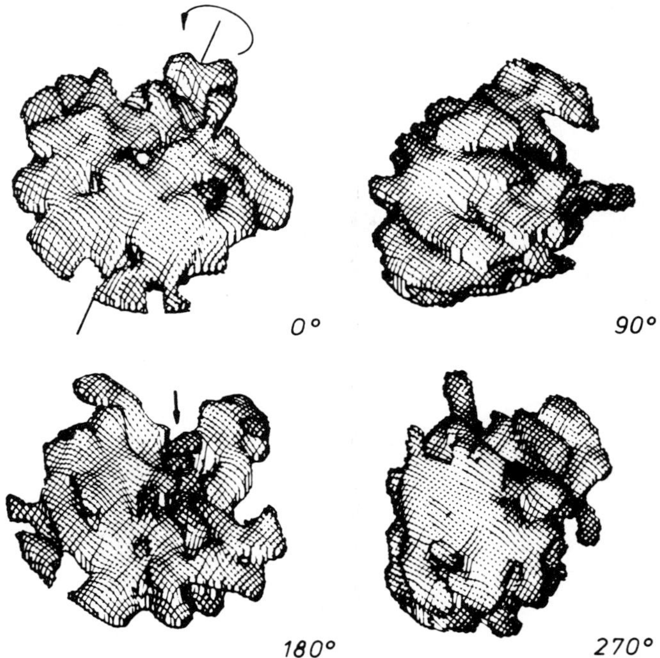

Fig. 109. Ribosomal structure. Shape of the 50 S ribosome subunit without "stalk". The different views are rotated in steps of 90° around the axis shown in the 0° view. The groove is marked by an arrow. (From HEGERL et al. 1984)

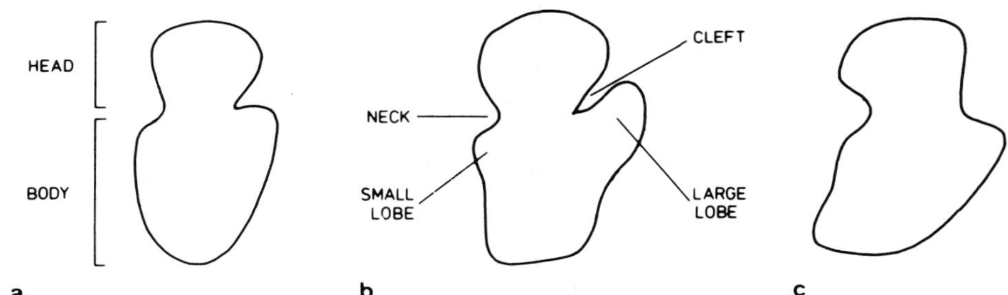

Fig. 110. Ribosome structure; three characteristic views of the 30 S subunit of the *E. coli* ribosome, giving the nomenclature of the typical structural features. (From STOEFFLER-MEILICKE & STOEFFLER 1984)

example, Fig. 113 demonstrates the positions of proteins in the 30 S subunit of the *E. coli* ribosome as obtained by neutron scattering used to measure distances between pairs of deuterated proteins. Measurements of fluorescence energy transfer have also been performed to estimate the separation between pairs of specific proteins. A common technique also used in this respect is the application of bifunctional reagents to covalently couple adjacent proteins. Various reagents spanning distances from 0.5 to 2.5 nm have been chosen. Moreover, the complete sequences of nucleotides of the 5 S,

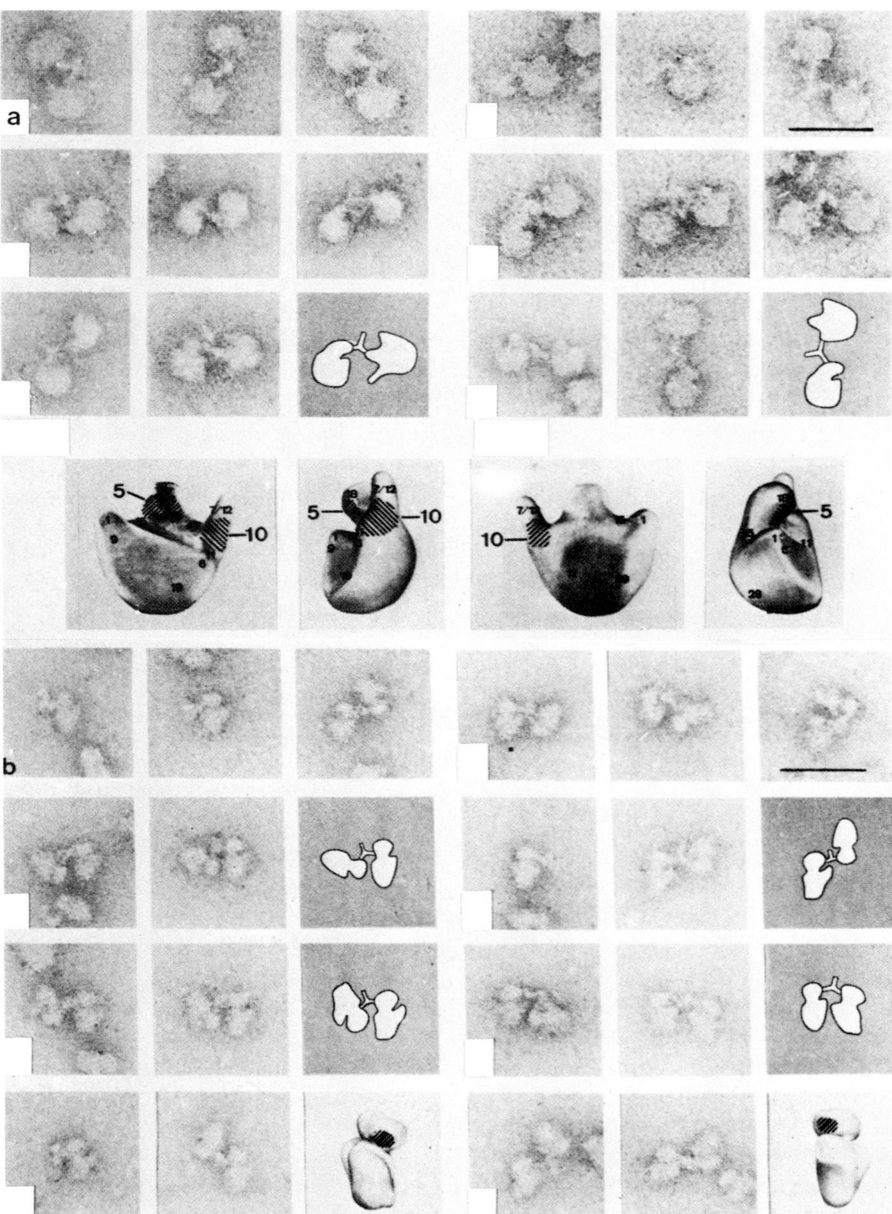

Fig. 111. Ribosomal structure. a, location of proteins on the 50 S subunit of *E. coli* ribosomes as deduced from immunoelectron microscopy. Bar: 50 nm. (From STOEFFLER et al. 1984); b, location of proteins on the 30 S subunit of *E. coli* ribosomes. Bar: 50 nm. (From BREITENREUTER et al. 1984)

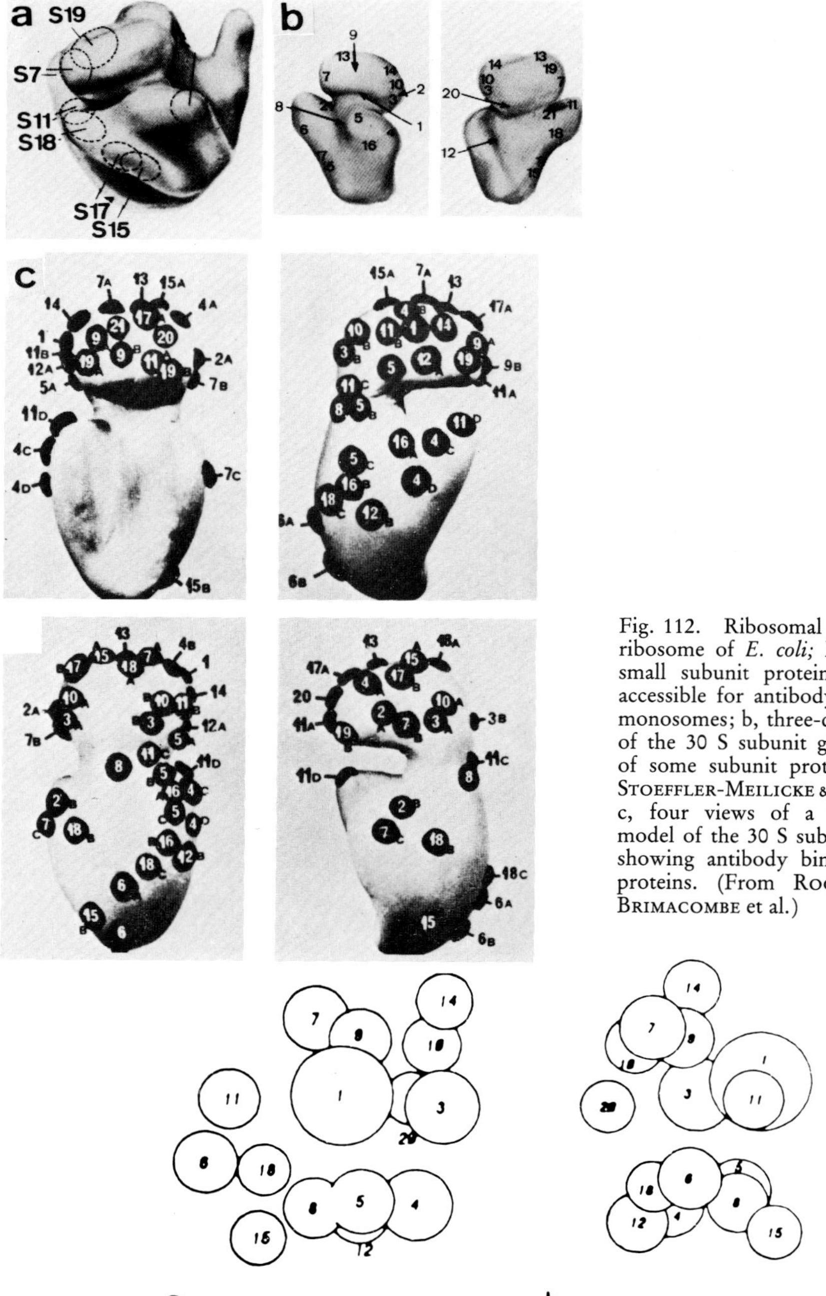

Fig. 112. Ribosomal structure. a, 70 S ribosome of *E. coli;* locations of those small subunit proteins which are not accessible for antibody binding in 70 S monosomes; b, three-dimensional model of the 30 S subunit giving the location of some subunit proteins. (a, b, from STOEFFLER-MEILICKE & STOEFFLER 1984); c, four views of a three-dimensional model of the 30 S subunit of ribosomes showing antibody binding sites for 21 proteins. (From ROGERS 1983, after BRIMACOMBE et al.)

Fig. 113. Ribosomal structure; neutron map of the positions of proteins in the 30 S subunit of *E. coli.* View (b) is rotated counterclockwise around a vertical axis in the plane of the page. (From STOEFFLER-MEILICKE & STOEFFLER 1984, after RAMAKRISHNAN et al., and MOORE)

16 S, and 23 S ribosomal RNA molecules are known, allowing the construction of models for the secondary structure of ribosomal RNA. Thus, the present view of the location of ribosomal components is a combined result of a variety of topographical methods. The description of ribosomal function is outside the scope of this book. However, it may be mentioned that the assembly process leading to the complete ribosome has been analyzed in considerable detail. Major aspects of these reactions have been summarized by INGRAHAM et al. (1983). Figs. 114 and 115 demonstrate the complexity of the assembly process.

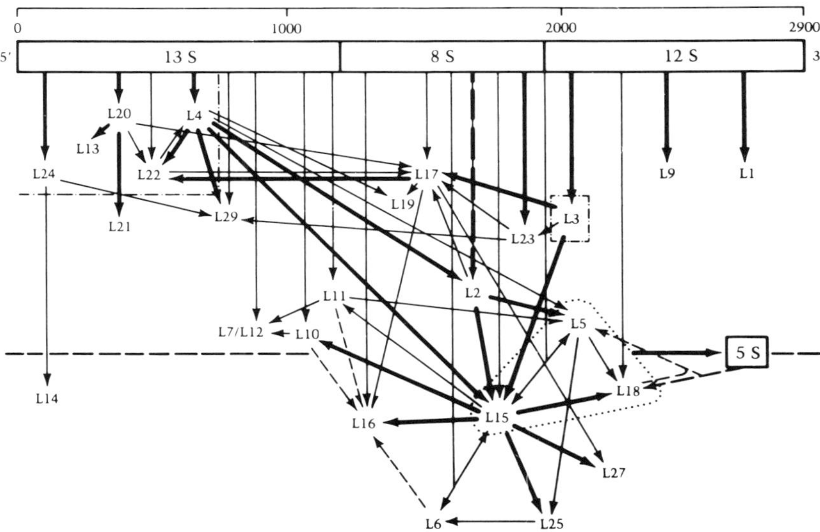

Fig. 114. Assembly map of the ribosomal 50 S subunit. The horizontal broken line separates proteins of the R I_{50} (1) particle from the ones added to give the R I_{50} (2) particle. Arrows indicate the facilitating effect on the binding of one protein by another. (From INGRAHAM et al. 1983, after NIERHAUS)

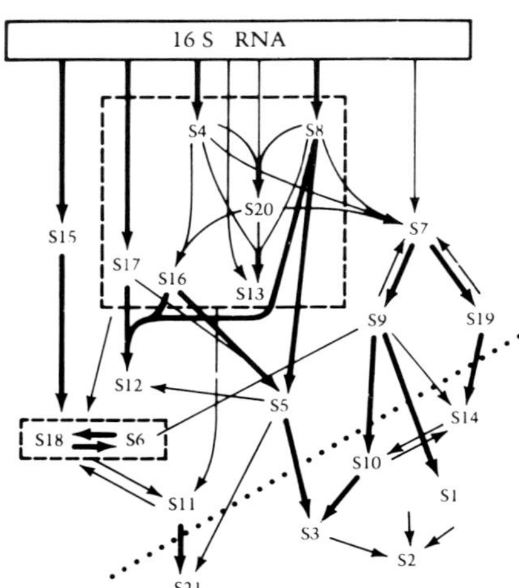

Fig. 115. Assembly map of the 30 S ribosomal subunit. Proteins above the dotted line are those that form the R I_{30} particles. The arrows indicate the facilitating effect on the binding of one protein by another. (From INGRAHAM et al. 1983; in part after HELD and after LAUGHREA)

5.8 The organization of the bacterial nucleoid

INGRAHAM et al. (1983) discussed the appropriate terms for the description of the genetic apparatus of bacteria. They proposed that "nuclear body" (or "nuclear region"), instead of "nucleus", should be used as the preferable term for what one sees in situ with preparations stained to reveal DNA in the light microscope, or with ultrathin sections viewed in the electron microscope. "Nucleoid" should be used to describe the structure one obtains after gentle lysis of the bacterial cell (Figs. 117 and 118). It represents a folded bacterial DNA molecule complexed with RNA and proteins. However, INGRAHAM et al. (1983) stated that, as information about nucleoids increases, it is likely that the term "nucleoid" will come to refer both to the in vitro and in vivo structure.

In light microscopic samples visualized under appropriate conditions, one to several nuclear bodies, depending on the rate at which the cells have been grown, can be seen. Each nuclear body contains the same haploid chromosome. In electron micrographs of ultrathin sections from samples prepared in the presence of OsO_4, the nuclear body is readily visible as a more or less distinct cell region ("OsO_4 nucleoid") containing complex fibrillar elements. In ultrathin sections cut from frozen-hydrated bacteria, the nuclear body appeared in exponentially growing bacteria as an ill-defined central region (Fig. 7). It was more clearly visible when bacteria were harvested for preparation from the stationary phase of growth.

The visibility of a distinct nucleoplasmic region in conventional ultrathin sections might be artificially caused by the destruction of the permeability barrier by fixation, thus creating a new ionic environment in the cell (WOLDRINGH 1973), or through the disruption of the association of transcription-translation complexes with DNA (DANEO-MOORE et al. 1980), or as a result of the loss of DNA-binding proteins (VALKENBURG & WOLDRINGH 1984). In any case, the nucleoplasmic region is devoid of ribosomes.

Not only from electron microscopy but also from other techniques indications have been obtained for the bacterial chromosome being attached to the cell membrane (DOYLE et al. 1983, ELLINGER et al. 1982, FIRSHEIN 1972, HEIDRICH & OLSEN 1975, HOROWITZ et al. 1979, OLSEN et al. 1974, RYTER 1968).

CAIRNS (1963 a, b) managed to demonstrate, by direct visualization from autoradiograms, that bacteria contain, per nuclear body, a single circular chromosome. In *E. coli,* the overall length of the chromosome was found to be about 1 mm. In Fig. 119, a bacterial chromosome which had been in the state of replication at the moment of preparation is depicted.

Considerable condensation is needed in order to place a bacterial chromosome within the cell body. This is assumed to be mediated by a variety of factors. EICKBUSH & MOUDRIANAKIS (1978) maintained that the double helix has an intrinsic potential to direct its own packaging into two different modes. The mode of DNA packaging was found to be determined by the electrostatic charge density and water activity of the immediate microenvironment of the helix. The two basic structures formed by both linear and covalently closed-circular DNA are: a left-handed supercoil characteristic of minimally charge-shielded DNA, and a smooth rod characteristic of fully charge-shielded DNA. The authors proposed that in the supercoil the double helix is overwound, while in the rod the helix is folded back and forth on itself, and they assumed that the observed structures of DNA in vitro mimic the packaging behavior of the DNA in vivo (s. also MEIJER et al. 1976 a, b, PETTIJOHN & SINDEN 1985, STONINGTON & PETTIJOHN 1971, WORCEL & BURGI 1972) where the DNA usually is complexed with proteins (s. below).

Bacterial DNA, not gently released from partially disrupted cells, when applied to a supporting film after spreading, shows only "naked" strands (Fig. 116). However, GRIFFITH (1976) succeeded

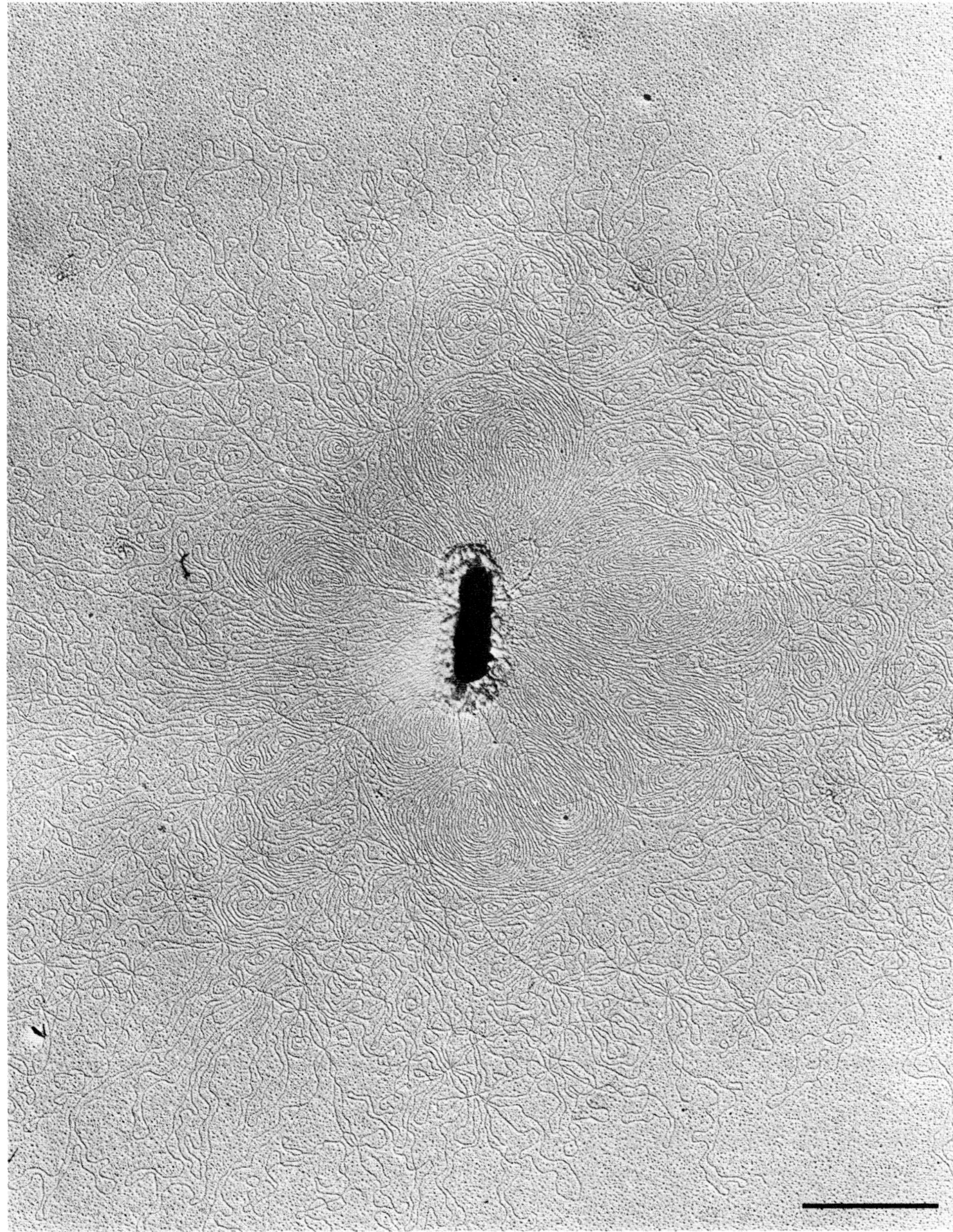

Fig. 116. Bacterial chromosome released from the cell by osmotic shock. Bar: 1 μm. (Original micrograph H. LÜNSDORF)

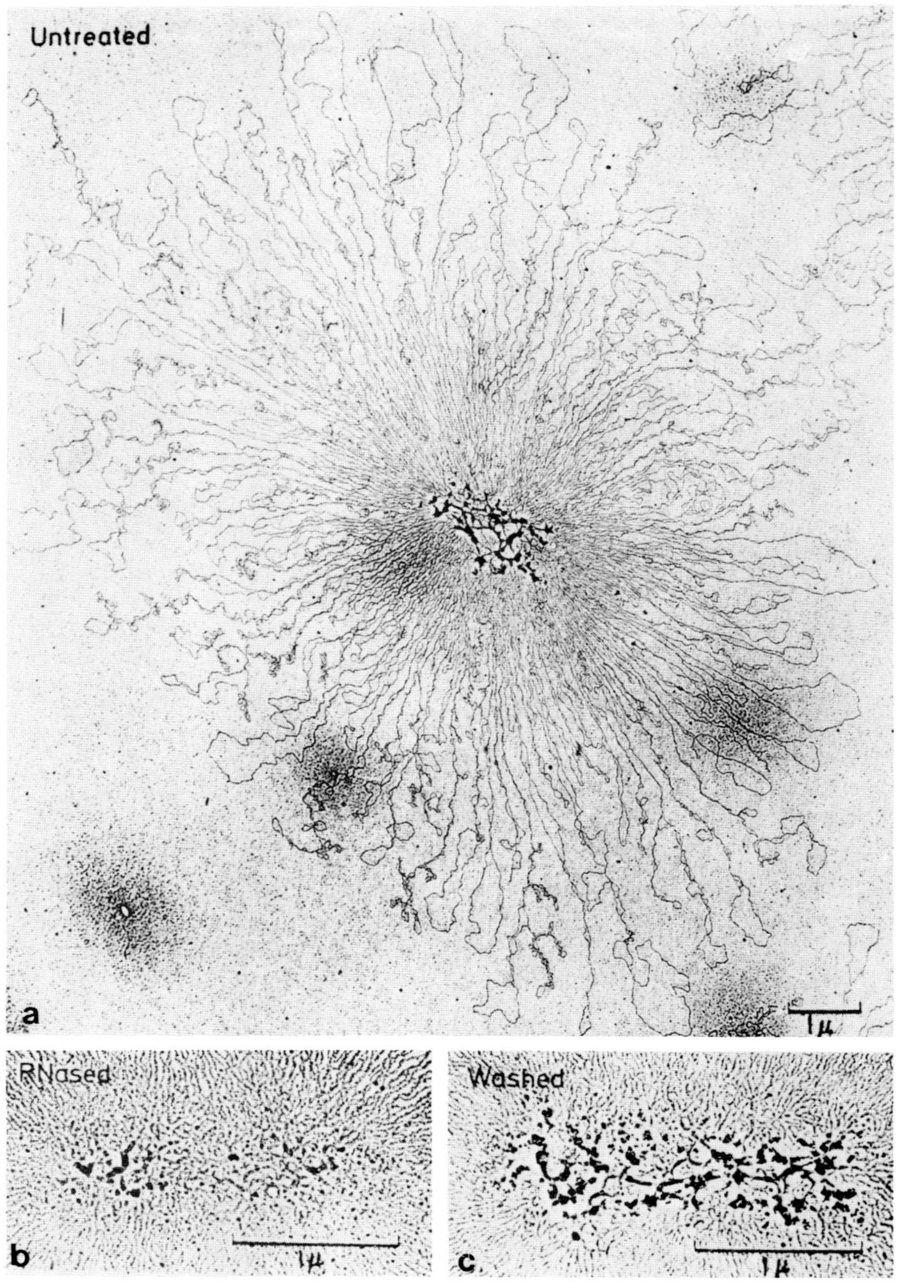

Fig. 117. a, the cell envelope-free nucleoid from *Escherichia coli;* loops can be seen; b, the material at the centre can be partly removed by treatment with RNase; c, control, not RNase-treated. (From ROGERS 1983)

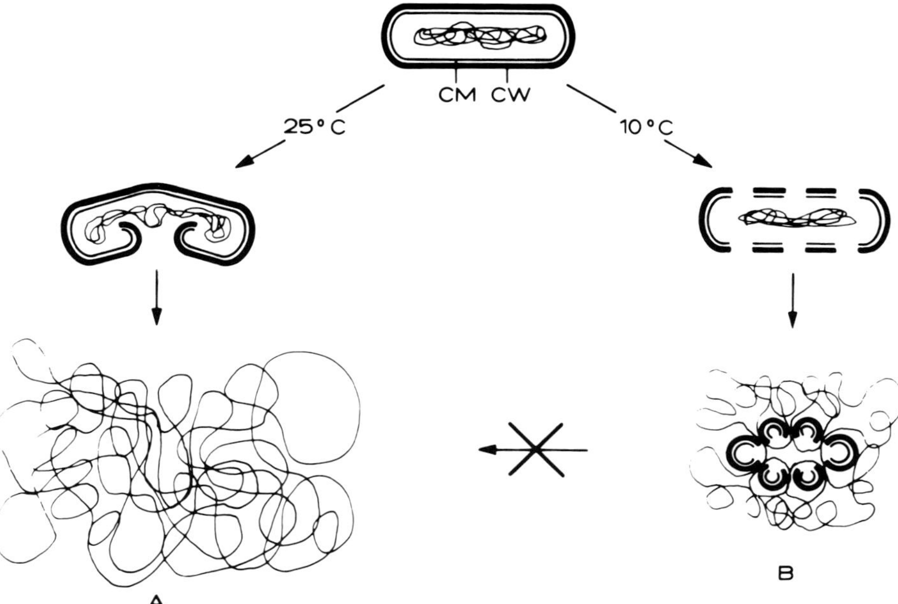

Fig. 118. Scheme representing the release of envelope-free (A) and envelope-bound (B) folded chromosomes from intact cells. – CM = cytoplasmic membrane; CW = cell wall. (From NANNINGA et al. 1984, after MEIJER et al.)

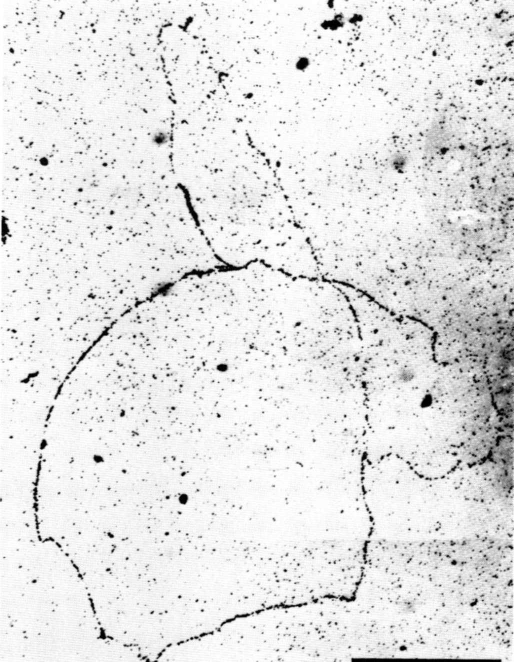

Fig. 119. Autoradiography of a chromosome of *Escherichia coli*; prepared from a culture that had been labelled for approximately 1.8 generations with [^{3}H] thymidine. Bar: 0.1 mm. (From CAIRNS 1963)

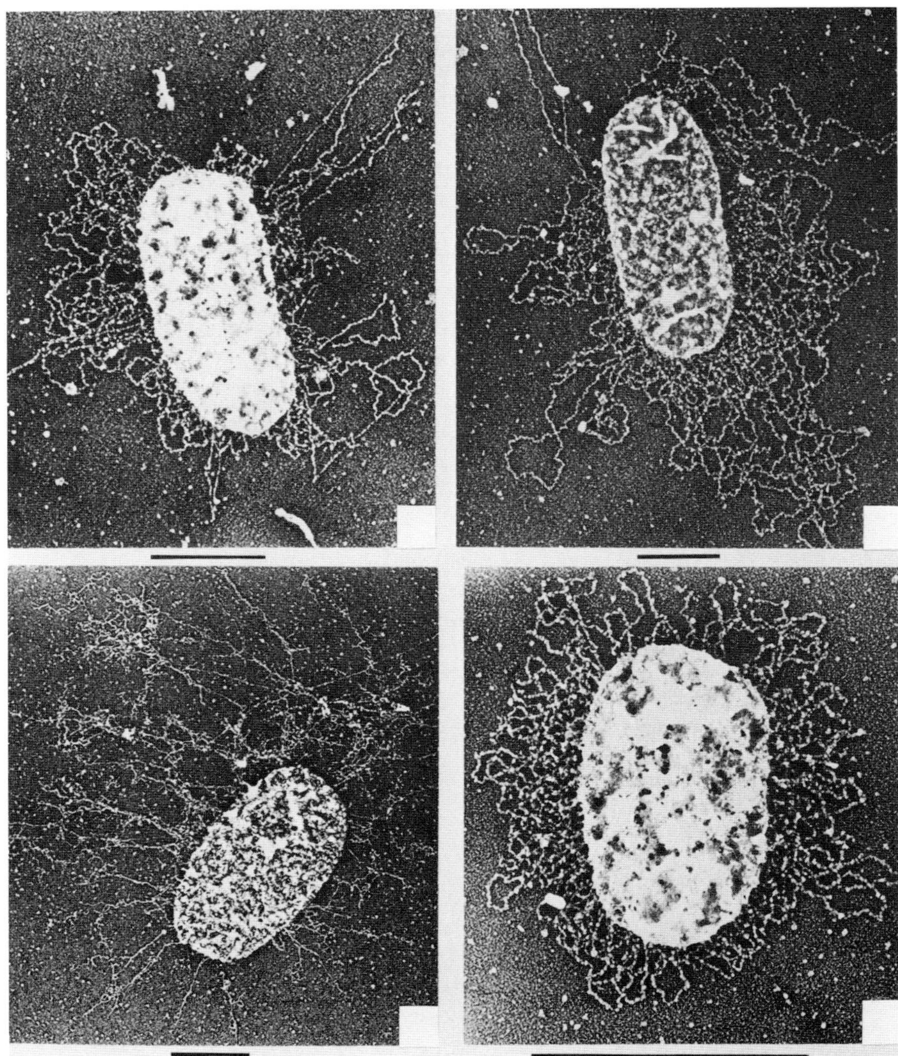

Fig. 120. Visualization of *E. coli* cells disrupted on the electron microscope supporting film. The DNA appears in regularly condensed chromatin-like fibers. Bars: 1 μm. (From GRIFFITH 1976)

in visualizing prokaryotic DNA in regularly condensed chromatin-like fibers (Figs. 120 and 121). These beaded fibers were detected only when the cells *(E. coli)* were gently lysed in 0.15 M NaCl solutions directly on the electron microscope-supporting film and if the dehydration steps began within 2 min of lysis. The fibers were found to have a diameter of 12 nm, with 13 nm repeating beaded substructure, and they were organized as loops attached to the cell envelope. GRIFFITH (1976) also discussed the factors that might stabilize DNA in a condensed state. He proposed that Mg^{2+}, polyamines, and perhaps a very heterogeneous class of weakly bound proteins might be

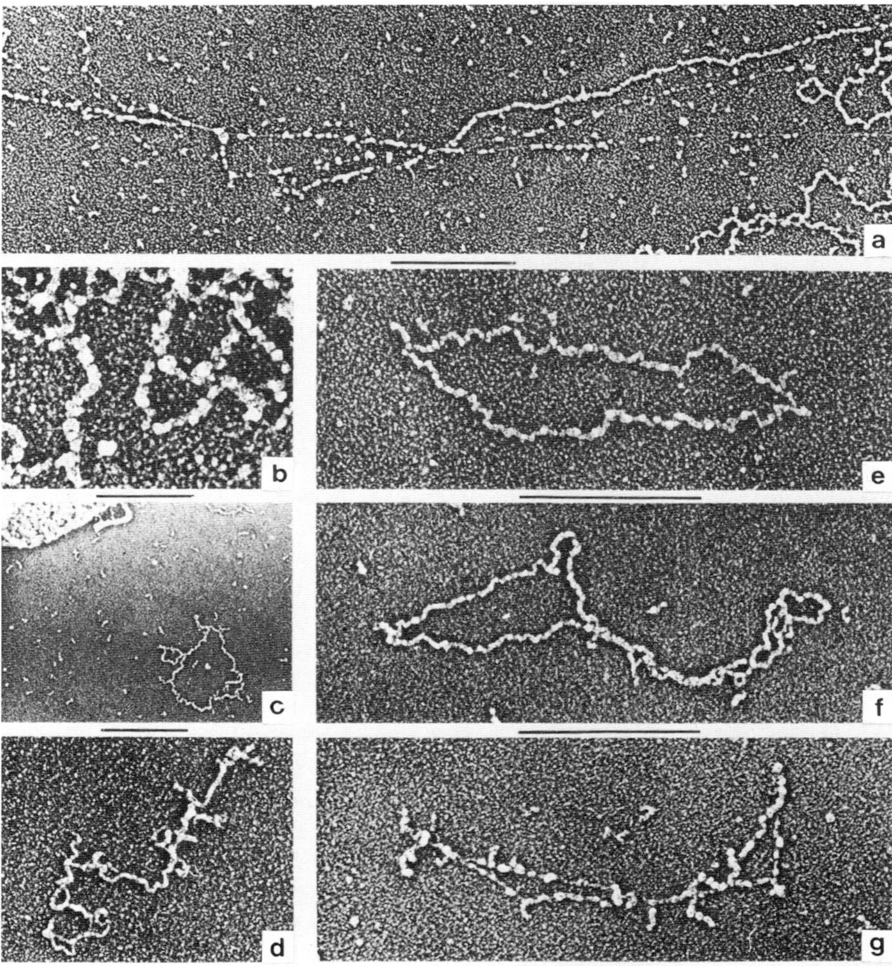

Fig. 121. *E. coli* and bacteriophage λ condensed DNA fibers. a, b, examples of 12 nm condensed fibers stretched during mounting (a) and unstretched (b); c–g, beaded nature of λ DNA prepared by disruption of *E. coli* cells. Bars: 0.5 μm except for b, which is 0.1 μm. (From GRIFFITH 1976)

responsible for this stabilization. SLOOF et al. (1983) isolated nucleoids from *Bacillus licheniformis* under low-salt conditions and without addition of detergents, polyamines or Mg^{2+}. These nucleoids were partially unfolded by treatments that disrupt protein-DNA interactions. The DNA was observed to be organized in independent, negatively supertwisted domains restrained by protein-DNA interactions. The number of these domains was estimated to be between 50 and more than 100. A major role for nascent RNA in restraining supertwisted DNA was not detected.

Small basic proteins have been discovered and purified from different prokaryotes. These proteins, in addition to being localized in the nuclear body in vivo, were found to promote the appearance of nucleosome-like structures when added to naked DNA. In *E. coli*, these proteins were given different names: HU (ROUVIÈRE-YANIV & GROS 1975), BH2 (VARSHAVSKY et al. 1977), NS (SURYANARAYANA & SUBRAMANIAN 1978), and DNA-binding protein II (GEIDER & HOFFMANN-BERLING 1981). They consist of two very similar polypeptide chains, probably arising from the duplication of a common ancestral gene. Molecular weights around 9000 D, 15000 D, 17000 D, and 30000 D have been determined. HASELKORN & ROUVIÈRE-YANIV (1976) detected that HU is highly conserved in prokaryotes, immunologically cross-reacting proteins being found in species of cyanobacteria, *Salmonella typhimurium* and *Bacillus subtilis.* Comparison of partial amino acid sequences indicated that the sequence variability between *E. coli* and cyanobacteria HU is as low as that found among eukaryotes for histones H2a and H2b. A certain microheterogeneity in the amino acid sequences of all HU proteins studied has been observed. Two HU subspecies, α and β, corresponding to the NS1 and NS2 proteins of MENDE et al. (1978), have been identified. Most native HU was found as an αβ-dimer or as tetramers (SURYANARAYANA & SUBRAMANIAN 1978).

From the functional point of view it appears that these proteins resemble eukaryotic histones. NS1 and NS2 bind preferentially to double-stranded DNA (ZENTGRAF et al. 1977) which they protect from nuclease digestion (VARSHAVSKY et al. 1977) and thermal denaturation (MIANO et al. 1982). Furthermore, these proteins were found to inhibit DNA synthesis (BERTHOLD & GEIDER 1976), but they have no effect on translation. Nevertheless the extent to which the analogy between prokaryotic and eukaryotic systems can be carried remains an open question. Some additional insight into structure-function relationships regarding the interactions of these proteins with DNA was obtained by PACI et al. (1984), using high-resolution ^{1}H NMR spectroscopy in a study of *E. coli* DNA-binding proteins NS1 and NS2, aimed at obtaining information on the tertiary and quaternary organization of these proteins. It was demonstrated that the tertiary and quaternary structures of NS are insensitive to extensive variations of ionic strength, unlike the situation with eukaryotic histones. The extensive tertiary and quaternary structures were found to involve hydrophobic interactions. The structures were lost upon heating, but readily reformed upon cooling. Molecules of the heterologous subunits (NS1 and NS2) were observed to dimerize preferentially in comparison to the self-aggregation of the homologous subunits.

Electron microscopy has revealed that a different type of *E. coli* protein, the single-strand-binding protein (SSB), when complexed with single-stranded DNA, also causes the creation of beaded fibers (Fig. 122). Detailed evaluations of these complexes combined with digestion experiments (CHRYSOGELOS & GRIFFITH 1982) suggested that SSB organizes single-stranded DNA in a manner similar to the organization of duplex DNA by histones (FELSENFELD 1978). SSB is not the only protein in *E. coli* that binds (cooperatively) to single-stranded DNA; the recA protein is also abundant. Both these proteins are involved in genetic recombination. Regarding functional aspects of these proteins, CHRYSOGELOS & GRIFFITH (1982) proposed that the bead- and linker arrangement of SSB-DNA complexes provides a mechanisms for extending the DNA while allowing other proteins access to a portion of the DNA. If another protein was to bind in a highly cooperative manner, beginning in the linker region, this might lead to release of the SSB. A further description of the details of these systems is outside the scope of this book; GEIDER & HOFFMANN-BERLING (1981), in a review on proteins controlling the helical structure of DNA, have summarized and discussed additional details such as purification and characterization of DNA-binding proteins, single-strand-binding proteins, double-strand-binding proteins, and have dealt with further characteristic aspects of DNA unwinding proteins, the recA protein, and DNA topoisomerases of types I and II.

Newly replicated bacterial DNA needs no special assembly process to become a chromosome (INGRAHAM et al. 1983). As the DNA is produced, it attracts all regulatory proteins, some of which

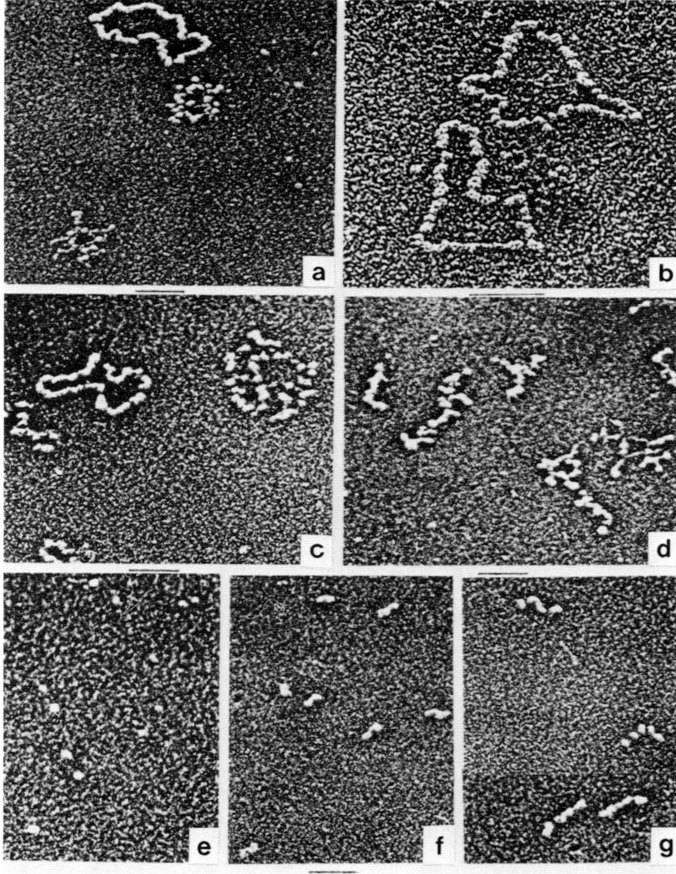

Fig. 122. Visualization of *E. coli* single-strand binding protein bound to bacteriophage fd single-strand DNA. The beaded organization of the complexes is apparent. Different aspects are obtained by variation of protein/DNA ratios (a–d), or by variation of the length of the DNA (e–g). Bars: 0.1 μm. (From Chrysogelos & Griffith 1982)

have been described above, that bind to DNA as their substrate, and some of them to their appropriate recognition sites, giving rise to the "maturation" of the chromosome by methylation, supercoiling, and folding. In addition, the DNA molecules are attached to the bacterial membrane by means that assure appropriate segregation of the chromosomes into daughter cells as they are formed (s. below).

Plasmids within bacterial cells have not been seen as distinct entities in ultrathin sections. They have only been visualized after extraction. Using conventional isolation procedures, plasmids appear as covalently closed or open circles of naked DNA. However, after gentle lysis plasmid pHG1 from *Alcaligenes eutrophus* exhibited a pearl necklace-like structure very similar to the complexes formed by bacterial chromosomal DNA and certain stabilizing proteins (Pape, personal commun.). Evidence has been obtained (Miller & Kline 1979) that, in the cell, RNA links F-plasmid DNA to the folded chromosome. This plasmid-RNA-chromosome complex was found to be alkaline labile. The chromosome-plasmid part of it was released by RNase treatment. Further indication for some mode of stable maintenance of plasmids within bacteria was obtained by FRIEDRICH et al. (1981) and HOGREFE et al. (1984). They observed that, even with curing agents, it was extremely difficult to eliminate megaplasmids of the type pHG from *Alcaligenes eutrophus* cells.

A different aspect of the organization of the bacterial nucleoid has been detected and described by MILLER et al. (1970) and HAMKALO & MILLER (1973) in their investigation, by electron microscopy, of fragile cells of *E. coli* after lysis. They adapted techniques developed for the visualization of the structure of active genes in eukaryotic cells for the observation of chromosomes of *E. coli*. It turned out that most of the *E. coli* chromosome is not genetically active at any one instant, that translation is completely coupled to transcription, and that 16S and 23S ribosomal RNA cistrons occur in tandem, in regions which are widely spaced on the chromosome. The authors made these interpretations from electron micrographs showing the extruded content of osmotically shocked *E. coli* cells. The content appeared as a mass of thin fibers with attached strings of granules with the size of bacterial ribosomes (Fig. 123). The fibers were shown, by treatment with DNase, to be the bacterial chromosome. The strings of granules were identified as ribosomes (polysomes) translating mRNA molecules at the time of isolation. The diameter of the fibers (around 4 nm) was probably caused, under the given conditions of gentle lysis, by proteins bound in vivo to the chromosomal DNA. Each polyribosome was seen to be attached to the chromosome at a site of an irregularly shaped granule 7.5 nm in diameter interpreted as an RNA polymerase molecule.

Inspection of electron micrographs obtained from ultrathin sections of bacterial cells, independent of the type of fixation, usually reveals a more or less pronounced nuclear region surrounded by the cytoplasm (EDELSTEIN et al. 1981, MOORE & HIRSCH 1973a, NANNINGA & WOLDRING 1981, RYTER & KELLENBERGER 1958, WOLDRING & NANNINGA 1976, 1985). No ribosomes, polysomes or

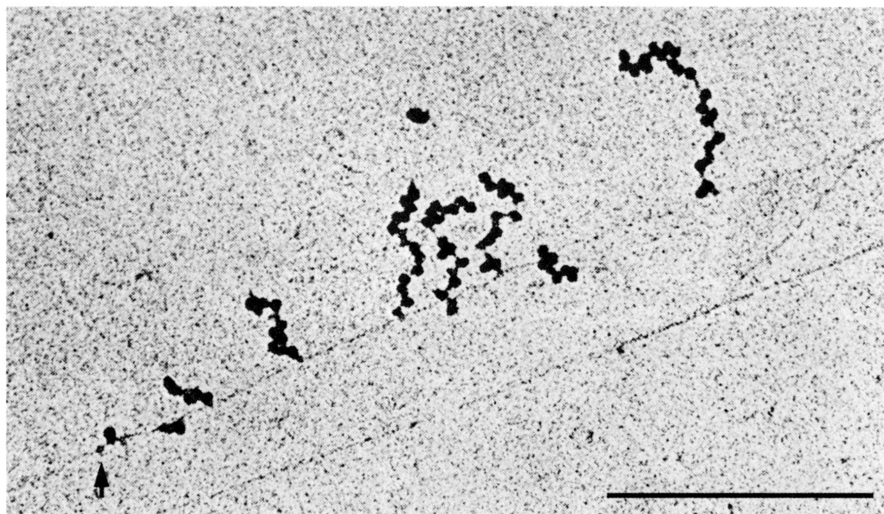

Fig. 123. Active region of *E. coli* genome. Structural operon, with the arrow indicating the presumptive initiation site for transcription. Bar: 0.5 µm. (From HAMKALO & MILLER 1973)

RNA polymerase molecules can be detected within the nuclear region. RYTER & CHANG (1975) demonstrated, by autoradiography on thin sections of *E. coli*, the absence of newly synthesized RNA in the nuclear region. Therefore, INGRAHAM et al. (1983) considered the possibility that newly synthesized RNA is made at the "interface" between the nuclear region and the cytoplasm. They proposed a model in which the RNA polymerase attaches to a promoter site "exposed" at the "interface" directed towards the cytoplasm. At a given instant most of the bacterial chromosomal DNA is "internal", i. e. within the nuclear region, and not exposed. However, one should assume that any part of the genome, due to thermal motion, would expose itself at the interface at relatively short intervals, probably seconds. In fact, CUMMINGS (1965) has demonstrated that the lactose operon is accessible to transcription throughout the normal cell cycle. These findings suggest that the polysomes attached to the DNA as shown by MILLER et al. (1970) are located, in vivo, in the cytoplasm, with the attachment site to the chromosome at the interface between nuclear region and cytoplasm as described above.

5.9 Non-flagellar bacterial appendages

5.9.1 Pili and fimbriae

Pili and fimbriae are non-flagellar, filamentous cell appendages which were first detected in *E. coli* (ANDERSON 1949) and in *Pseudomonas aeruginosa* (HOUWINK & VAN ITERSON 1950) by electron microscopy (BIEBRICHER 1984a, b). These structures are widely distributed among Enterobacteriaceae, Pseudomonadaceae and a variety of other bacteria including methanogens (DODDEMA et al. 1979). The terms "fimbriae" (DUGUID et al. 1955) and "pili" (BRINTON 1959) were introduced for these appendages, and these terms have been used by several authors with different meanings. JONES & ISAACSON (1983), in a review on proteinaceous bacterial adhesins and their receptors, stated that the terms "fimbriae" and "pili" have usually been used denoting morphology, but not function. They suggested that the terms pili and fimbriae be applied to appendages involved in gene transfer and bacterial attachment to surfaces associated with colonization, respectively. It is suggested that these terms should come into common use, in contrast to alternative nomenclatures such as "conjugative pili" and "somatic pili".

5.9.1.1 Pili

Pili are plasmid-coded. Besides several other properties such as resistance against certain antibiotica, colicin- or hemolysin production, and interaction with male-specific bacteriophages (BRINTON & BEER 1967), the respective plasmids endow the ability for conjugation, i. e. genetic transfer from one to another bacterium by direct cell-cell contact. The prototype of the pili is the F-pilus (Fig. 124) in *E. coli K-12* (BRINTON 1971, TOMOEDA et al. 1973, FIVES-TAYLOR 1978). The plasmid encoding its genetic information has been termed F-factor; it is either carried freely in the cytoplasm of F^+ cells, or it is inserted into the bacterial chromosome of Hfr cells. For conjugation (Fig. 125) of these donor cell types with the recipient cell type F^-, the presence of F-pili is necessary (CURTISS 1969). BRINTON (1965) has discussed the transport of bacterial DNA through an axial canal within the F-pilus ("F-pili conduction model") whereas MARVIN & HOHN (1965) and CURTISS et al. (1969) postulated that the F-pili, after making contact with the F^- cell, are retracted thus finally producing direct envelope–envelope contact between the conjugating cells ("F-pili retraction model"). Probably, though challenged, the latter of the two models might be closer to reality

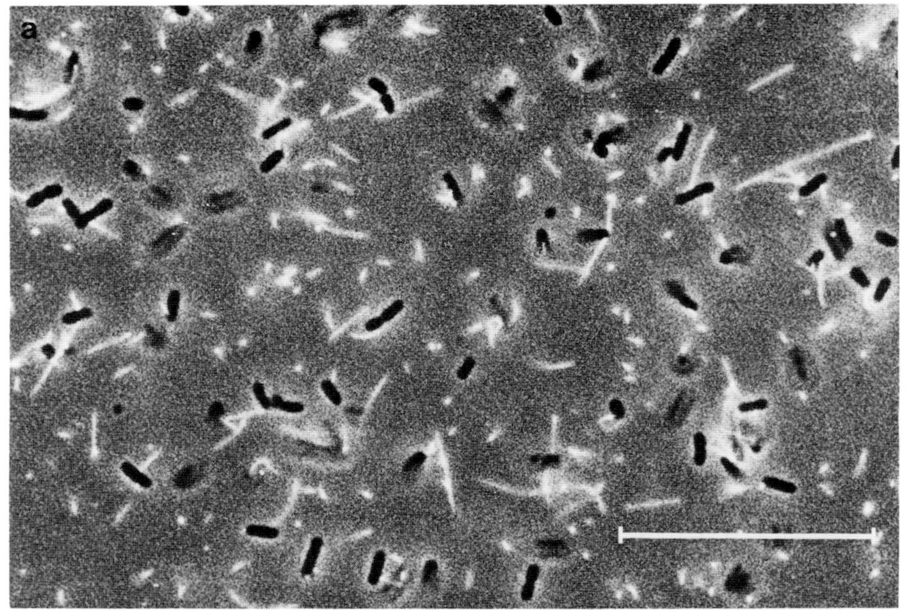

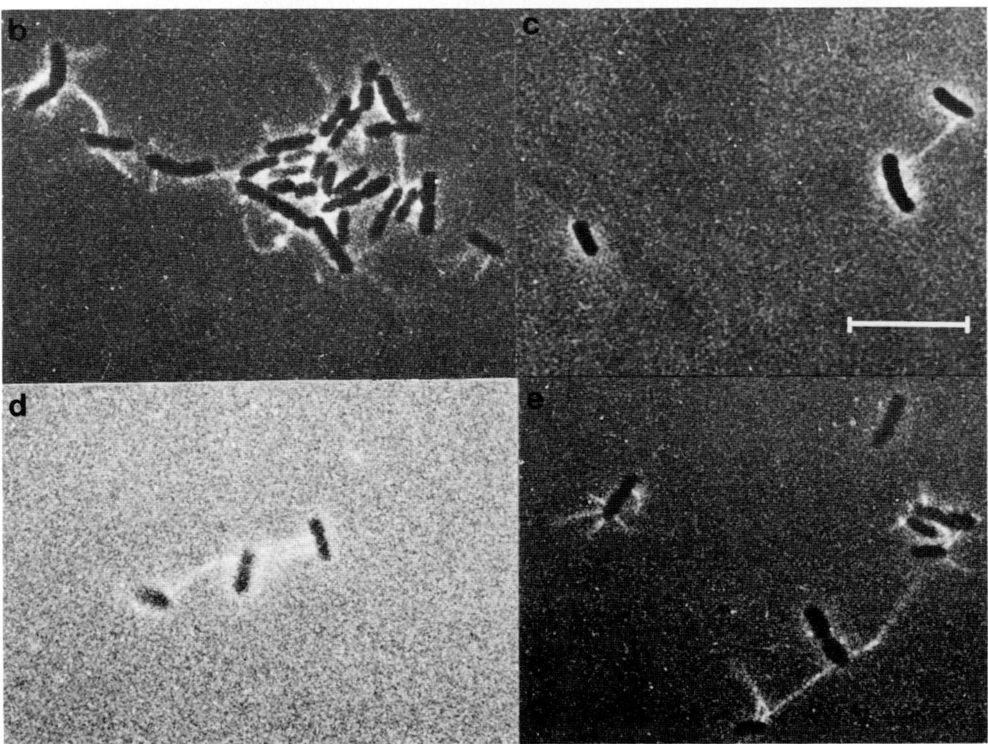

Fig. 124. a, *Escherichia coli K-12 Hfr H;* visualization of the F-pili with fluorescence-labelled bacteriophages (RITC-*MS 2* system). Bar: 20 µm; b–e, labelling of F-pili as in a, but conjugation system *E. coli Hfr H* × *N 10 F⁻*. Bar: 10 µm. (From JARCHAU 1985)

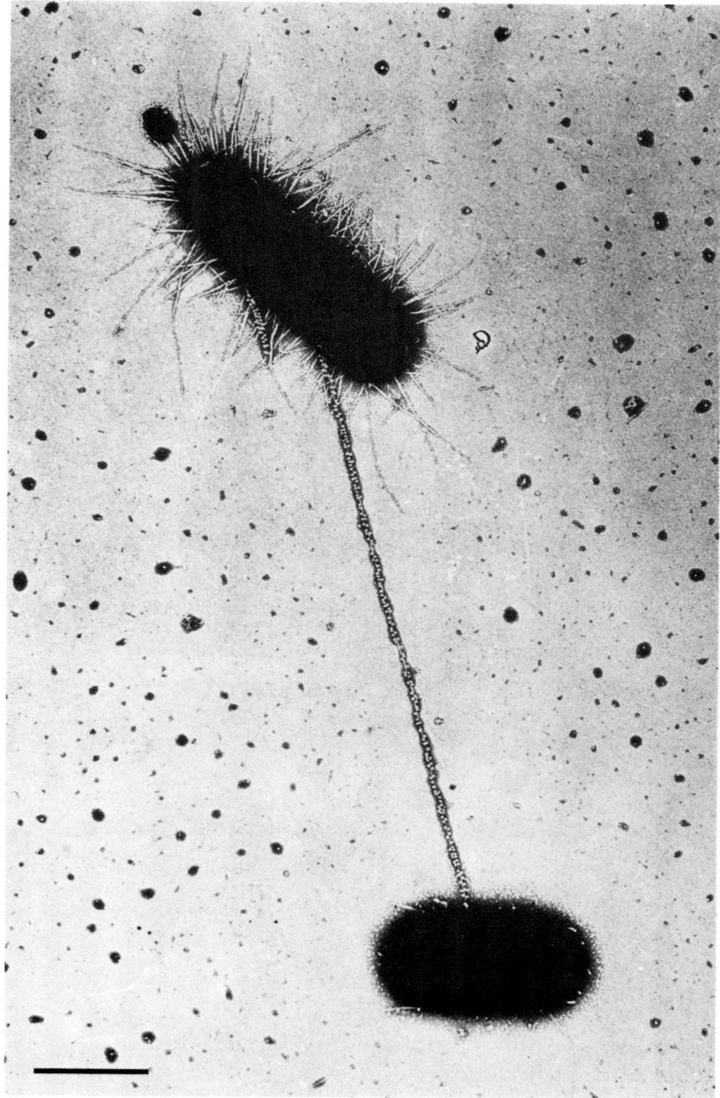

Fig. 125. Two *Escherichia coli* cells joined by a pilus during conjugation. Bacteriophages attached to the F-pili. Bar: 0.5 µm. (From STRYER 1981, after BRINTON & CARNAHAN)

(ACHTMAN & SKURRAY 1977, MANNING & ACHTMAN 1979, WILLETTS & SKURRAY 1980). F-pilus retraction might be an important factor for infection by male-specific bacteriophages (CRAWFORD & GESTELAND 1964, BRINTON et al. 1964, CARO & SCHNÖS 1966).

The subunit of the F-pili (the F-pilin) has a molecular weight of 11 200 D; it has been shown to be a phosphoglycoprotein whose phosphorylation and glycosylation vary depending on the bacterial

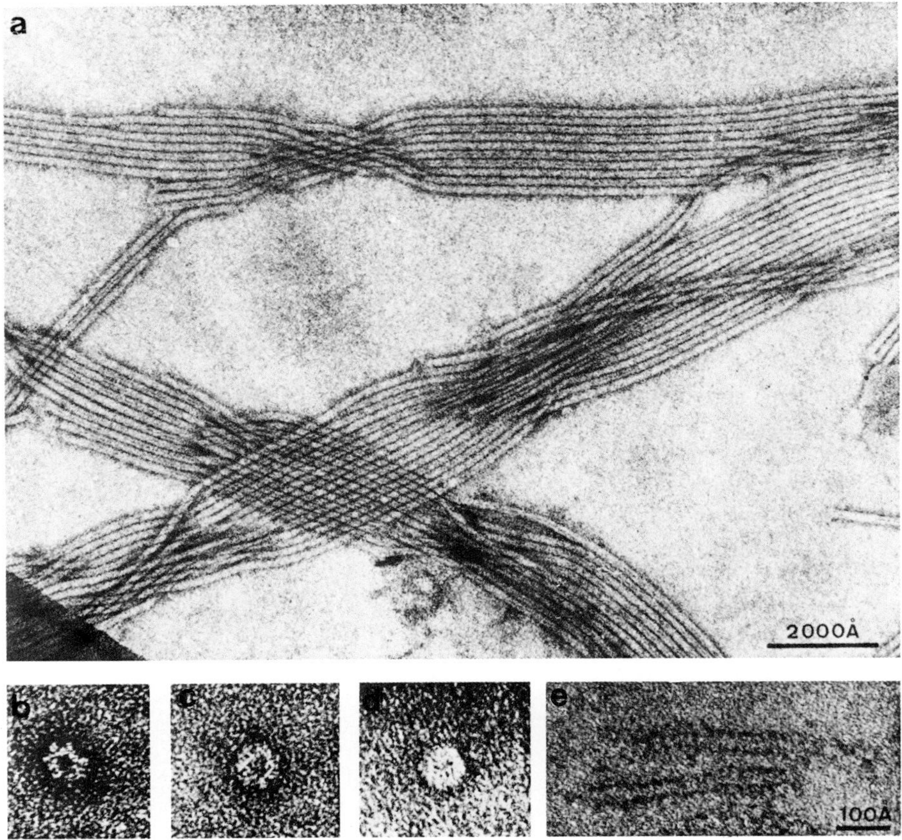

Fig. 126. Structure of *E. coli* F-pili. a, bundles of parallel F-pili obtained by adhesion onto mica, and negatively stained; b–d, fragments of negatively stained F-pili standing perpendicular to the supporting film; e, ultrathin section of F-pili. The plane of sectioning is nearly parallel to the axis of the pili. (From FOLKHARD et al. 1979)

strain. A "propilin" with a higher molecular weight has been detected (JARCHAU 1985), indicating that there might be processing and modification reactions taking place finally leading to F-pilin, a situation typical for exported proteins.

The F-pilus, according to electron microscopic (Fig. 126) and X-ray diffraction studies, is a tube with an outer diameter of 8 nm and a wall thickness around 3 nm, resulting in a diameter for the axial hole or canal of 2 nm. The F-pilus wall is made up of 4 helices wound around a common axis, consisting of F-pilin. Presumably, hydrophobic interactions are responsible for the stabilization of the structure. The synthesis of F-pili is brought about by the action of at least 13 genes located on the *tra* operon of the F-factor (MANNING & ACHTMAN 1979, WILLETTS & SKURRAY 1980). As F-pili are polymers made up of only one species of polypeptide chain, only one of these genes (either the *traA* cistron, MINKLEY et al. 1976, or the *traJ* cistron, MANNING & ACHTMAN 1979) is the gene which codes for the structure. The remaining genes are responsible for processing, export of the protein, polymerization of the F-pilus, depiliation, or the synthesis of a hypothetical "base structure" for the F-pilus in the cell envelope. F-pilin synthesis and utilization have been studied by SOWA et al. (1983). Their data are consistent with a model in which pilin subunits are assembled transiently into F-pili, conserved by retraction (NOVOTNY & FIVES-TAYLOR 1974), and made available for subsequent reassembly by depolymerization into F-pilin subunits and their storage within the cell, probably in the cytoplasmic membrane (MOORE et al. 1981). The existence of an F-pilin pool in the outer membrane has also been discussed (BEARD & CONOLLY 1975). WOROBEC et al. (1983) and FROST et al. (1983) have performed studies on the localization of the major antigenic determinant of *EDP 208* pilin, indicating that the N-terminal dodecapeptide is an important antigenic determinant on that pilus protein. Additional results have shown that the C-terminal fragment (residues 5 to 12) contains one or two additional antigenic sites.

5.9.1.2 Fimbriae

Fimbriae usually are shorter than F-pili, but their number per cell may be very high. They may be inserted peritrichously or polarly. The prototype of the fimbriae is the "1-pilus" (Fig. 127) observed in several *E. coli* strains (BRINTON 1959, 1965, SWANEY et al. 1977). These structures have been reported for many bacterial species, both Gram-negative and Gram-positive (Fig. 128). Their diameters as measured from electron micrographs of negatively stained samples are between 3 and 7 nm, depending on the bacterial strain. Similar to the F-pilus, they are hollow cylinders made up of helically arranged subunits with molecular weights ranging between values below 6000 D and around 17000 D. The diameter of the axial canal was seen to be 2 nm in "1-pili" of *E. coli* (BRINTON 1965, MITSUI et al. 1973), or less in fimbriae of *Rhizobium lupini* (MAYER 1969). "1-pilin", the subunit of "1-pili" of *E. coli*, is a glycoprotein containing hexose residues. The genes coding for the structure and the expression of the subunits of fimbriae in *E. coli* were found to be located on the chromosome; several cistrons participate in coding of their structure and synthesis. The structural gene for "1-pilin" has not yet been identified. The regulation of the expression of the "1-pili" in *E. coli* occurs by phase variation (EISENSTEIN 1981): the cell can show two different phenotypes, with or without fimbriae, and the variation of the two phases occurs spontaneously and by accident. The situation might be similar to that in *Salmonella* flagellar phase variation where the existence of an invertable insertion element was postulated (JARCHAU 1985; further references therein). Alternative explanations for the "1-pili" phase variation might involve the control of the *"pil"* genes by the metastable *flu* gene, or some influence of cyclic AMP regulating the expression of "1-pili" by regulation of transcription. "Propilin" type polypeptides for "1-pilin" have not yet been detected. The existence of a substantial subunit pool was made unlikely by experiments using chloramphenicol (JARCHAU 1985). MCMICHAEL & OU (1979) could not detect "1-pilin" in the cytoplasmic membrane of *E. coli.* As first shown by BRINTON (1965), depolymerized "1-pili" of *E. coli* are capable of spontaneous repolymerisation interpreted to be a self-assembly process. "Depiliation" for fimbriae has not yet been investigated in detail but there are no indications for retraction. However, observations in experiments with conjugating soil bacteria indicated a decomposition of polar fimbria-like structures into bundles of very thin filaments during a process which finally led to wall–wall contact of bacteria in a "star"-like cell aggregate (MAYER 1971). A model for this process has been proposed, together with additional fine structure data on the intact tubular structures (HEUMANN & MARX 1964), and biochemical data on their subunits (MAYER & SCHMIDT 1971). Insertion sites of fimbriae have been seen in several electron microscopic studies, and the existence of annular base structures for fimbriae in the cell envelope (Figs. 128 and 129) has been demonstrated in some cases (HAMILTON et al. 1975, MAYER 1979, WELLS et al. 1983).

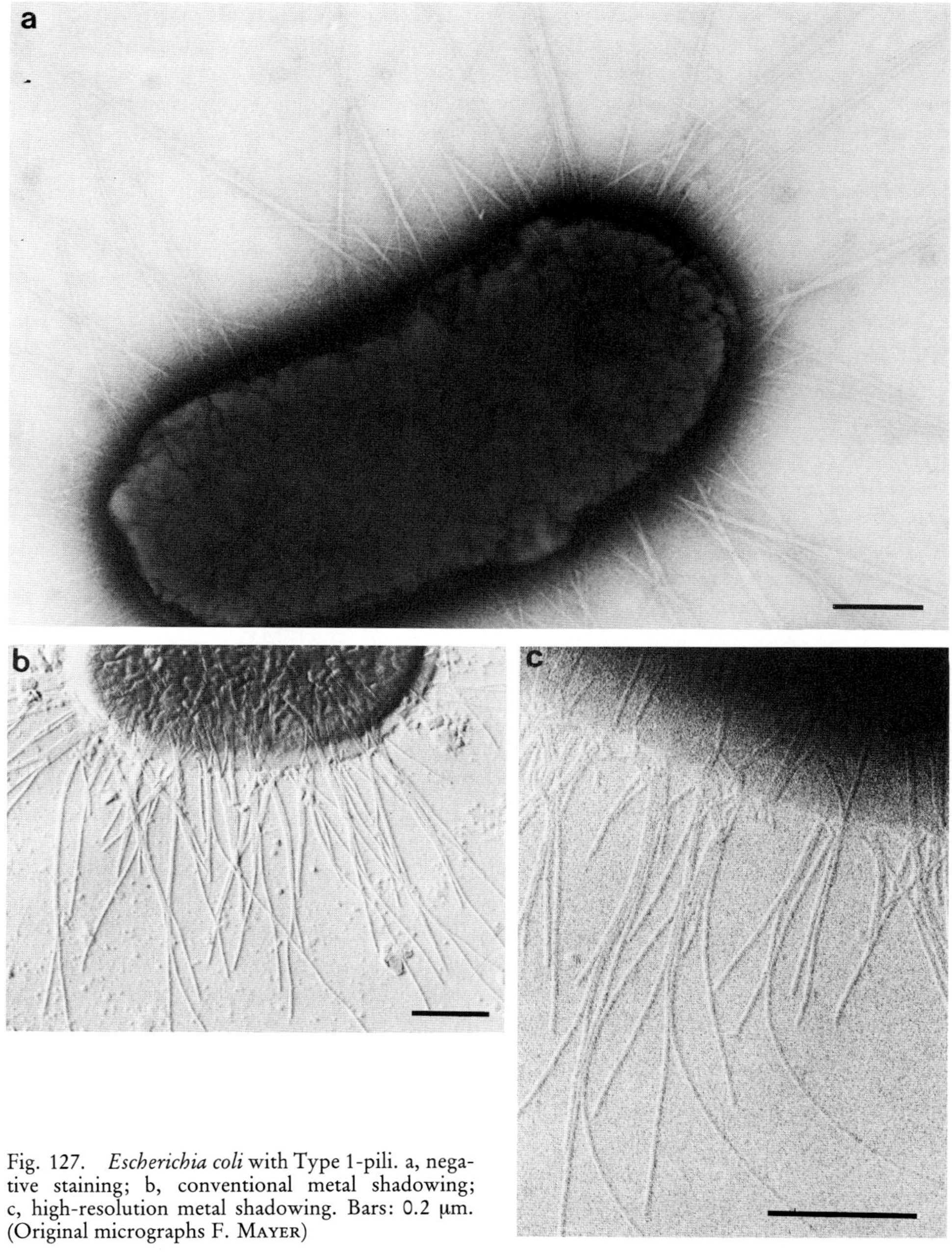

Fig. 127. *Escherichia coli* with Type 1-pili. a, negative staining; b, conventional metal shadowing; c, high-resolution metal shadowing. Bars: 0.2 μm. (Original micrographs F. MAYER)

Numerous studies have indicated that bacterial fimbriae of different types are involved in the process of bacterial adhesion not only to other bacteria of the same strain as demonstrated in the "star" formation, but also to eukaryotic cells and other surfaces as well. These findings have been

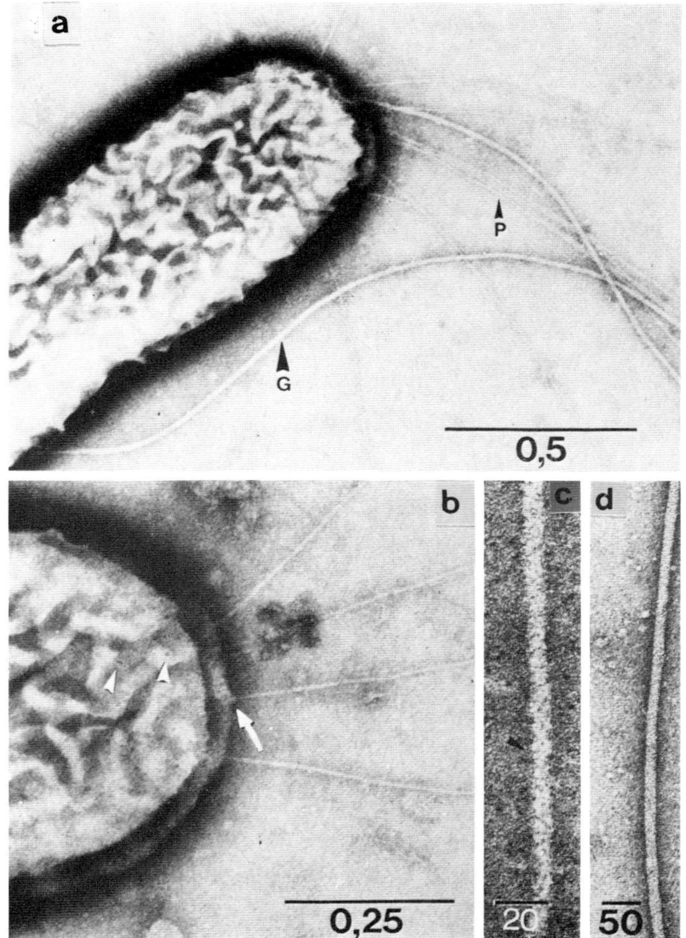

Fig. 128. a, flagella (G) and pilus-like appendages (P) at a cell of the hydrogen-oxidizing bacterium *Pseudomonas taeniospiralis* strain *2 K 1;* b, arrow points to differentiations located at the basis of the pilus-like structures; arrowheads: annular pore-like structures; c, higher magnification of part of a pilus-like structure; d, part of a flagellum. (From MAYER 1979)

extensively described and reviewed (DUGUID et al. 1966, DUGUID & GILLIES 1957, DUGUID & OLD 1980, SALIT & GOTSCHLICH 1977). Different mechanisms of fimbriae-mediated adhesion have been detected by means of hemagglutination (HA): mannose-sensitive HA mediated by common "1-pili" and mannose-resistant HA mediated by fimbriae isolated from several *E. coli* strains. These features indicate that HA is brought about by different interactions and several target cell specificity patterns. METT et al. (1983) have found that mannose-resistant adhesion to tissue-culture cells can be classified into different subtypes. These adhesions appear to be carbohydrate-mediated. KORHONEN et al. (1983) carried out a molecular characterization of type 3 fimbriae of *Klebsiella* sp. and their role in bacterial adhesion to plant roots. They proposed that type 3 fimbriae are involved

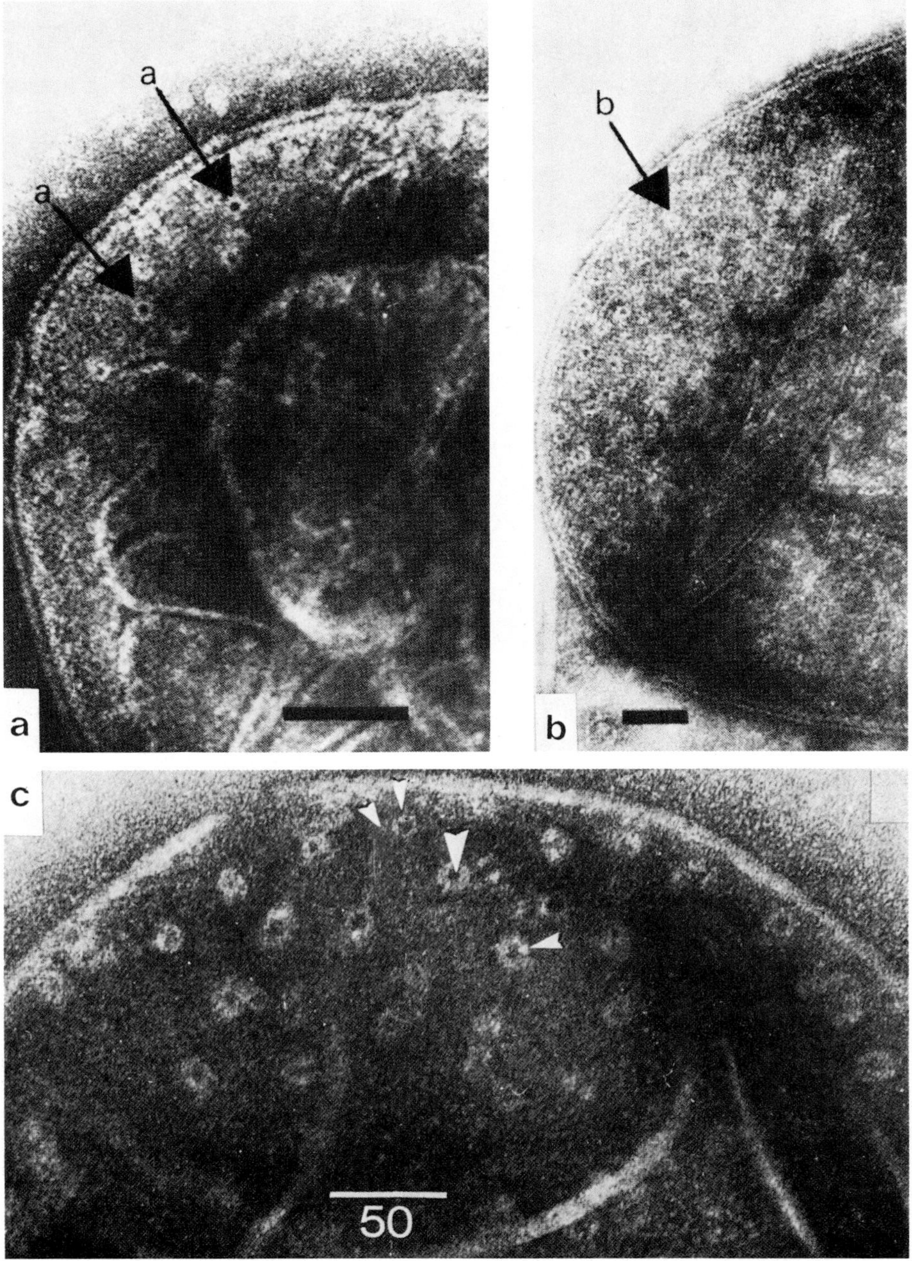

Fig. 129. a, b, annular pore-like structures (arrows, "a" and "b") in the envelope of intact cells of *Pseudomonas avenae*. Bars: 0.1 μm. (From WELLS et al. 1983); c, annular pore-like structures (arrowheads) in the envelope of the hydrogen-oxidizing bacterium *Pseudomonas taeniospiralis* strain *2 K 1;* these structures are assumed to be insertion aggregates for pilus-like appendages. (From MAYER 1979)

in the establishment of the plant–bacterium association concerning nitrogen fixing *Klebsiella* strains living on the roots of non-leguminous plants. Thus, adhesion mediated by fimbriae is a counterpart to non-fimbrial adhesion brought about by cell surface carbohydrates and by microbial exopolysaccharides of different kinds.

5.9.2 Bacterial spinae

Bacterial spinae are tubular, pericellular, non-prosthecate rigid cell appendages (Fig. 130) that are produced by certain Gram-negative bacteria (McGregor-Shaw et al. 1973, Moll & Ahrens 1970). Spinae are comprised of a single protein moiety, spinin, that is able to self-assemble in vitro

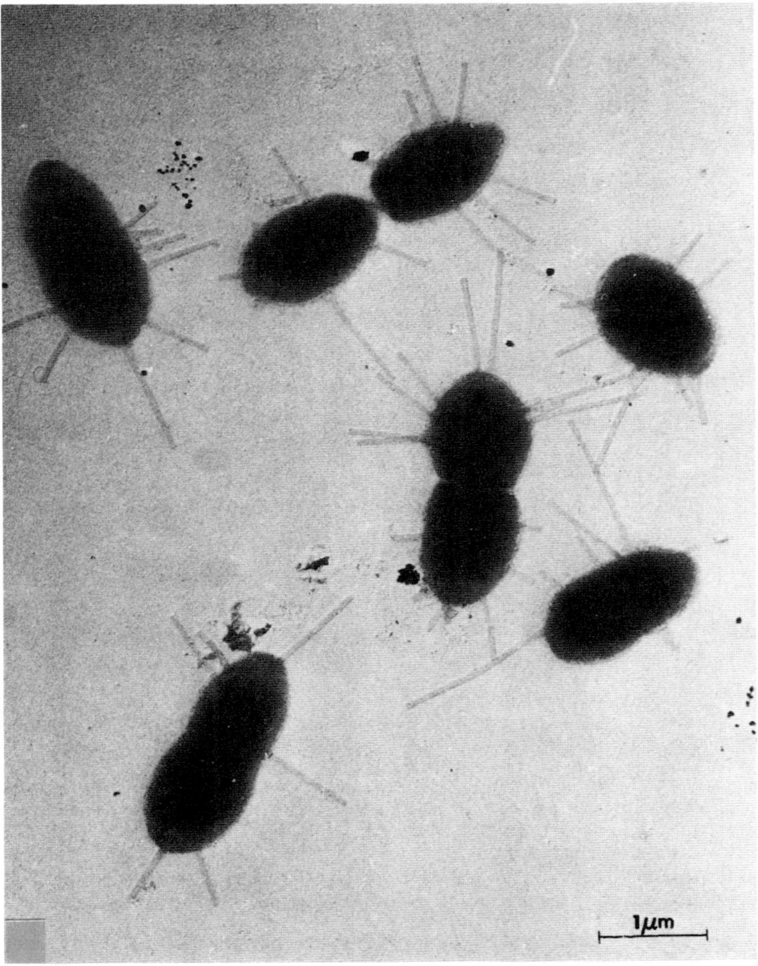

Fig. 130. Isolate from sample containing decaying marine algae; echinuliform (non-prosthecate) appendages (spinae). (From McGregor-Shaw et al. 1973)

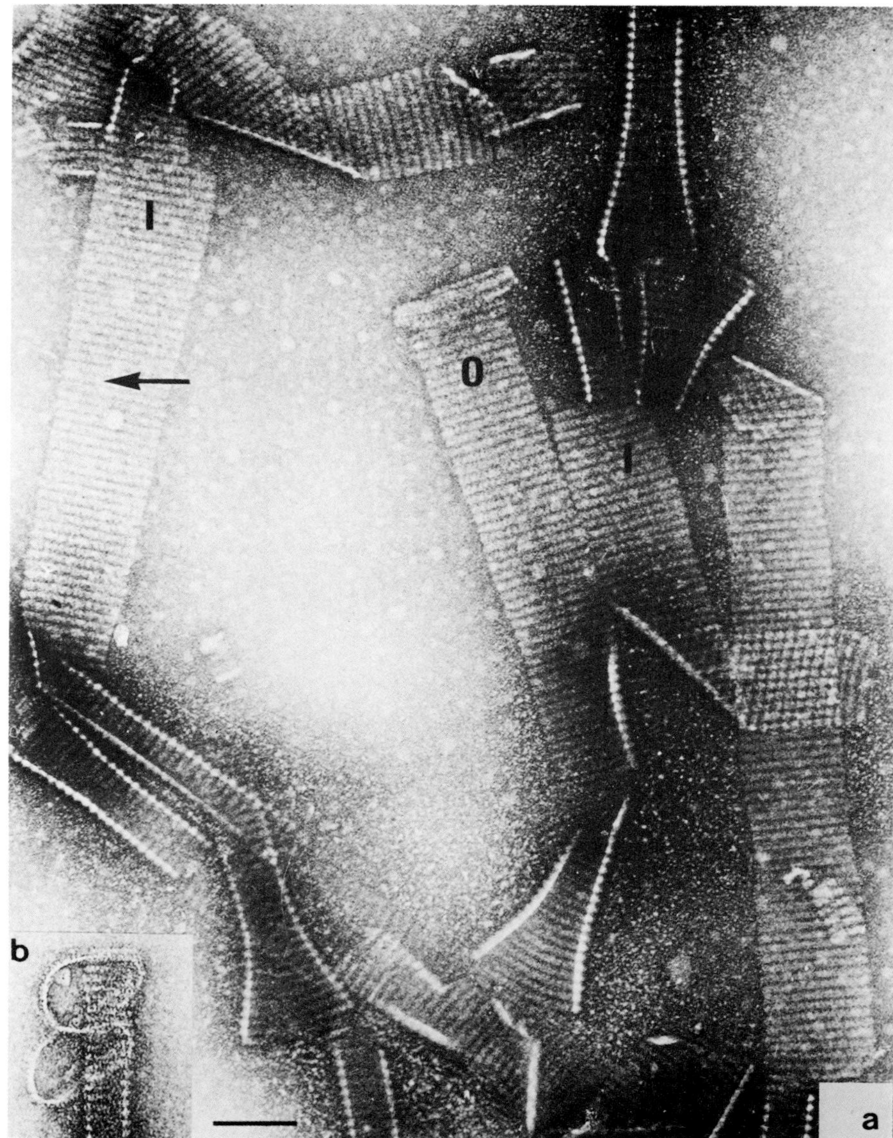

Fig. 131. Structure of bacterial spinae. a, b, purified preparation of spinae from the marine pseudomonad *D 71*, negatively stained. Many spinae have collapsed and split open to form ribbons. In different ribbons, inner (I) and outer (O) surfaces are presumed to be uppermost. Note the "paired line" substructure in the ribs (arrowed) and the variable orientation of oligomers in the centre ribbon. Insert (b) shows a spina that has slightly unwound to reveal the wall structure. Bar: 0.1 µm. (From EASTERBROOK et al. 1976)

into filaments (COOMBS et al. 1975, EASTERBROOK & COOMBS 1976, EASTERBROOK et al. 1976). In vivo, these filaments make up the wall of a hollow tube, 60 nm in diameter and 5 nm thick, which is expanded at its zone of contact with the surface of the cell envelope, reaching a maximum diameter of 120 nm (Fig. 131). Detailed studies of spinin have shown that its molecular weight is 19000 D. In its native state it exists almost entirely in the β-conformation. About 10 spinin molecules were found to be assembled in a morphological subunit with dimensions 11.5 by 5.6 nm which in turn is a constituent of a helically wound thread cross-linked between successive turns (EASTERBROOK 1974, EASTERBROOK et al. 1973). Even though bacterial spinae have been known for quite some time, their function is not yet clear. Similarities have been seen to exist between spinin and outer membrane proteins. The possible significance of this deserves further investigation to provide insight into the function of spinae. Possibly they might help the cell to adjust to certain environmental conditions. It has been shown by EASTERBROOK & ALEXANDER (1983) for the marine pseudomonad *D71* that this organism can be maintained in either the unspined or spined state depending on the salinity, pH, and temperature. Shifting bacteria from conditions restrictive for spination to those permissive for spination resulted in the production of spinae after a short lag period. The percentage of spined bacteria in a population increased from 0 to nearly 100% in one generation time. Concomitantly, the average number of spinae per bacterium increased to reach a maximum after 6 hours. Analysis of initiation and growth of spinae revealed that the attachment sites of the spinae, the spina bases, were first observed at either the pole or in a central location at sites where division constriction had either occurred or was about to occur. In comparing the initiation of spina bases and the regulation of the insertion of peptidoglycan components and outer membrane proteins into the bacterial envelope, it has been seen that all three types of constituents are inserted at a similar location and time. The stability of spinae attachment to the surface of the cell argues for some kind of interaction between spinae and membrane proteins and/or peptidoglycan. Spinae in the pseudomonad mentioned above attained full length, usually 1 to 3 μm, in about 15 min and grew by the addition of spinin subunits at their distal end. It was assumed that the rate-limiting step of growth of spinae is the establishment of the base structure. Once this exists, the secretion of spinin would occur in the region of the membrane circumscribed by the spina base, into the lumen of the growing spina. Inhibition studies have shown that whereas there is no significant pool of protein the mRNA is stable and is present in bacteria 30 min before spina structures are detectable.

5.10 Structural and functional aspects of bacterial motility

5.10.1 Bacterial motility mediated by flagella

5.10.1.1 Structural aspects

Many bacteria swim actively with the aid of flagella (Figs. 132 and 133) which are tubular helical appendages attached polarly, bipolarly, or peritrichously (ARAGNO & SCHLEGEL 1977, 1981, ARAGNO et al. 1977, ASTBURY et al. 1955, BODE 1973, CHAMPNESS 1971, HOUWINK & VAN ITERSON 1950, KERRIDGE et al. 1962, LEIFSON 1960, LOWY & HANSON 1965, LOWY & SPENCER 1968, DE PAMPHILIS & ADLER 1971a, PIETSCHMANN 1942, SCHMITT 1972, SIMON et al. 1978, SMITH & KOFFLER 1971, TAUSCHEL & DREWS 1970). Electron microscopic studies intended to reveal the presence and type of insertion of bacterial flagella (KODAKA et al. 1982, MAYFIELD & INNIS 1977) should be performed with great care as it has been shown that flagellar filaments tend to break during the preparation procedures. Moreover, it has been found that flagella may be present in certain growth phases but absent in others. Flagella are about 11 to 20 nm in diameter (Fig. 134) and can be up to 20 μm long.

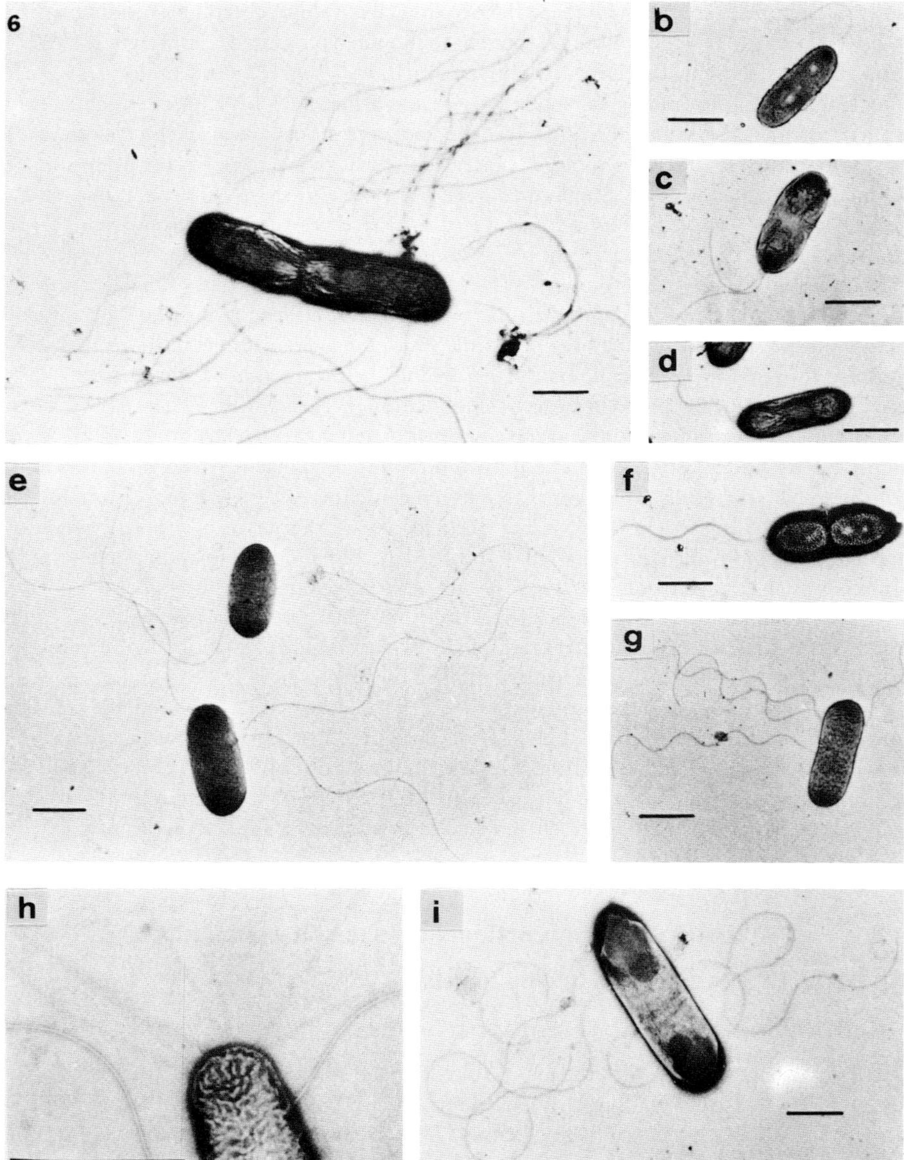

Fig. 132. Flagellation of hydrogen bacteria. a, *Alcaligenes eutrophus H 16;* b, *Pseudomonas pseudoflava;* c, *P. pseudoflava;* biflagellated cell; d, *P. flava;* e, *P. facilis;* f, *Alcaligenes paradoxus;* g, *A. paradoxus* with "curly" flagella; h, *A. paradoxus* with fimbriae; i, strain *MA 2.* Bars: 1 µm. (From Aragno & Schlegel 1981)

A number of mutants from several well investigated bacterial strains *(Escherichia coli, Salmonella typhimurium)* with differently shaped flagella have been isolated. Some had straight flagella which entirely lacked the helical form (O'Brien & Bennett 1972, Shirakihara & Wakabayashi 1979,

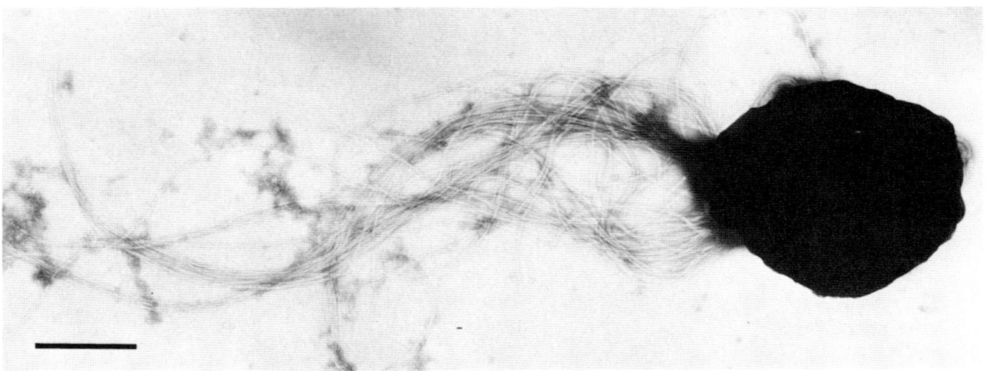

Fig. 133. Flagellation of the methanogen *Methanococcus jannaschii* isolated from a submarine hydrothermal vent. Bar: 0.5 µm. (From JONES et al. 1983)

SUZUKI & IINO 1977), others were "curly", with a tighter helical structure than the wild type. It has been found that the conditions used for culturing bacteria can also affect the form of the flagella although the form remains, in principle, strain-specific (SILVERMAN & SIMON 1972). For example, flagella with a normal wave length can become "curly" when the pH of the medium is decreased (KAMIYA & ASAKURA 1976). CALLADINE (1975, 1982), in reports on the construction of bacterial flagellar filaments and on aspects of their conversion to different helical forms, has demonstrated that a mechanically bistable but otherwise linear-elastic subunit can build filaments which are not only helical but also possess the observed polymorphic properties existing under certain conditions. He has also discussed the problem which concerns the hydrodynamic performance of a helical filament able to switch wave-form when subjected to mechanical overload.

The bacterial flagellum is composed of a distal filament (a semi-rigid helical propellor) which has an empty core (SLEYTR & GLAUERT 1973), and which can be "normal" or "complex" (SCHMITT et al. 1974) or even "sheathed" (Fig. 136) (GLAUERT et al. 1963, NOVICK & TYLER 1985), a short proximal "hook" (a flexible coupling) (ABRAM et al. 1965, 1966, 1970, LOWY 1965, RASKA et al. 1976, WAGENKNECHT et al. 1981), and a rod and set of rings (COHEN-BAZIRE & LONDON 1967, HOENIGER et al. 1966, VAN ITERSON et al. 1966, DE PAMPHILIS & ADLER 1971 b, 1971 c) embedded in the cell wall and cytoplasmic membrane (Fig. 137) (a reversible rotary motor, s. below).

As compared to "normal" flagellar filaments, "complex" filaments (Fig. 135) exhibit additional band-like helical structures made up of subunits, surrounding a central tubular core. "Sheathed" flagella are enveloped by an additional tube-like structure originating, according to electron microscopic images, from the cell envelope.

The flagellar filament has been found to be made up of subunits (flagellin) (Fig. 134) that vary in molecular weight from 33000 D to 60000 D according to the bacterial species. The amino acid composition of several flagellins has been determined; cysteine and tryptophan are never found, and the amounts of proline, tyrosine and histidine are very low.

Flagellin monomers are polypeptides about 4.5 µm in diameter, with a rather elongated form (BODE et al. 1972). The filaments are made up of the flagellin monomers (KOBAYASHI et al. 1959, KOFFLER & KOBAYASHI 1957, TAUSCHEL 1971) arranged in hexagonal packing, leaving a central hole or tube sometimes filled with material probably representing flagellin monomers (INGRAHAM et al. 1983; further references therein).

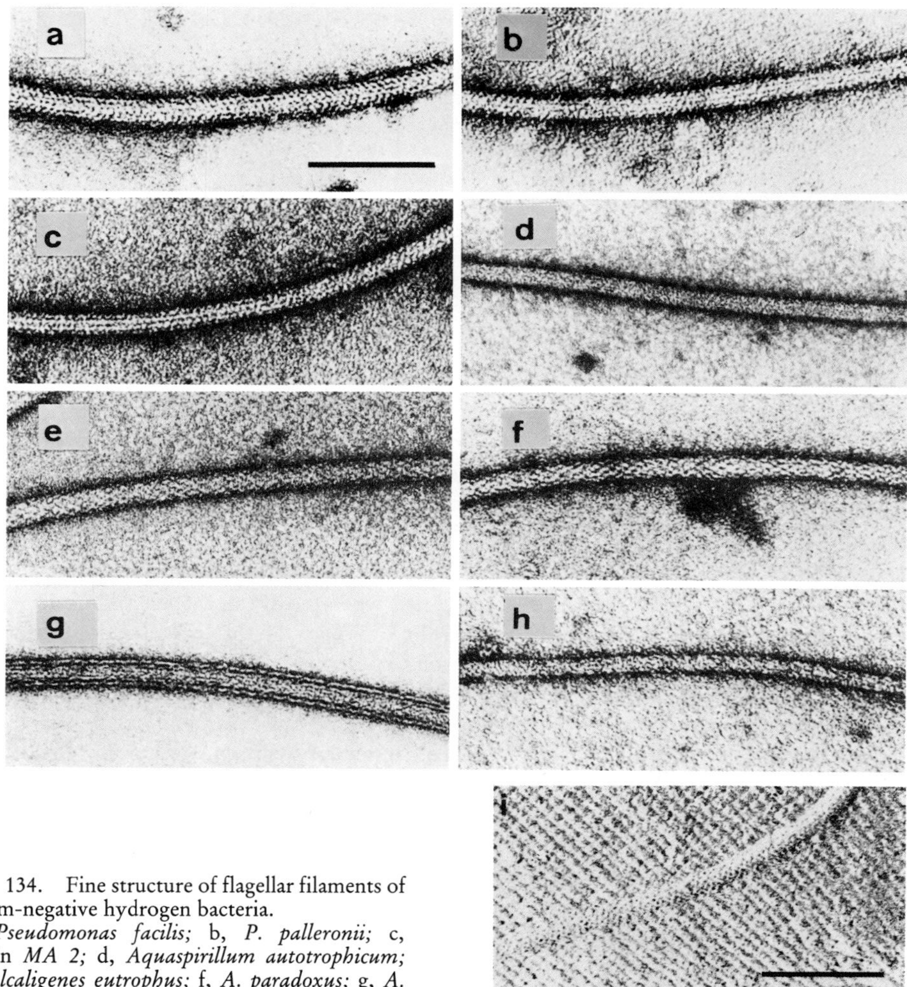

Fig. 134. Fine structure of flagellar filaments of Gram-negative hydrogen bacteria.
a, *Pseudomonas facilis;* b, *P. palleronii;* c, strain *MA 2;* d, *Aquaspirillum autotrophicum;* e, *Alcaligenes eutrophus;* f, *A. paradoxus;* g, *A. ruhlandii;* h, *Pseudomonas pseudoflava.* All micrographs identical magnification (bar: 100 nm). (From ARAGNO et al. 1977); i, freeze-etching preparation of *Clostridium aceticum;* cell wall with periodic surface structure and a flagellum with clearly visible flagellin molecules. Bar: 100 nm. (From BRAUN et al. 1981)

The longitudinal arrangement of the flagellin molecules, as shown by light-optical diffraction and filtering, and by direct visualization in metal-shadowed preparations, is helical. For example, the subunits of the filaments of *Salmonella typhimurium* flagella are packed in a surface lattice which consists of 11-, 1-, 5- and 6-start helical lines; the 11-start lines run almost longitudinally, and there are in the order of 11×500 subunits in one complete turn of the flagellar helix.

Filaments can be partially degraded into wavy fibers (Figs. 138 and 139) by a variety of compounds such as urea, acid, alkali, alcohol, and other chemicals, or by treatments such as freezing and thawing or ultrasonication. The fibers presumably persist since the forces between the flagellin monomers are greater in longitudinal direction than in lateral direction. More rigorous

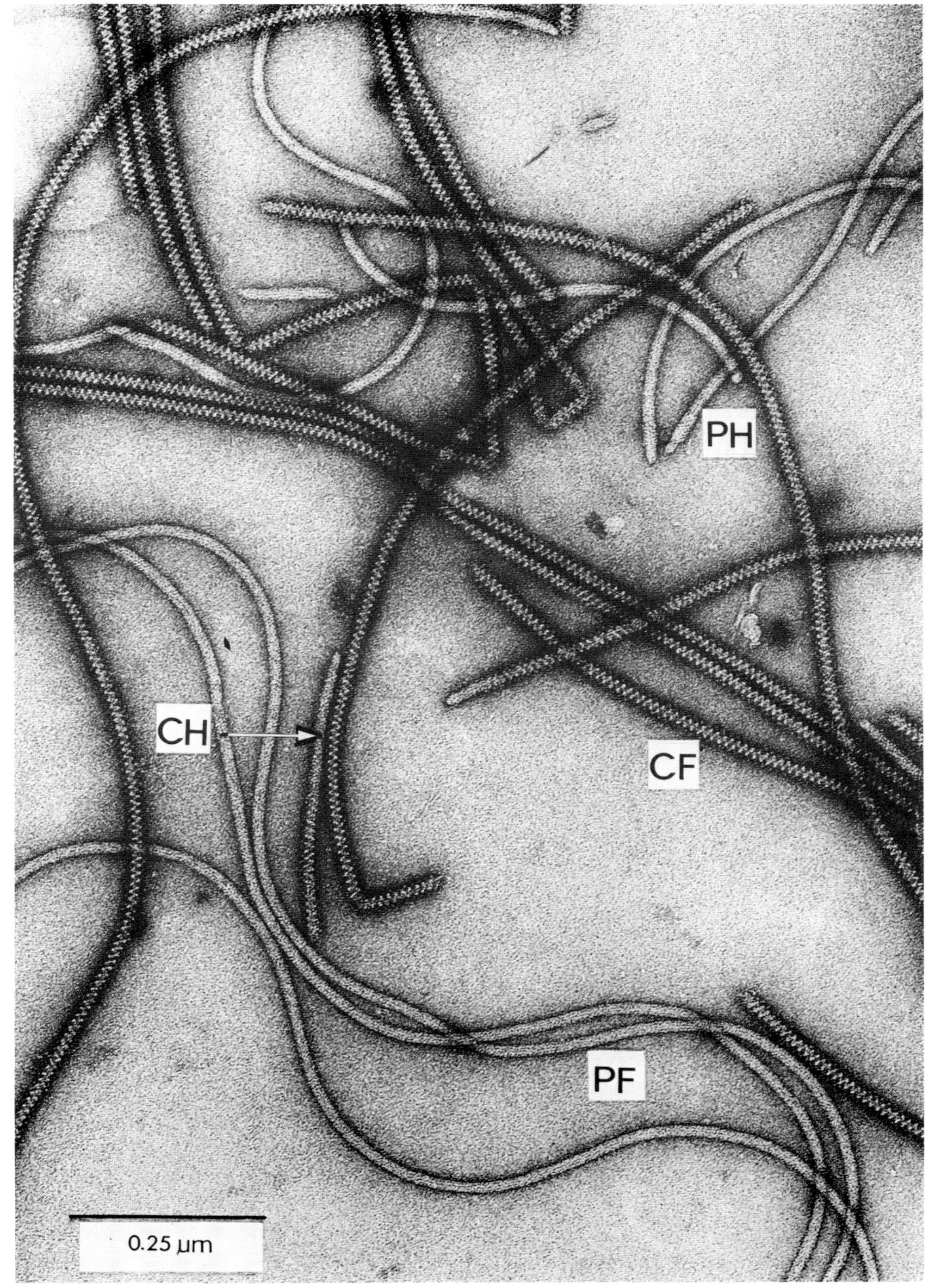

Fig. 135. Structure of "plain" (PF) and "complex" (CF) bacterial flagella of *Pseudomonas rhodos 9-6*, some with their hooks (PH and CH, respectively) attached. (From SCHMITT et al. 1974 a)

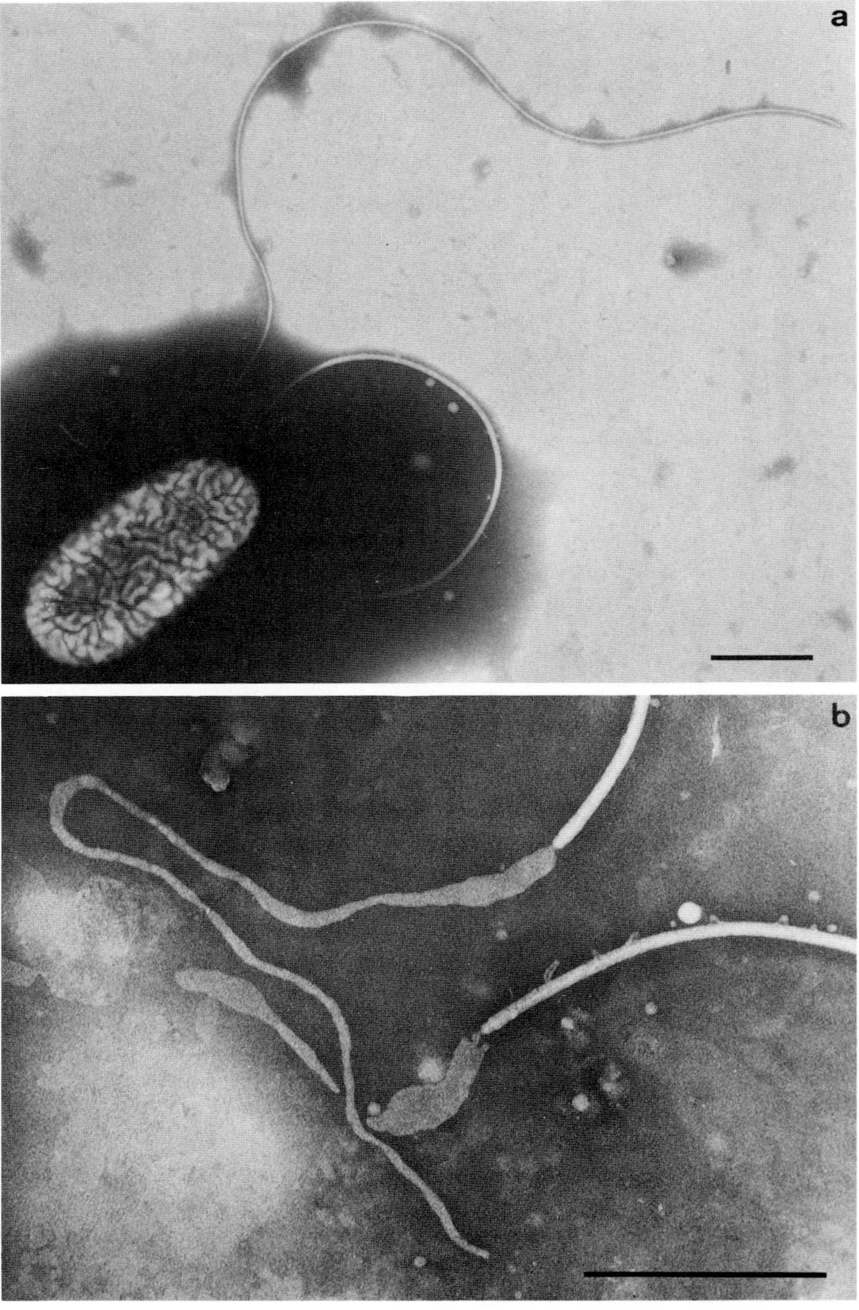

Fig. 136. Unidentified soil bacterium with sheathed and unsheathed flagella. a, thick (sheathed) and normal flagellum; b, during negative staining, the sheath is disintegrating. Bars: 0.5 μm. (Original micrograph F. MAYER)

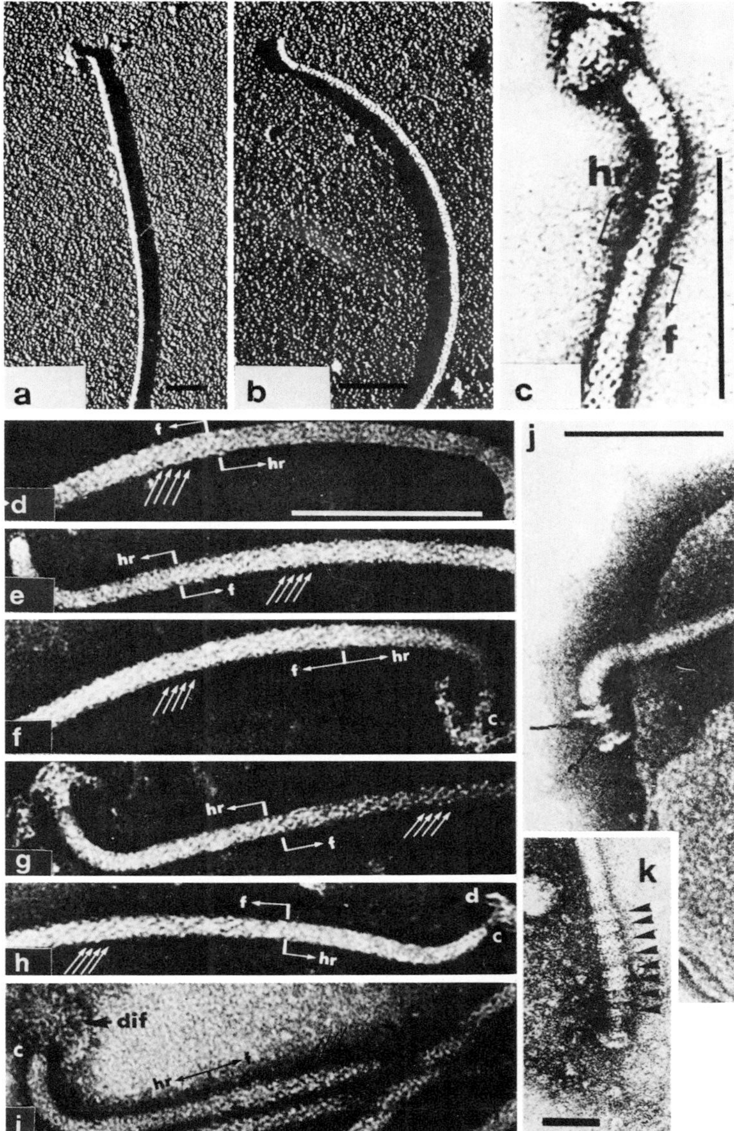

Fig. 137. Flagellar insertion: hook structrures. a–i, *Bacillus stearothermophilus 194;* the micrographs show the different regions of the flagella: the proximal hook region (hr) and the filament (f); the length of the hook region may vary; note a constriction (c) at the end of the hook and the material attached to the base of the flagella, especially the disc-shaped (d) structure, and the diffuse (dif) appearance of some attached material. Bars: 0.1 µm. (From ABRAM et al. 1966); j, flagellum and membrane fragments liberated by enzymatic lysis of *Rhodospirillum molischianum.* The basal organelle consists of two pairs of discs connected by a collar to each other and to the flagellar hook. Bar: 0.1 µm. (From COHEN-BAZIRE & LONDON 1967); k, separated flagellar hook from *Clostridium thermoautotrophicum* with unusual ringlike structures (arrowheads). Bar: 50 nm. (From WIEGEL et al. 1981)

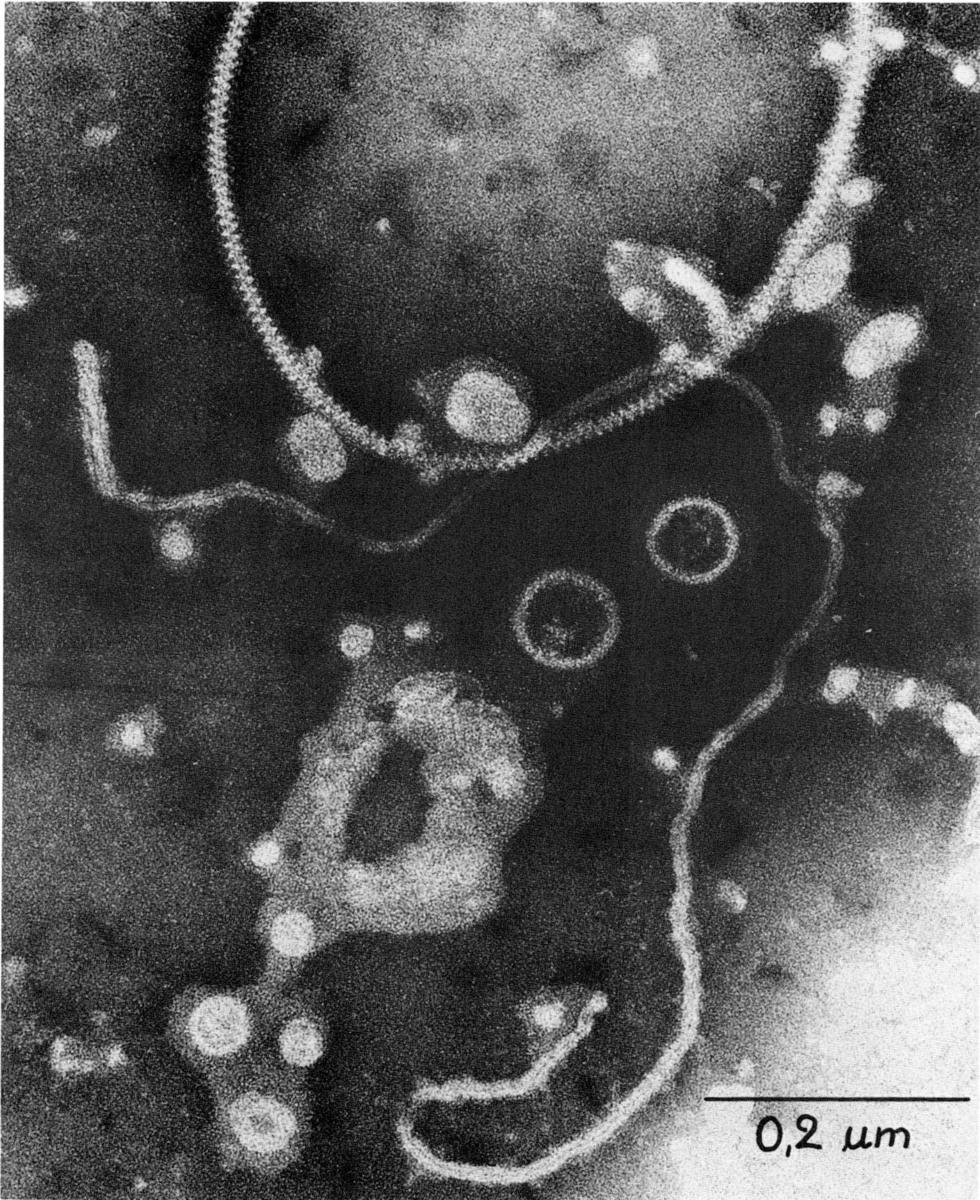

Fig. 138. "Complex" flagella of *Pseudomonas rhodos* after treatment with 6 M urea. Portion of an intact filament and a naked core with a partially disintegrated hook at one end. The overall structure of the core appears "soft", probably owing to the lack of the helical outer component. A central tube of 2.5 nm width is discernible. Remnants of the decomposed helical outer structure appear as flat aggregates. (From SCHMITT et al. 1974)

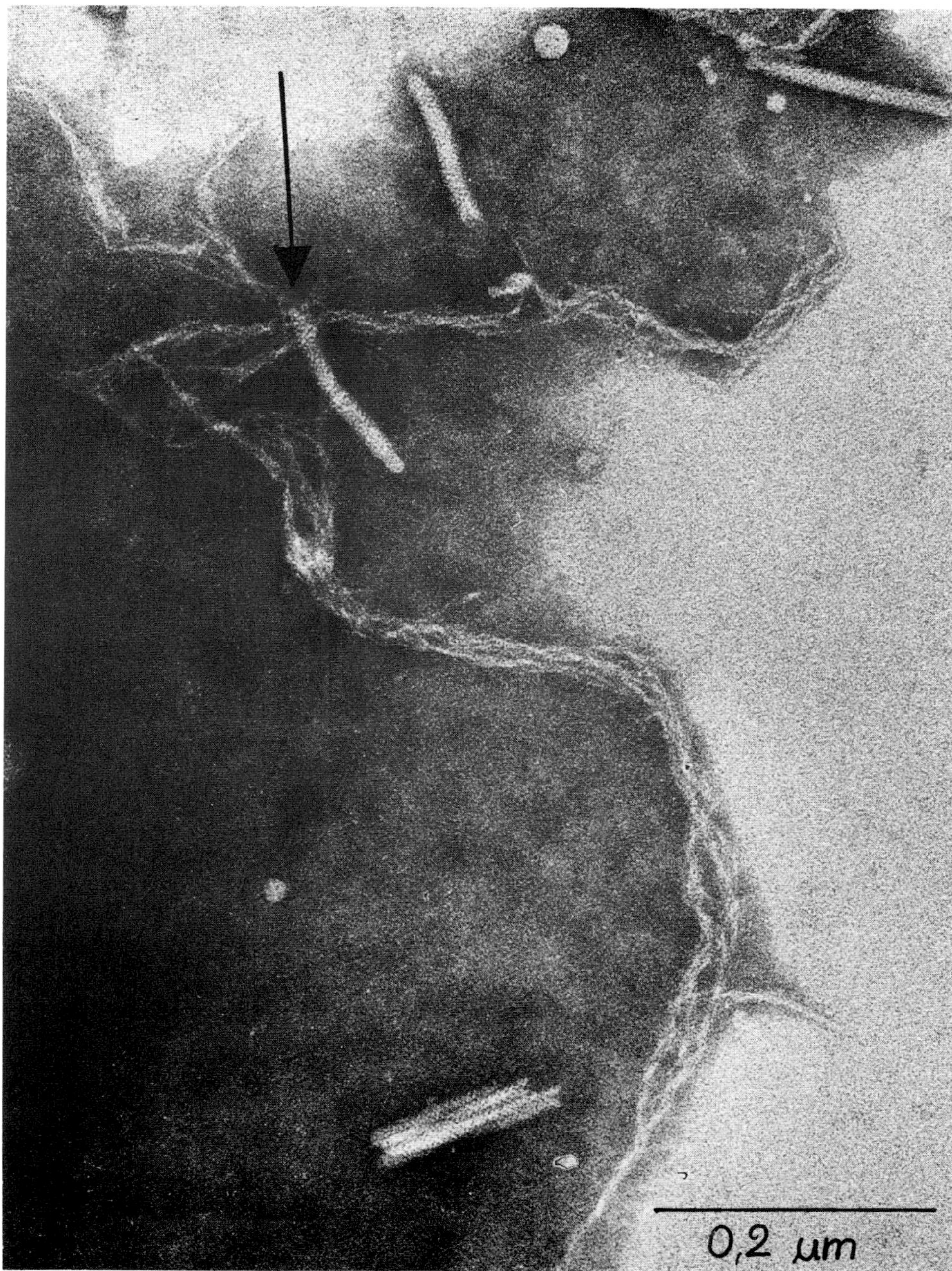

Fig. 139. "Complex" flagella of *Pseudomonas rhodos* after prolonged treatment with 6 M urea. Disintegration of internal core into coiled bundles of eight to ten fibers probably corresponding to rows of subunits originally arranged in large-scale helices. Arrow marks apparent origin of one bundle at a flagellar hook, which under these conditions is not visibly decomposed. (From SCHMITT et al. 1974)

conditions have been found to lead to subunits with the hook still preserved. Flagellin monomers obtained from such treatments can be induced to reassociate to give flagella-like structures.

The conditions for in vitro-reassembly have been determined for several flagellin species (ABRAM & KOFFLER 1964, ASAKURA 1970). They were found to be as diverse as those for disaggregation. In reassociation experiments using different pH values, straight filaments, after disaggregation into subunits, could be converted into a helical type of filament. In many cases, especially at low salt concentrations in the reassembly solution, "seeding" with fragments of filaments was necessary for in vitro polymerization. The in vitro "growth" is assumed to occur by self-assembly, a prócess similar to crystallization, but from one end only. Co-polymerization of monomers from different strains of a bacterium was shown to build different helical forms. KAGAWA et al. (1981) reported that hook structures attached to cell bodies of *flbC* mutants of *E. coli* may act as a seed for flagellar filament formation in solutions with added monomer flagellin, and that such cell bodies developed motility. Similar findings have been obtained by KAGAWA et al. (1983) during studies on the motility development of *Salmonella typhimurium* cells with *flaV* mutations (corresponding to *flaC* mutations in *E. coli*) after the addition of exogenous flagellin.

The growth of flagellar filaments (EMERSON et al. 1970, IINO 1969) in vivo has been shown to occur from their distal poles by the incorporation of amino acid analogues (p-fluorophenylalanine) or by the addition of radioactively labelled amino acids to the culture medium, followed by autoradiography. Some experimental evidence supports the view that flagellin necessary for distal growth of the filament is not simply secreted by the cell but is passed up the central flagellar cavity. This finding might explain that the central canal is sometimes observed to contain material (s. above).

IKEDA et al. (1983) have studied a non-motile mutant of *Salmonella typhimurium* which has very short flagella due to a mutation in the structural gene of flagellin (H2). They demonstrated that addition of ammonium sulfate to the culture medium caused precipitation of large amounts of protein identified as flagellin. This mutant flagellin appeared to be secreted in the monomeric form, in an amount comparable to the amount in the flagellar filaments of wild-type bacteria. The mutant flagellin did not polymerize in vitro. These results suggested that the mutant bacteria secreted flagellin because the flagellin polymerized poorly and therefore could not be trapped at the tip of the flagellar filament. IKEDA et al. (1983) suggested in addition that this short-flagella mutant should be useful for the study of the mechanism of flagellin transport. INGRAHAM et al. (1983) discuss further aspects of flagellin synthesis and assembly of flagellar components (Fig. 140) (further references therein).

Attached to the proximal end of the filament is the "hook" (Fig. 137) which connects the filament with the "basal body" (INGRAHAM et al. 1983; further references therein). The hook, usually with a bent shape, is between 100 and 900 nm long depending on the bacterial species concerned. It consists of one type of polypeptide which has been demonstrated to be different from the flagellin composing the respective filament, as revealed by immunological techniques. Examples are known where the hook appears to be complexed with additional periodically attached structures of yet unknown function (WIEGEL et al. 1981).

Chemical work with hooks has been facilitated by the isolation of polyhook mutants, i. e. mutants that cannot terminate hook formation and therefore form extended hook structures. In electron micrographs they appear as curly appendages 1 to 2 µm long. The "rod", connected to the hook, usually is about 27 nm long. It has been found to bear, in the samples obtained from *E. coli* and *S. typhimurium*, two pairs of rings, and in *B. subtilis* one pair, corresponding in its location to the inner pair of rings in *E. coli* and *S. typhimurium*. The system of rod and rings has been called the

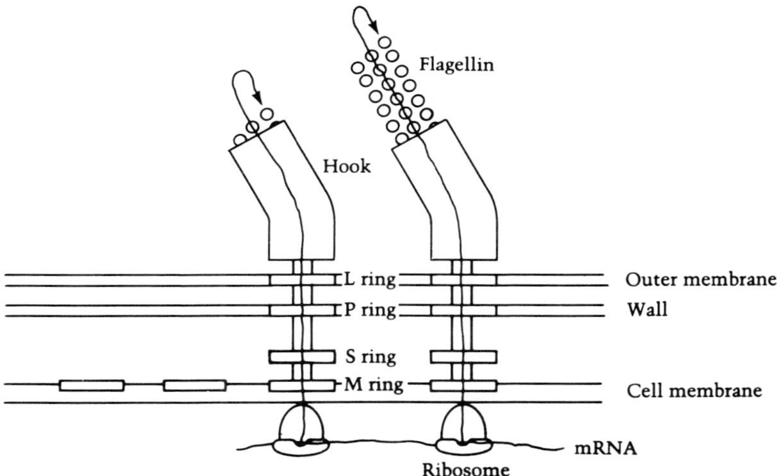

Fig. 140. Assembly of the flagellum in a Gram-negative bacterium. (From INGRAHAM et al. 1983; in part after DE PAMPHLIS)

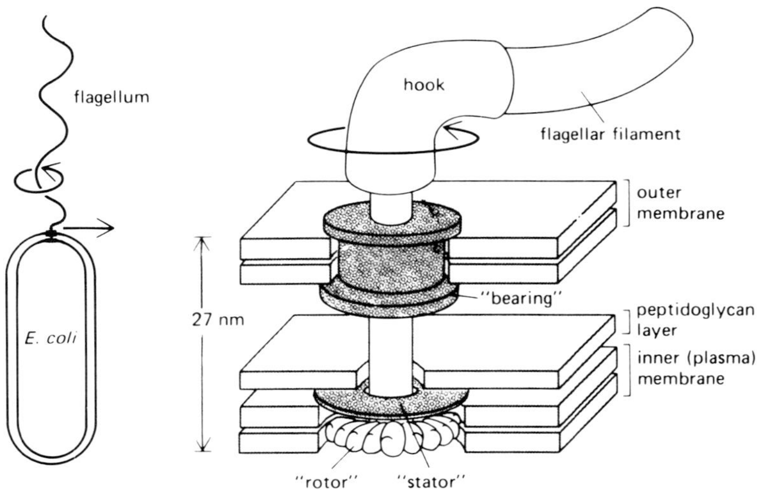

Fig. 141. Schematic drawing of the flagellar rotary motor of *E. coli.* The parts of the assembly that remain stationary are shaded, while the rotating parts of the flagellum are white. (From ALBERTS et al. 1983)

"basal body" of the bacterial flagellum (Fig. 141). The inner pair of rings appears to be the prerequisite for flagellar motion. The innermost ring (M) is associated with the cytoplasmic membrane; the other ring of this pair (S), according to BERG (1975) is mounted in the cell wall in Gram-positive bacteria. The M-ring is assumed to be the "rotor"; it is fixed to the rod (the "driveshaft"), and it rotates freely in the cytoplasmic membrane. The S-ring, being not in contact with the rod, is postulated to be the "stator". "Stator" and "rotor" act together producing rotation

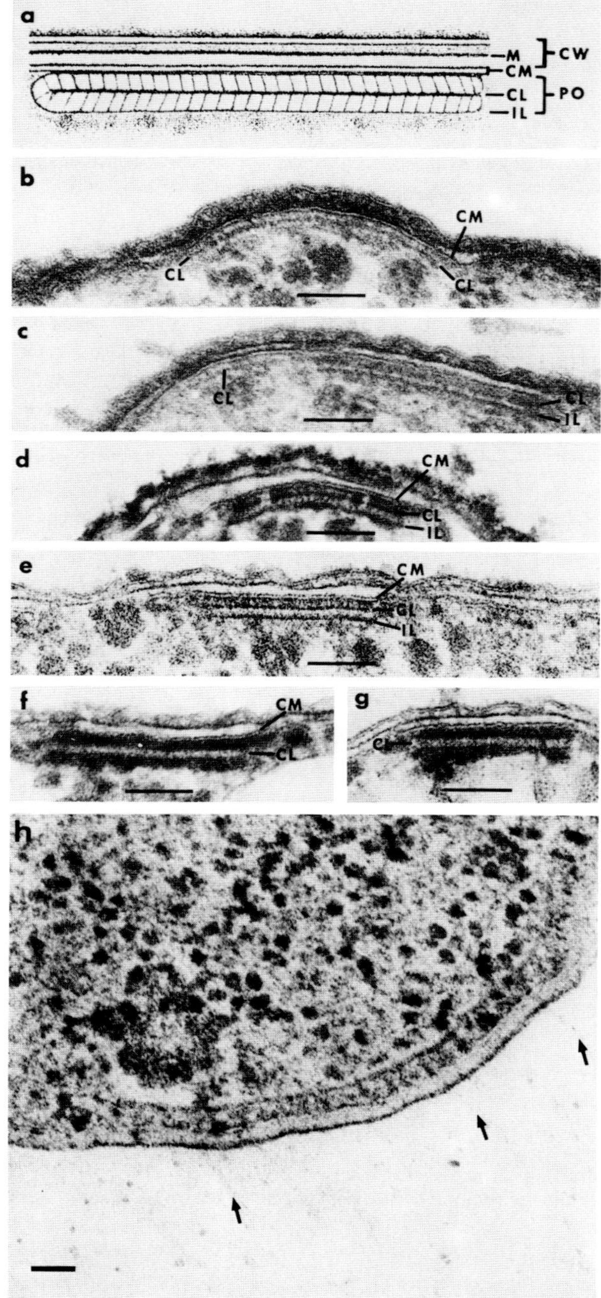

Fig. 142. Flagellar apparatus; the polar organelle. a, schematic drawing (*Sphaerotilus natans*); b–g, ultrathin sections; d, treated for demonstration of ATPase; e, treated for cytochrome c oxidase in the presence of KCN; f, g, treated for cytochrome c oxidase; activity is demonstrated by heavy osmiophilic deposits. CL = central layer; CM = cytoplasmic membrane; CW = cell wall; IL = inner layer; M = murein sacculus; PO = polar organelle. Bars: 0.1 µm. (From TAUSCHEL 1985). h, polar organelle in *Methanococcus jannaschii;* indications for the presence of flagella (arrows). Bar: 50 nm. (From JONES et al. 1983)

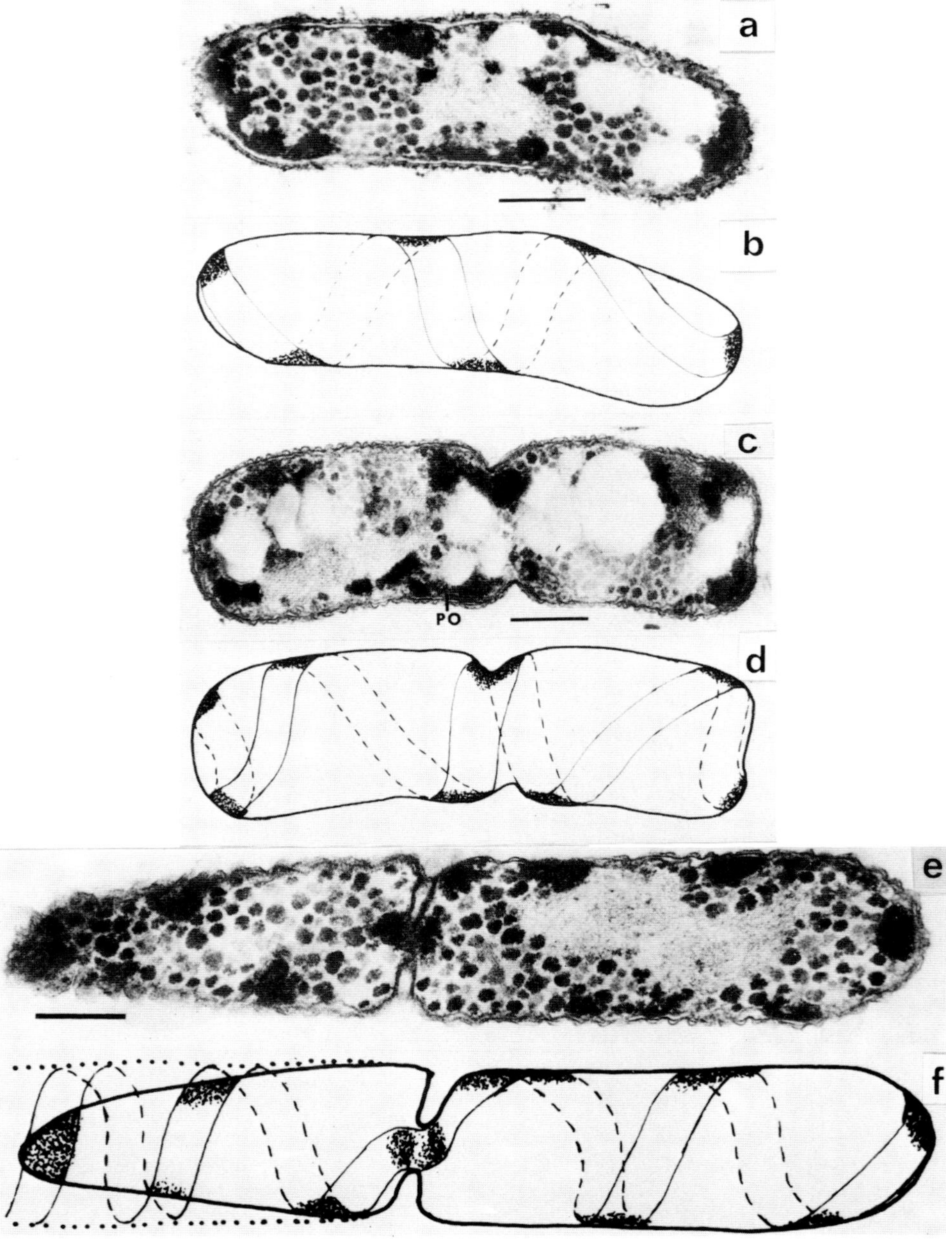

Fig. 143. The polar organelle (PO) in *Sphaerotilus natans*. Cells treated for ATPase activity indicating the localization of the polar organelles in ultrathin sections. a, b, the ribbon-like polar organelle forms a helical structure; c, d, cell division also involves the polar organelle; e, f, fission of the cell with concommitant division of the polar organelle. Bars: 0.5 μm. (From TAUSCHEL 1985)

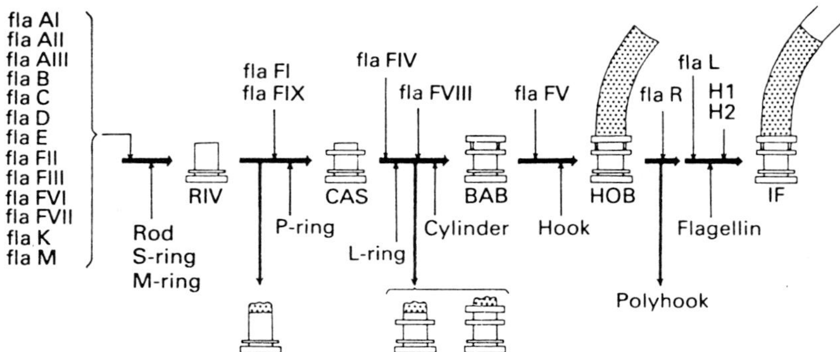

Fig. 144. Diagram suggesting some of the functions for the flagellar gene products in flagellar morphogenesis in *Salmonella*. – RIV = rod-inner ring complex; M-ring is in the cytoplasmic membrane, S-ring is immediately above it. P-ring is associated with the peptidoglycan, L-ring is associated with the lipopolysaccharide. BAB = basal body; CAS = RIV-P-ring complex; HOB = hook-basal body complex; IF = completed flagellum. (From Rogers 1983, after Iino)

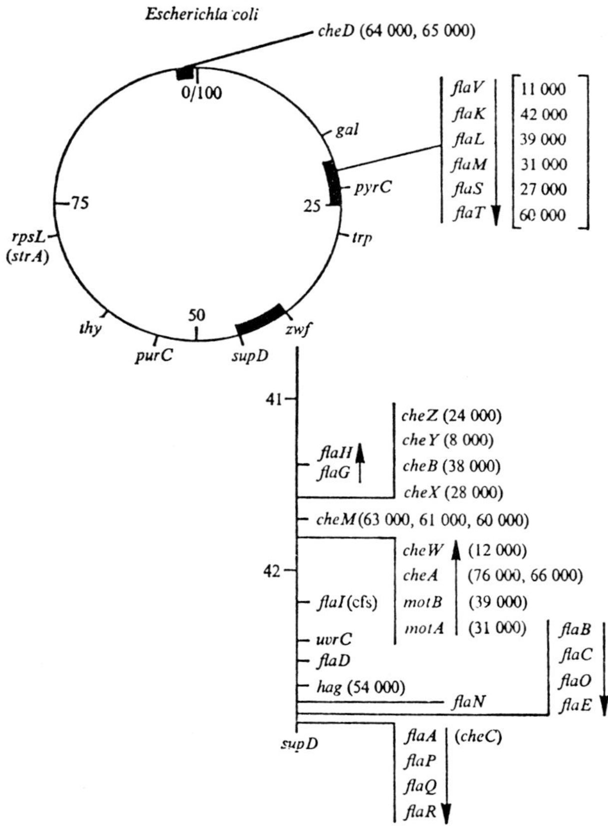

Fig. 145. The *Escherichia coli* chromosome with regions of flagellar related genes indicated together with the molecular weights of gene products specified. (From Rogers 1983, after Simon et al.)

of the rod (and, hence, of the hook and the filament) relative to the cell body. The torque for flagellar rotation is generated between M- and S-rings. Unlike filament and hook, the "basal body" contains at least 9 (or even up to 13) different types of polypeptides. Face views of inner ring pairs have exhibited periodic substructures interpreted to be molecular components of the rotary motor. The additional (outer) pair of rings observed in Gram-negative bacteria is connected to the peptidoglycan layer (P-ring) and the outer membrane or lipopolysaccharide layer (L-ring). This pair of rings fulfills functions as bearings for the flagellar rod passing through the cell wall.

In addition to the flagellar apparatus described so far, a rather complex structure closely associated with the basal regions of flagellar tufts has been described. It has been termed the polar organelle (Figs. 142 and 143) (ALAM & OESTERHELT 1984, ROWLES et al. 1976, TAUSCHEL & DREWS 1969). It is a conspicuous structure not present in non-flagellate bacteria. It consists of a central layer linked both to an inner layer and to the cytoplasmic membrane by spoke-like structures. In the absence of further structural details, it has been discussed (ALAM & OESTERHELT 1984) that this organelle might be involved in some regulation of the flagellar motors of halobacterial flagella. TAUSCHEL (1985) has performed cytochemical studies on *Sphaerotilus natans* and *Rhodopseudomonas palustris* in order to further characterize this structure. He has applied cytochemical

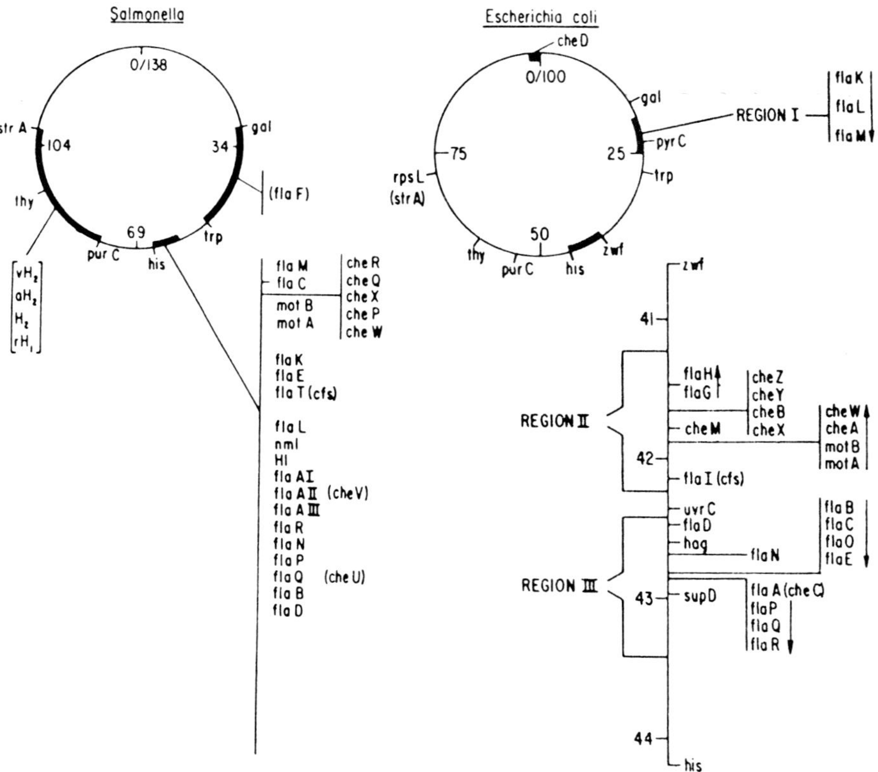

Fig. 146. Distribution of flagellar genes on the genetic maps of *Salmonella* and *E. coli*. (From ROGERS 1983, after SILVERMAN & SIMON)

techniques for the demonstration of ATPase and cytochrome c oxidase activities. Each enzyme was found to be located in a specific area of the organelle, the ATPase being associated with the inner bounding layer, and the cytochrome c oxidase with the fibrillar spokes. Characteristic deposits of lead phosphate were seen at areas which usually altered from side to side along the length of longitudinal cell sections (Fig. 143). In some instances, these deposits were found to be located adjacent to the polar organelle. TAUSCHEL (1985) suggested, as a simple interpretation of his findings, that the polar organelle forms a continuous, helically shaped, ribbon-like structure underlying the cytoplasmic membrane of the cell. As seen from electron micrographs of cells dividing by binary fission, this ribbon divides concomitantly with cell fission, and is passed on to each daughter cell as postulated earlier (MURRAY & BIRCH-ANDERSEN 1963).

Flagella formation is under genetic control (Figs. 144 to 146). The *fla* genes of *E. coli* and *S. typhimurium* are similarly organized on the genetic maps indicating that the final step of flagellar morphogenesis of both bacteria occurs by the same mechanism. SILVERMAN & SIMON (1977) have published a diagram depicting the distribution of flagellar genes on the genetic maps of both bacteria. The complexity of flagella-related genes on the *E. coli* chromosome is demonstrated in Fig. 145 (SIMON et al. 1978), and suggestions of some of the functions of the products of the flagellar genes in *Salmonella* have been given by IINO (1969, 1977) as seen from Fig. 146.

5.10.1.2 Functional aspects

The flagellar filament is a passive structure. It is driven by a rotary motor (Fig. 148) located in the cell envelope (s. above), a feature supported by a wealth of evidence (BERG 1974, BLOCK & BERG 1984, BLOCK et al. 1982, HYAMS & BORING 1975, MITCHELL 1984, SILVERMAN & SIMON 1974a, TAYLOR & KOSHLAND 1974). The way in which a bacterium swims depends on the direction of rotation of its flagellar motor(s), and the control of this direction is the basis of the chemotactic

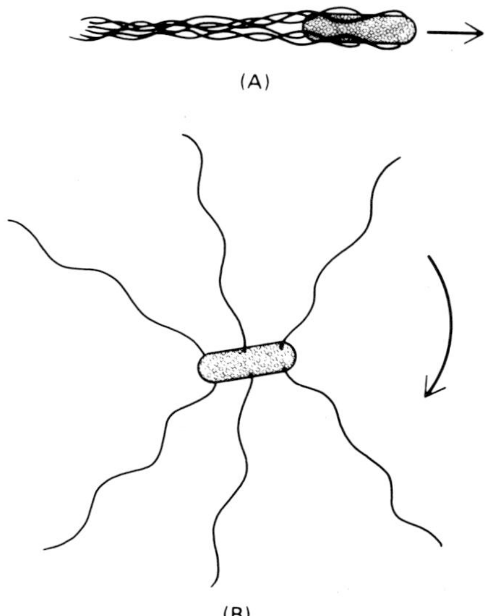

Fig. 147. Schematic drawing showing the positions of the flagella on *E. coli* during swimming. When the flagella rotate counterclockwise (A), they are drawn together into a single bundle, which acts as a propeller to produce smooth swimming. When the flagella rotate clockwise (B), they fly apart and produce tumbling. (From ALBERTS et al. 1983)

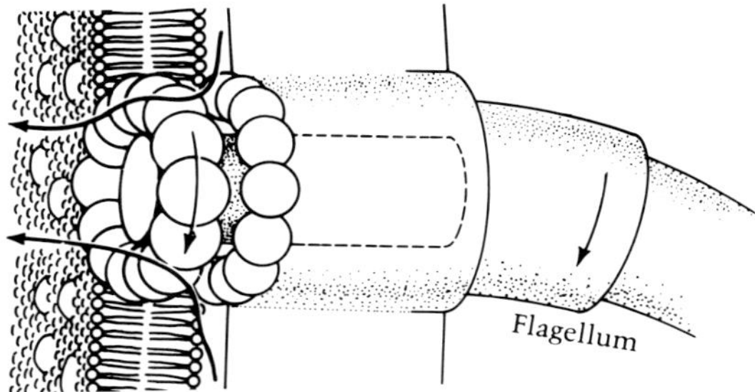

Fig. 148. View of the flagellar motor and its components. (From INGRAHAM et al. 1983, after HINKLE)

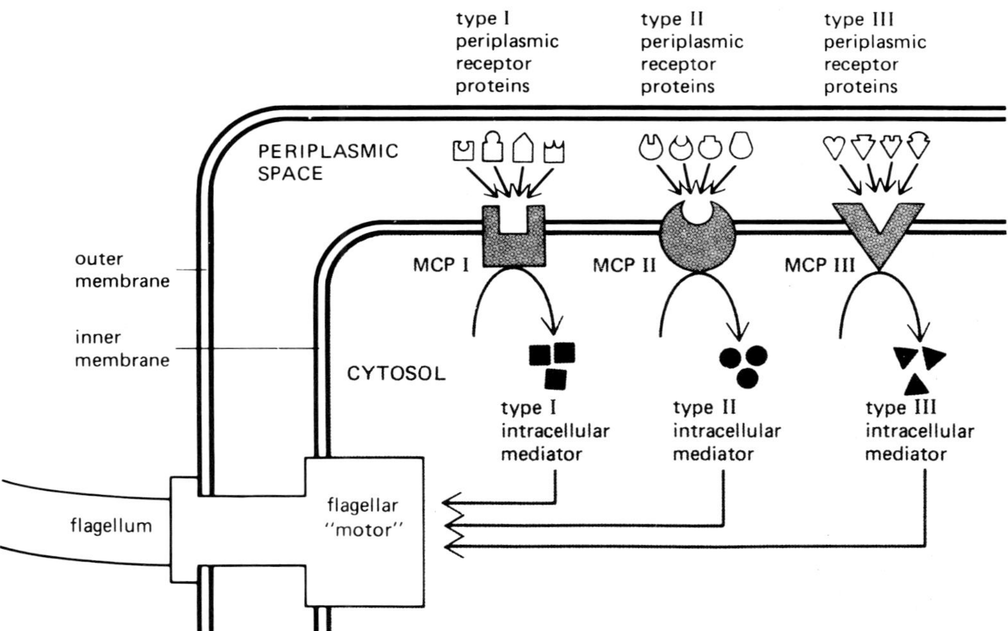

Fig. 149. The steps in signal transduction during bacterial chemotaxis. Chemical attractants (not shown) bind to specific receptor proteins in the periplasmic space. The receptors then interact with one of the three methyl-accepting chemotaxis proteins (MCPs) in the cytoplasmic membrane. This activates the MCPs to produce an intracellular mediator that causes the flagellar "motor" to continue to rotate counterclockwise. Some chemical attractants directly bind to and activate MCP; in such cases the MCP acts as both receptor and transducer. (From ALBERTS et al. 1983)

response (LARSEN et al. 1974). For *E. coli* ist has been shown that the cell is propelled by about six flagellar filaments arising at random points on the surface of the cell (Fig. 147). Each filament is powered by a rotary motor. During counterclockwise rotation, the filaments work together in a bundle that drives the cell steadily forward – the bacterium "runs"; when the motors turn clockwise, the bundle flies apart, and the motion of the bacterium is erratic – the cell "tumbles" (Fig. 147). Runs and tumbles occur in an alternating sequence, each run constituting a step in a three-dimensional walk. Chemotaxis of bacteria is made possible by the existence of receptors at the cell periphery which are able to sense attractants or repellants (Fig. 149). The bias of the flagellar motors (the fraction of time spinning counterclockwise) increases in proportion to the rate of change of receptor occupancy. A model developed by KOSHLAND (1977, 1979) assumes that different flagellar motors on the same cell reverse synchronously. An alternative model for the coordination of flagella has been deviced by ISHIHARA et al. (1983). These authors have used filamentous, non-septate *E. coli* cells which had been produced by introducing a cell division mutation into a strain that was polyhook but otherwise wild type for chemotaxis. Markers for its flagellar motors were attached to antihook antibodies. The markers were driven alternately clockwise and counterclockwise, at angular velocities comparable to those observed when wild-type cells are tethered to glass. The directions of rotation of different markers on the same cell were not correlated; reversals of the flagellar motors occurred asynchronously. The bias (s. above) of the motors changed with time. Variations in bias were correlated, provided that the motors were within a few micrometers of one another. Thus, although the directions of rotation of flagellar motors are not controlled by a common intracellular signal, their biases are. From these results it appears that the signal has a limited range. These experiments have confirmed and extended the work of MACNAB (1983) who observed asynchronous motion of flagellar filaments on cells of *S. typhimurium* of normal size.

BERG et al. (1982), in a report on dynamics and energetics of flagellar rotation in bacteria, have described results from studies on the energetic properties of the flagellar motor in a motile *Streptococcus* strain which lacks endogenous energy reserves and is susceptible to various ionophores. They summarized the findings as follows: "The motor is driven by a proton flux induced by a transmembrane pH (GLAGOLEV & SKULACHEV 1978) or electrical potential gradient. In a metabolizing cell, protons move inward, and the sense of rotation is determined by chemotactic signalling: the motor turns alternately clockwise and counterclockwise; chemoattractants induce a counterclockwise bias. In a starved cell, when protons move inward in response to an artificially imposed pH or electrical potential gradient, the motor turns clockwise in the absence of chemoattractants, counterclockwise in their presence. The rotation rate is linearly proportional to the protonmotive force, suggesting that a constant number of protons is transported per revolution. Assuming 100% efficiency, this number is estimated to be about 300. Experiments in which the electric potential gradient is abolished by adding K^+ and valinomycin to the external medium show that a pH shift of 0.1–0.3 units is sufficient to support rotation. Thus, the motor turns even when the potential energy of a proton due to the gradient is less than its thermokinetic energy. The motor can be driven by an artificially induced proton flux of either polarity. In a wild-type cell, the sense of rotation is clockwise whether protons move inward or outward. When a chemoattractant is added, the motor goes mostly counterclockwise when protons move into the cell, mostly clockwise when they move out, although this clockwise bias is not as strong as it is in the absence of a chemoattractant. In one generally non-chemotactic mutant *(SM197)*, the motor always turns counterclockwise when protons move into the cell, clockwise when they move out. Experiments with other mutants suggest that chemoattractants may enhance counterclockwise rotation without reference to the sign of the proton flux". A detailed model involving a protonmotive redox loop has been presented to account for the rotation of the M-ring relative to the S-ring under the influence of a proton gradient (BERG et al. 1982).

Bacterial motility mediated by flagella is an excellent example for a complex problem which can only be understood in detail by a combination of physical, chemical, and genetic approaches, with cytological data obtained by electron microscopy providing the basis for such studies.

5.10.2 Motility of spirochetes and structurally related bacteria

Spirochetes and structurally related bacteria are unique as far as cytology and type of motility are concerned (CANALE-PAROLA 1978, 1981, GOULBOURNE & GREENBERG 1981). Usually, they exhibit a helical shape which varies, depending on the specific strain, in wave length, amplitude, the regularity of their cell coiling, and the form of their cell ends (blunt, tapered, pointed). Divers spirochetes show a wide range of cell lengths and diameters. The helical shape of the cells is maintained by the peptidoglycan layer. This has been shown by treatment with agents that affect peptidoglycan (either its integrity or its synthesis) resulting in loss of the helical shape. Loss of the helical shape was also observed in some mutants, and some of them were motile, though markedly less than the wild-type strains. Cells of *Treponema pallidum* and some other *Treponema* strains have been reported to have a flat wave form rather than a helical shape. HOVIND-HOUGEN (1976),

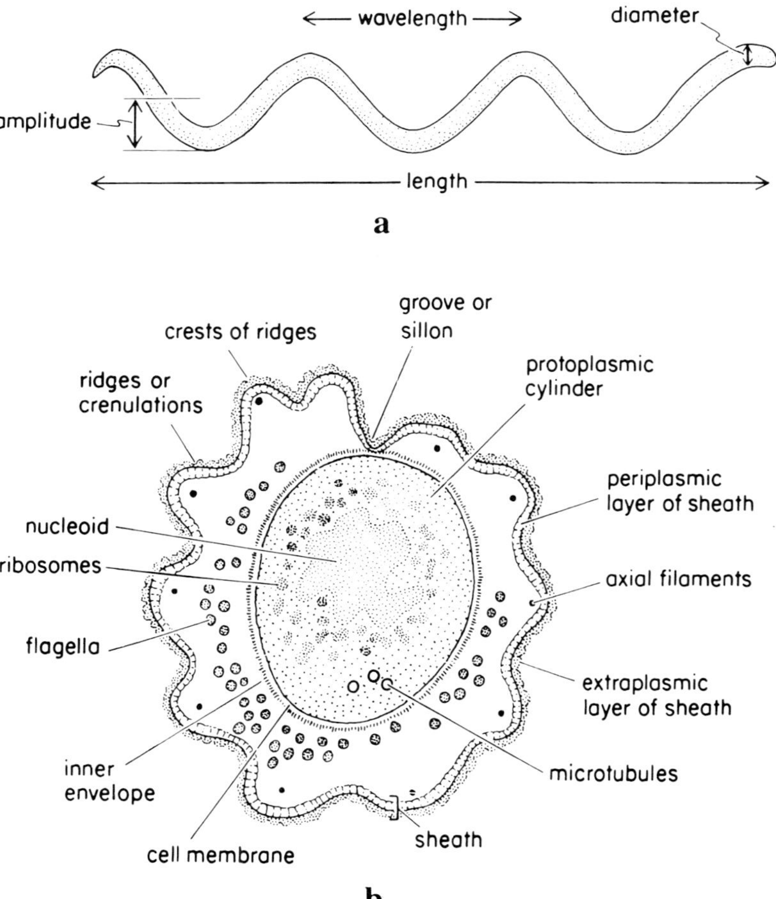

Fig. 150. Diagram of morphological terminology of spirochetes. a, by light microscopy; b, by electron microscopy; transverse section. (From MARGULIS et al. 1981)

however, has demonstrated by electron microscopy, that cells of *T. pallidum* and *T. pertenue* have a helical shape.

During movement spirochetes, though retaining their basic typical shape, assume a variety of modified shapes caused by contractions, undulations, flexions, wave propagations, and vibration of the cells.

A unique feature of spirochetes are filamentous structures wound around the helical protoplasmic cylinder; the name "periplasmic fibrils" has been proposed for them instead of the terms "axial filament" or "axial fiber" (Figs. 150 and 151). Electron microscopy has revealed that individual periplasmic fibrils are similar to flagella in fine structure, composition and behaviour upon treatment with degrading agents. They consist of a filamentous portion, a hook and an anchoring apparatus with several discs or rings. Some biochemical evidence suggests that periplasmic fibrils of spirochetes might be more complex than the flagella of most bacteria when the number of different types of polypeptides constituting the fibril structure is compared to the number of types of proteins in regular flagella. Each periplasmic fibril is inserted at one end only of the cell, but it extends along most of the cell length; thus, the fibrils inserted at the opposite cell poles overlap in the central portion of the cell. This has been demonstrated for *Treponema* and *Leptospira* strains.

Electron microscopy has revealed that the outer sheath or outer cell envelope observed in spirochetes as a very typical and unique structural feature is a three- or possibly five-layered membrane which completely surrounds the cell including the periplasmic fibrils. Kaiser & Doetsch (1976) have suggested that this sheath might act as a semipermeable membrane. It is not known whether this sheath is attached to the protoplasmic cylinder or not. In a model for

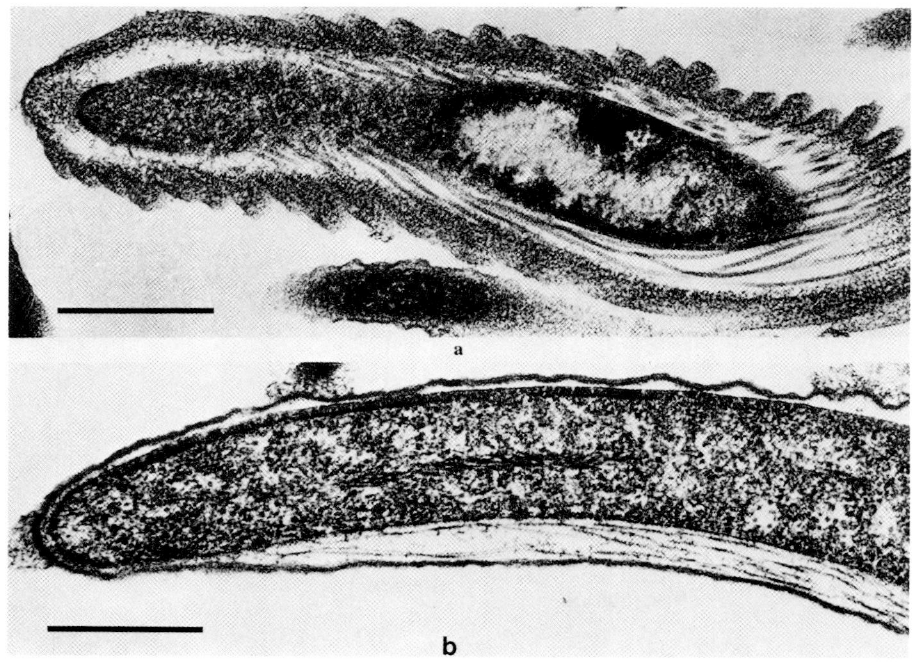

Fig. 151. Longitudinal sections through a, *Pillotina* sp.; b, *Hollandina* sp. Bars: 0.5 µm. (From Margulis et al. 1981)

spirochetal propulsion, the sheath is assumed to be free to rotate about the cell body. However, various models and theories aimed at interpreting the mechanism of the three major types of spirochetal movement, i. e. translational motion, rotation about the longitudinal axis, and flexing have been published. As flexes are followed by changes of translational direction, GREENBERG & CANALE-PAROLA (1977) suggested that flexes may correspond to the tumbles or twiddles of flagellated bacteria. This view is supported by the finding that several spirochetes have the ability for chemotactic behaviour. Some spirochetes, e. g. *Spirochaeta plicatilis,* have been reported to creep or crawl on solid surfaces. CANALE-PAROLA (1978), in a review on motility and chemotaxis of spirochetes, has described and discussed several theories developed to explain the various aspects of spirochetal locomotion. The reader is referred to this review as a detailed discussion of all published models for this specific kind of bacterial movement is outside the scope of this book.

5.10.3 Gliding, and other types of active bacterial locomotion

Besides motility mediated by flagella, gliding (Figs. 152 and 153) is the other major type of locomotion used by bacteria (BURCHARD 1981, REICHENBACH & DWORKIN 1981 a, b) in order to change the local environment or to form complex aggregates in a tendency towards formation of colonies or fruiting bodies. This type of motility has been examined and reviewed by several authors. Obvious locomotory organelles have, however, not yet been described with one possible exception (PATE & CHANG 1979). These authors claim that gliding motility in prokaryotic cells is driven by rotary assemblies in the cell envelopes (s. below). Gliding has been defined by PRINGSHEIM (1951) as "a translocation along solid bodies ... during which no wriggling, contraction or peristaltic alterations are visible, the change of shape being restricted to bending ... Gliding movements are not always regular, but intermittent and hesitant, with frequent changes in direction". It has been shown that solid supports are not absolutely required for gliding. *Oscillatoria* can glide through a viscous medium, and gliding of bacteria at water-air interfaces has been observed. Depending on the type of bacterium, gliding can be a locomotion in a direction parallel to the long axis of a filament *(Flexibacter),* or in a direction perpendicular to the long axes of chains of cells (trichomes of *Simonsiella*). With some exceptions such as *Simonsiella* which grows and glides as a ribbon-shaped trichome and which has a morphologically distinct, concave, ventral surface (Fig. 154), typical gliding bacteria are axially symmetrical, not only in form but also in their ability to glide. During gliding, rotation around the long axis was found to be common in cyanobacteria and *Beggiatoa,* with a species-specific sense of rotation. *Flexibacter polymorphus* has helical filaments, and the sense of rotation conforms with the sense of the helix. It has been noted in cine films and by scanning electron microscopy that gliding bacteria etch agar surfaces. Trails appear as furrows or grooves, the gliding tracks of *Simonsiella* being replicas of the concave ventral surface of the filaments. Since most of the gliders lack agarase activity, the grooves may not generally reflect the enzymatic degradation of agar. Instead, they might be caused by local displacement or dehydration of the agar surface. REICHENBACH (1965) has suggested that "trail-following", i. e. gliding along such grooves produced by other bacteria of the same strain, would facilitate the communal behaviour of myxobacteria such as the formation of multicellular swarms and aggregation culminating in fruiting body formation (GRÄF 1965, REICHENBACH 1974). Extracellular slime has been detected marking the paths of myxobacteria and *Cytophaga;* it has also been observed as a collapsed sheath left behind gliding oscillatorian cyanobacteria. Consequently, directed and active slime secretion has been proposed to provide the propulsive force for gliding. Secretion of slime would require envelope pores. In those species where they have been found, the pores did not appear to

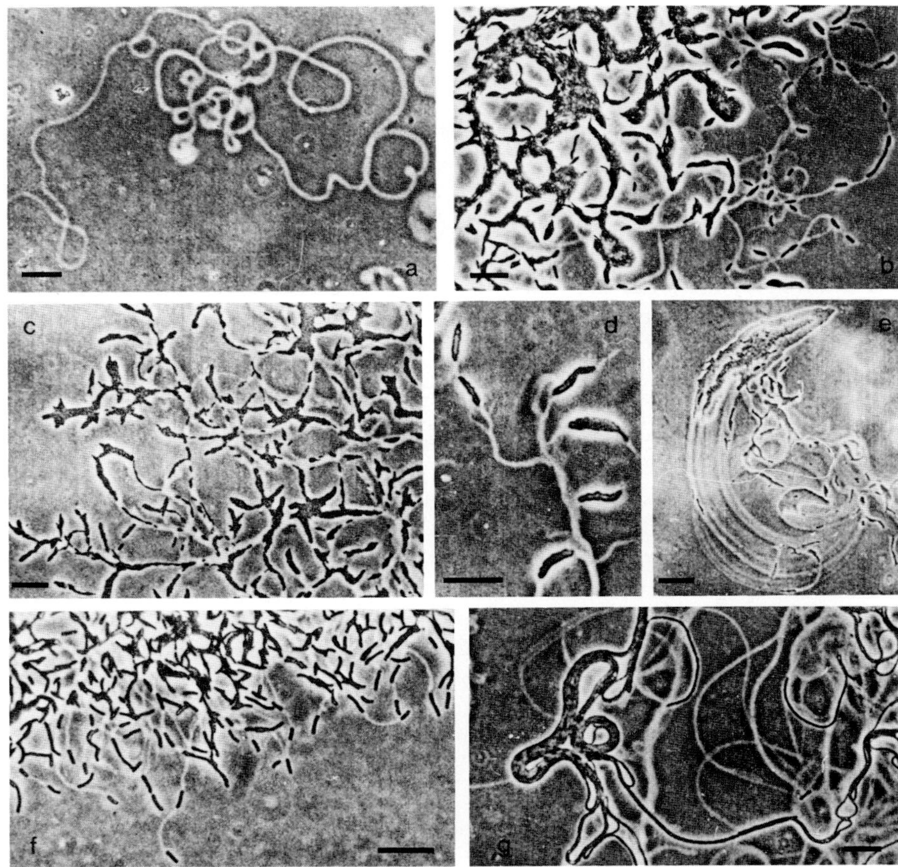

Fig. 152. Gliding bacteria; slime tracks on agar. a, *Nannocystis exedens* (Myxobacterales); b, *Myxococcus fulvus;* c, *Cytophaga* sp.; d, *Flexibacter elegans* (Cytophagales); e, *Vitreoscilla stercoraria* (Leucotrichales); f, *Lysobacter* (Cytophagales); g, *Herpetosiphon giganteus* (Leucotrichales). Bars: 20 μm. (From REICHENBACH & DWORKIN 1981 a)

perforate every layer of the envelope, nor have slime fibers been observed associated with such pores or channels. Pores have not been detected in all gliders. These pores that have been visualized lacked any orientation suggestive of unidirectional slime production. A pore system would also require a mechanism for reorientation to account for reversal of the gliding direction. JAROSCH (1962) has stated that slime is sectreted over the entire surface of the gliding cell, and additional findings indicated that slime is released ahead of and lateral to gliding flexibacters. It has been calculated that the volume of slime that would have to be extruded to propel the cyanobacterium *Symploca muscorum* at its average speed of 3 μm/sec would be equivalent to several cell volumes per second. These and other observations taken together, provide evidence against the theory that gliding is brought about by direct slime extrusion. That slime is however required for gliding is suggested by observations that cyanobacteria producing no detectable amounts of slime are unable to glide unless their trichomes are juxtaposed to one another. It has been proposed that slime may

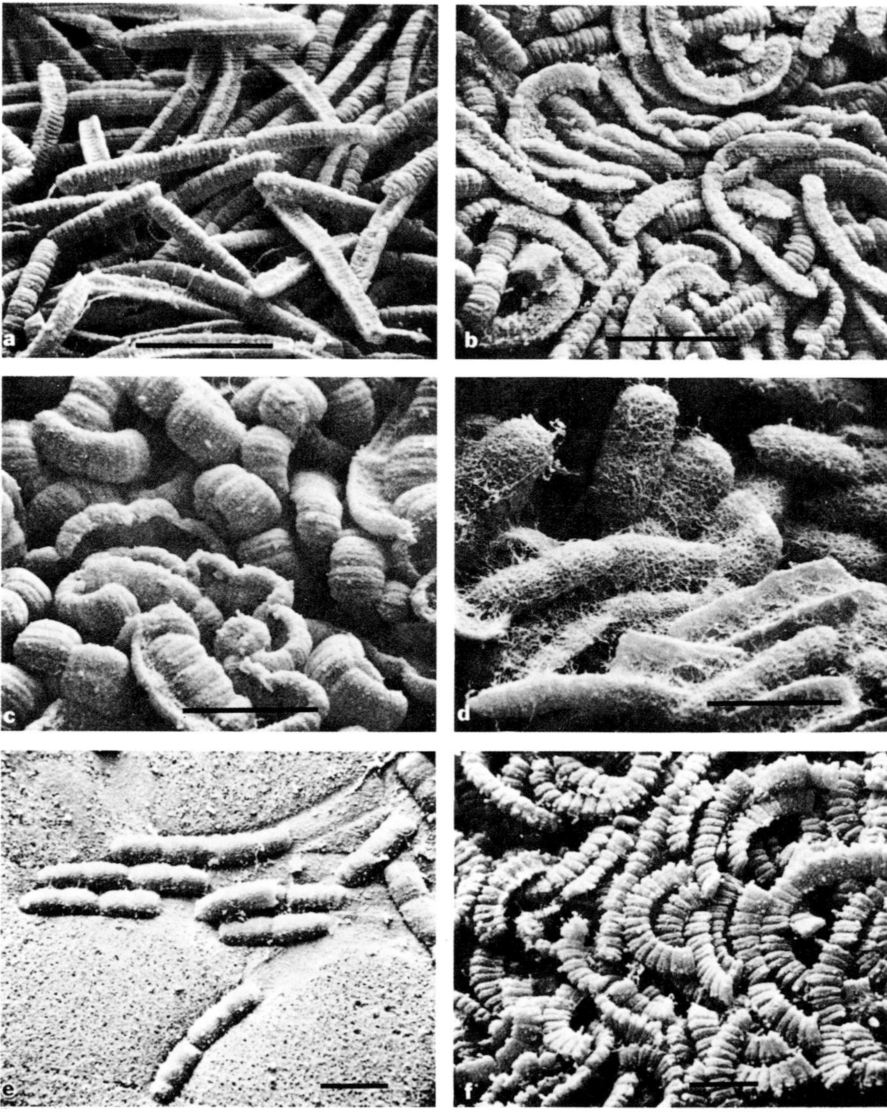

Fig. 153. Gliding bacteria; members of the family Simonsiellaceae. a, *Simonsiella crassa;* b, *S.* sp.; c, d, *S. steedae*; e, as c, gliding tracks; f, *Alysiella filiformis*. Bars: 10 μm. (From KUHN 1981)

mediate adsorption of the glider to the substratum. Viscous slime isolated from *Flexibacter* was found to be a linear colloid with properties of a temporary adhesive which could serve to hold the cell against its substratum but permit translational movement on that substratum. These properties of slime might also permit travelling waves of local and temporary trichome separation from the substratum by gliding *Oscillatoria*, as observed by AMBROSE (1961). Polysaccharide slimes, being hygroscopic, may play an additional role in protecting gliding bacteria from dehydration. Mainte-

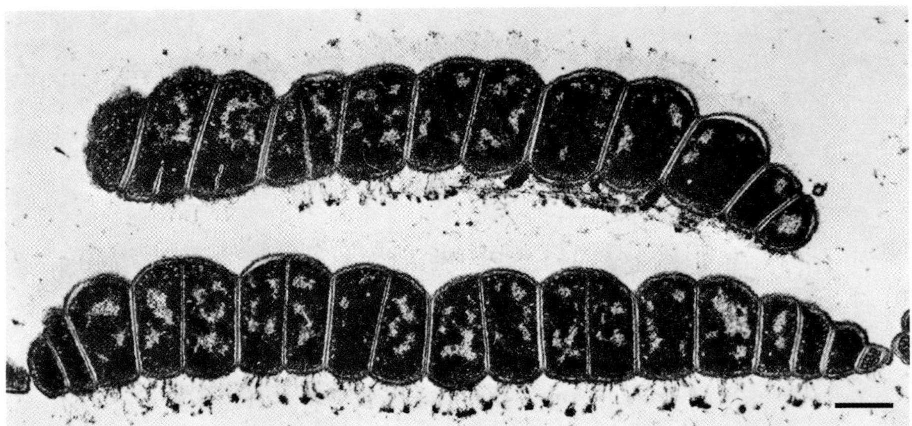

Fig. 154. Longitudinal section of multicellular filaments of *Simonsiella steedae* showing the dorsal-ventral differentiation; the dorsal capsule is undifferentiated; the ventral capsule contains fibrils emerging from the cells at right angles. Bar: 0.5 μm. (From KUHN 1981, after PANGBORN et al.)

nance of an aqueous environment adjacent to the cell might facilitate functioning of the gliding mechanism.

As the theory of directed slime extrusion through envelope pores has not withstood experimental control, other ultrastructural features in the cellular organization of gliding bacteria have been discussed to be the structural basis for gliding and related movements. Fimbriae were thought to be mechanically involved in gliding; however, DOBSON et al. (1979a, b) have reported that some motile strains of *Myxococcus* lack fimbriae. It has been demonstrated that fimbriae, though not required for gliding, are necessary for the swarming or social type of motility. In the absence of fimbriae, cells glide as individuals only. Fimbriae were also found to mediate adhesion to surfaces on which the cells glide. These filamentous structures may also be involved in twitching motility, a form of surface-associated, intermittent, jerky movement that does not regularly follow the cell long axis. This translocation, as shown for *Pseudomonas aeruginosa* bearing retractile polar fimbriae, presumably occurs by outgrowth of the appendages, their adhesion to the surface of the substratum, and retraction or contraction, resulting in the cell being pulled towards the adhesion site. This type of movement which has been observed in a variety of bacteria definitely is not gliding. Specific envelope structures have also been discussed as being directly related to gliding, together with a flexibility of the envelope layers. However, flexibility has not been found to be a prerequisite for gliding as demonstrated in *Flexibacter BH 3* which is rigid, and other bacteria such as *Chondromyces apiculatus* and some cyanobacteria. There have been reports on convolutions in the envelopes of gliding bacteria. It has been discussed that gliding "is produced by regular wave motion in the outer membrane which in turn is produced by a directional sequence of "make-and-break" interactions between the flexible outer membrane and the more rigid peptidoglycan layer" (HUMPHREY et al. 1979). However, not all gliders demonstrate obvious envelope convolutions. Envelope fibrils, i.e. fibrils associated with cell envelope layers of gliding bacteria, have been described and discussed (BURCHARD 1981; further references therein) as the structural basis for gliding, with the most extensive characterization of such structures coming from studies on *Oscillatoria princeps*. A parallel array of 6 to 9 nm diameter fibrils in a defined layer of the cell envelope was seen with several electron microscopic techniques. The fibrils were found to be arranged as parallel, right-handed helices. The fibrils could be eliminated by pepsin treatment. Trypsin treatment of trichomes resulted in gliding motility becoming jerky and irregular, before it finally stopped. The fibrils might have contractile properties, and fibrillar contractions may account for the "wave-like distorsions" in the fibrils and the closely associated outer membrane. Propagation of these distorsions may result in forces applied against the substratum which might be responsible for translocation. GLAGOLEVA et al. (1980) have proposed that these fibrils may be attached to membrane-bound basal bodies similar to those that drive bacterial flagella. Resulting fibrillar torsions would thus be driven by basal body rotation relative to the cell envelope. This hypothesis raises the possibility of a functional similarity between oscillatorian envelope fibrils and the axial fibrils that appear to serve as motility organelles of spirochetes. Like gliders, some spirochetes are able to translocate on solid surfaces by "creeping and crawling" (s. above), and they are able to flex. In addition, some structural similarities have been found between the cytoplasmic fibrils present in the gliding *Myxococcus xanthus* and those found in a spirochete of the genus *Treponema.*

The existence of rotary assemblies, i. e. ring-like structures associated with the envelope layers of *Cytophaga johnsonae* and *Flexibacter columnaris,* led to the proposal of a unified theory for prokaryotic motility (PATE & CHANG 1979). In flagellated bacteria, rotation of the basal body relative to the envelope drives the flagellar filament. Gliding bacteria would be propelled by the spinning of the rotary assemblies against the substratum. Non-gliding mutants of gliders lack these assemblies. Inhibitors of gliding motility blocked latex sphere movements caused by the assemblies. These structures have been isolated by addition of KI, a salt known to solubilize eukaryotic actomyosin, followed by dialysis against a buffer selected to keep actin in fibrillar form. The rings were found to be approximately 20 nm in diameter and 10 nm thick. Some of them were arranged in two-ring stacks. Experiments designed for the isolation of similar structures from other bacteria have been performed, and comparable assemblies, though variable in size and morphologically irregular, have been obtained from *Cytophaga U67* and *Myxococcus xanthus.* However, the functional relevance of these structures for gliding motility has not been settled.

Goblet-shaped structures described to be present in hexagonal arrays on the external surface of *Flexibacter polymorphus* have been suggested to be extensions of rotary assemblies that facilitate movement in a marine environment. Outer membrane-goblets are, however, not unique of gliding bacteria. Therefore, their functional significance appears to be doubtful. Cytoplasmic filaments similar to microfilaments in eukaryotic cells have been found in some gliding bacteria. *Myxococcus xanthus* cells have been observed to contain bundles of filaments 4 to 5 nm in diameter, closely and longitudinally associated with the inner face of the cytoplasmic membrane, and exhibiting a periodic substructure. One end of the bundle appeared to be in contact with the cell envelope, with the other end terminating in the cytoplasm. It may be that these filaments can only be visualized when at least one point of envelope attachment breaks, permitting contraction. Assuming two attachment points in the living cell, these structures may be contractile organelles for gliding motility in an inchworm-like fashion. Such a gliding mechanism might be possible if longitudinal contractions and elongations were postulated to be of a very small degree and/or high frequency. Cine films of gliding *Oscillatoria* do not indicate significant changes suggestive of cyclic contractions. Rod-shaped or cylindrical cytoplasmic elements described as filaments have also been detected in some other bacteria, especially in more than 10 strains of myxobacteria. However, they were reported to penetrate the cell envelope, emerging extracellularly as bundles of flagella. Therefore, again, functional relevance for a gliding motility is rather doubtful. To date, there has been no solid evidence in support of a concept that motility of (enveloped) gliding bacteria is driven by interactions of eukaryotic-like actin and myosin. On the contrary, DWORKIN et al. (1983) and KELLER et al. (1983) have presented theoretical and experimental support for a known model that gliding motility of *Myxococcus xanthus* is driven by a surface-tension gradient at the interface that surrounds the cell, thus providing the driving force for active and regulated movement. It was assumed to be mediated by a surfactant secreted in a polarized and reversible way. These authors stress the point that gliding of myxobacteria may differ substantially from that of other gliding bacteria, a view shared also by other authors. BURCHARD (1981; further references on gliding motility therein) has discussed the surface-tension hypothesis very critically and has concluded that many attributes of gliding bacteria are difficult to explain with such a model.

A surface-associated movement that has been called gliding has also been observed among mycoplasmas (BREDT 1973, MANILOFF 1979, WALL et al. 1983), though these bacteria, according to 16 S ribosomal RNA sequencing data, may have evolved from ancestral clostridia and are not classified among gliding bacteria by traditional taxonomy. Mycoplasmas lack a typical envelope; instead, they are bounded by a stabilized lipoprotein unit membrane. Moving mycoplasmas exhibit an anterior-posterior axis and a specialized terminal region (a tip) that might be important in adhesion to the substratum. An extracellular mucoprotein layer, particularly thick in the tip region, may serve the same function in moving mycoplasmas as does extracellular slime for enveloped gliding bacteria. Other types of movement have also been observed in mycoplasmas and the related spiroplasmas, such as rapid, active, and reversible changes of cell shape, "swimming" in viscous media, whirling, spinning, undulating, and flexing. Fibrils similar to those described in *Myxococcus xanthus* have been seen in *Spiroplasma citri,* and a "skeleton" of thin fibers, morphologically similar to those of rabbit muscle F actin, have been found in *Mycoplasma pneumoniae*. Indications for actin-like protein in extracts of *Mycoplasma pneumoniae* have been obtained (NEIMARK 1977). Filaments were observed to be formed from components in these extracts upon treatment with ATP-Mg^{2+}. These filaments could be decorated with heavy meromyosin. Labelling experiments with antiserum raised against rabbit muscle actin had positive results, labelling being particularly intense at the tips.

Besides ultrastructural studies, biochemical and physiological investigations, and the analysis of mutants are necessary approaches for a better understanding of gliding and related types of motility both in mycoplasmas and in enveloped bacteria.

5.11 Bacterial endospores, exospores, cysts, myxospores, and other persistent states

Some bacteria are capable of producing cell states or differentiations which are resistant against unfavourable environmental conditions (BEAMAN et al. 1982, BEWLEY 1979) such as unphysiologically high temperatures, lack of water, radiation, mechanical stress or chemical influences. These special cell states originate from vegetative cells and can easily be visualized under the light microscope due to optical properties different from those of vegetative cells. Such states have been found as endospores, exospores, cysts, myxospores or other persistent cell forms (Fig. 160) (ARONSON & FITZ-JAMES 1968, FITZ-JAMES & YOUNG 1969, FREHEL & RYTER 1969, GOULD & HURST 1969, HANSON et al. 1970, HOLT & LEADBETTER 1969, MURRELL et al. 1969, NORRIS 1969; WALKER 1970).

Endospores are spores formed within a vegetative cell (Figs. 155–157); typical examples are the spores of bacilli and clostridia, i. e. Gram-positive bacteria. However, endospores have also been found in *Sporomusa* (MOELLER et al. 1984), a genus of Gram-negative anaerobic bacteria (Fig. 158).

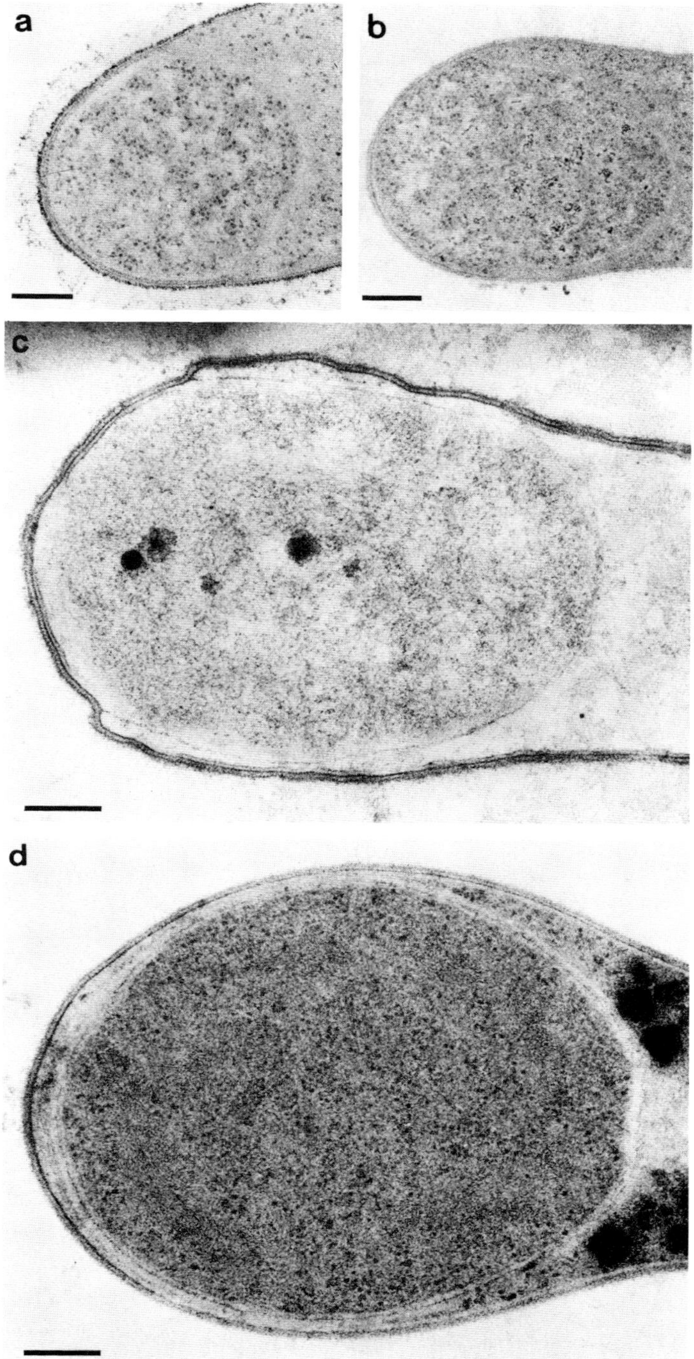

Fig. 155. Early stages in spore formation in *Clostridium formicoaceticum*. a–c, growing septum; d, after end of septum growth. Bars: 0.2 μm. (Original micrographs F. MAYER)

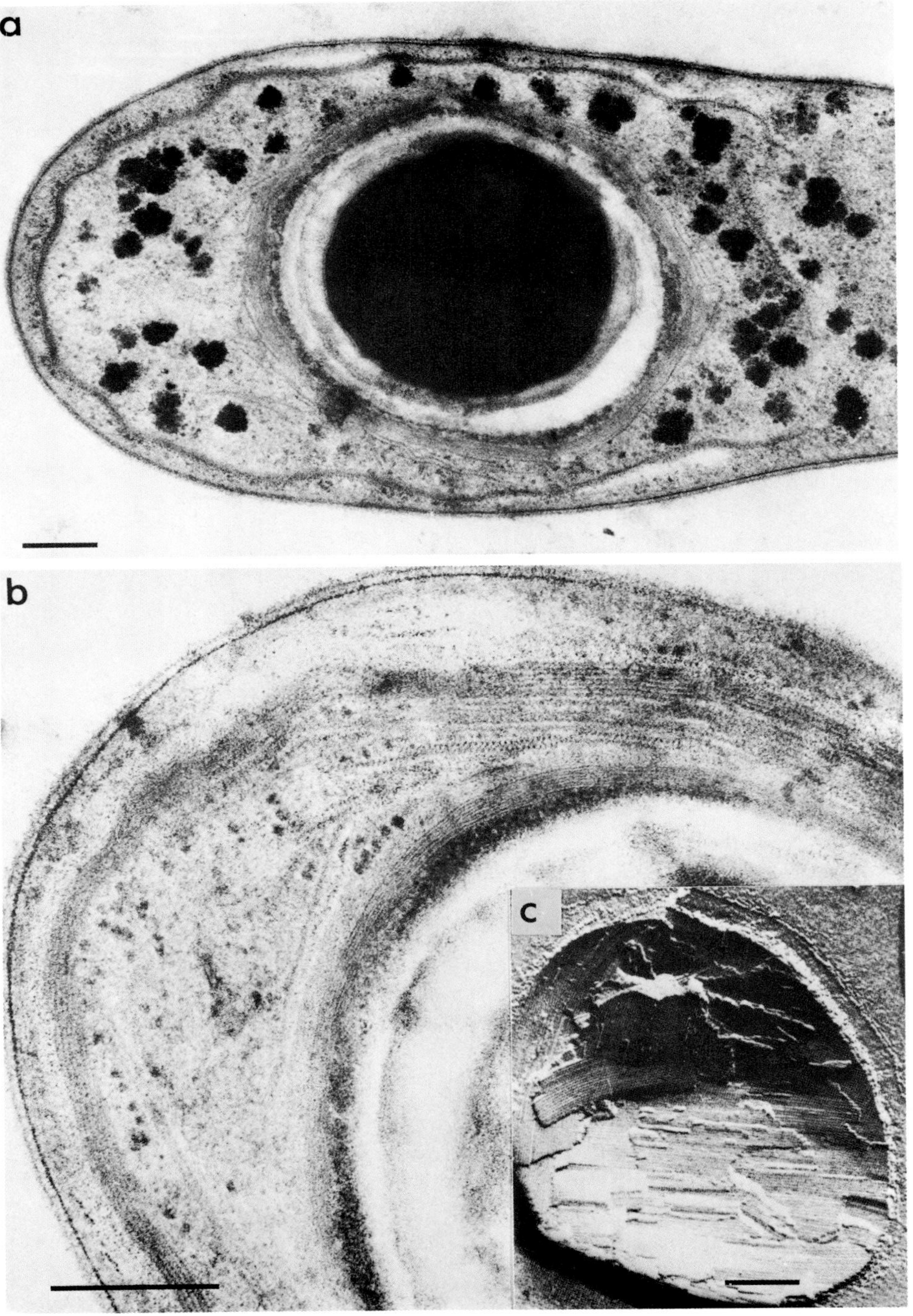

Fig. 156. Structure of the spore of *Clostridium formicoaceticum*. a, late stage of spore formation; b, thin section through spore layers; c, freeze-etching sample of a spore showing fibrillar components in the spore layers. Bars: 0.2 µm. (Original micrographs F. MAYER)

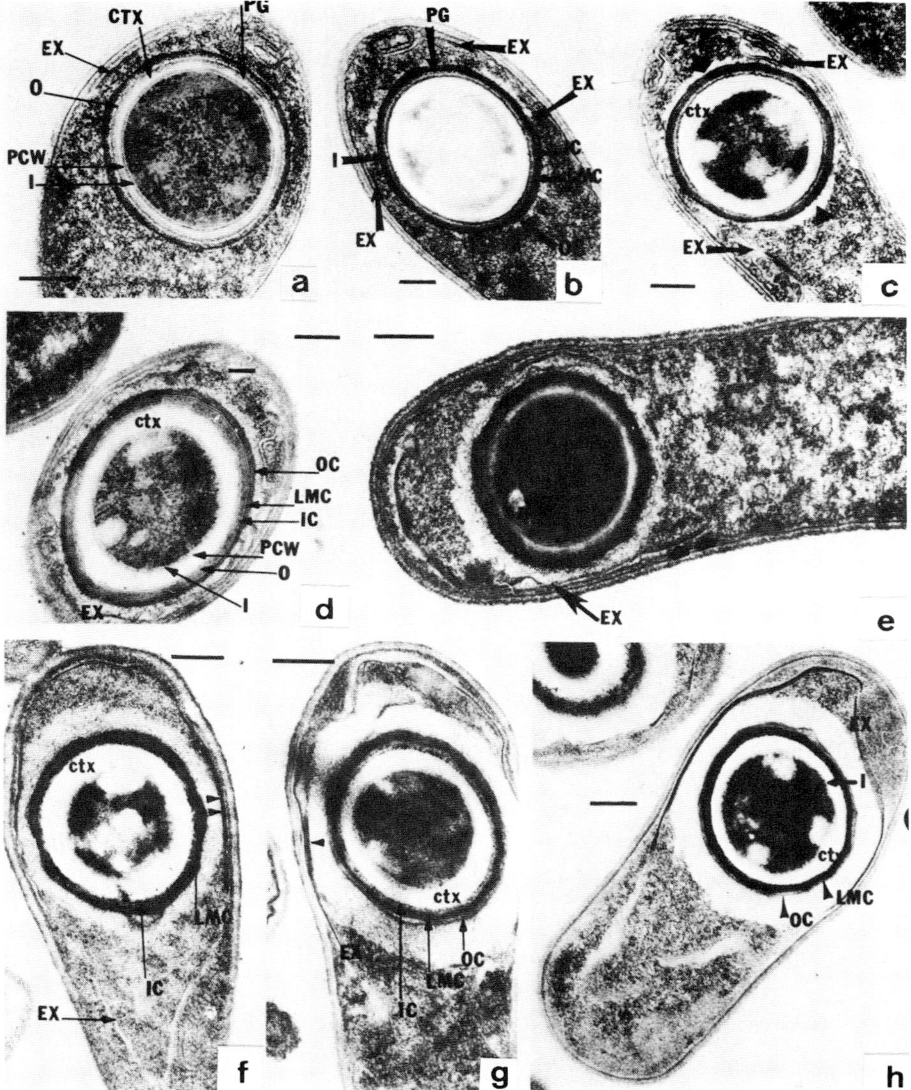

Fig. 157. Ultrastructure of *Bacillus sphaericus* spore formation; late stages in sporulation shown in a sequence of stages (a–h). – CTX, ctx = cortex; EX = exosporium; I = inner forespore membrane; IC = inner coat; LMC = lamellar midcoat; O = outer forespore membrane; OC = outer coat; PCW = primordial cell wall; PG = peptidoglycan layer. Bars: 0.25 µm. (From Holt et al. 1975)

Motile spores have been noted in *Kineosporia aurantiaca*, a genus of the order Actinomycetales.

The formation of exospores has so far only been observed for the methane-utilizing bacterium *Methylosinus trichosporium* (Schlegel 1981) and in the genus *Rhodomicrobium* (Pfennig & Trueper 1974). These spores originate from a vegetative cell by a budding process. Their phys-

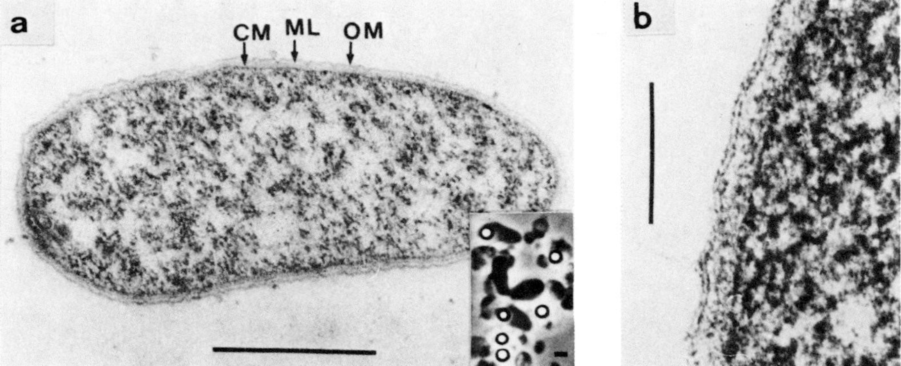

Fig. 158

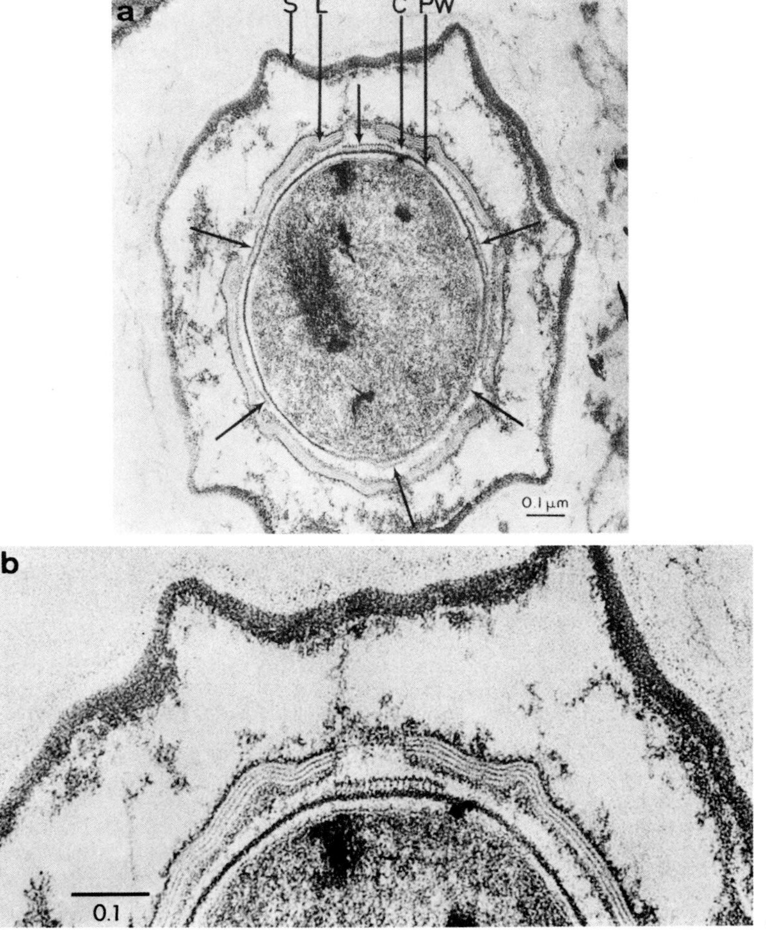

Fig. 159

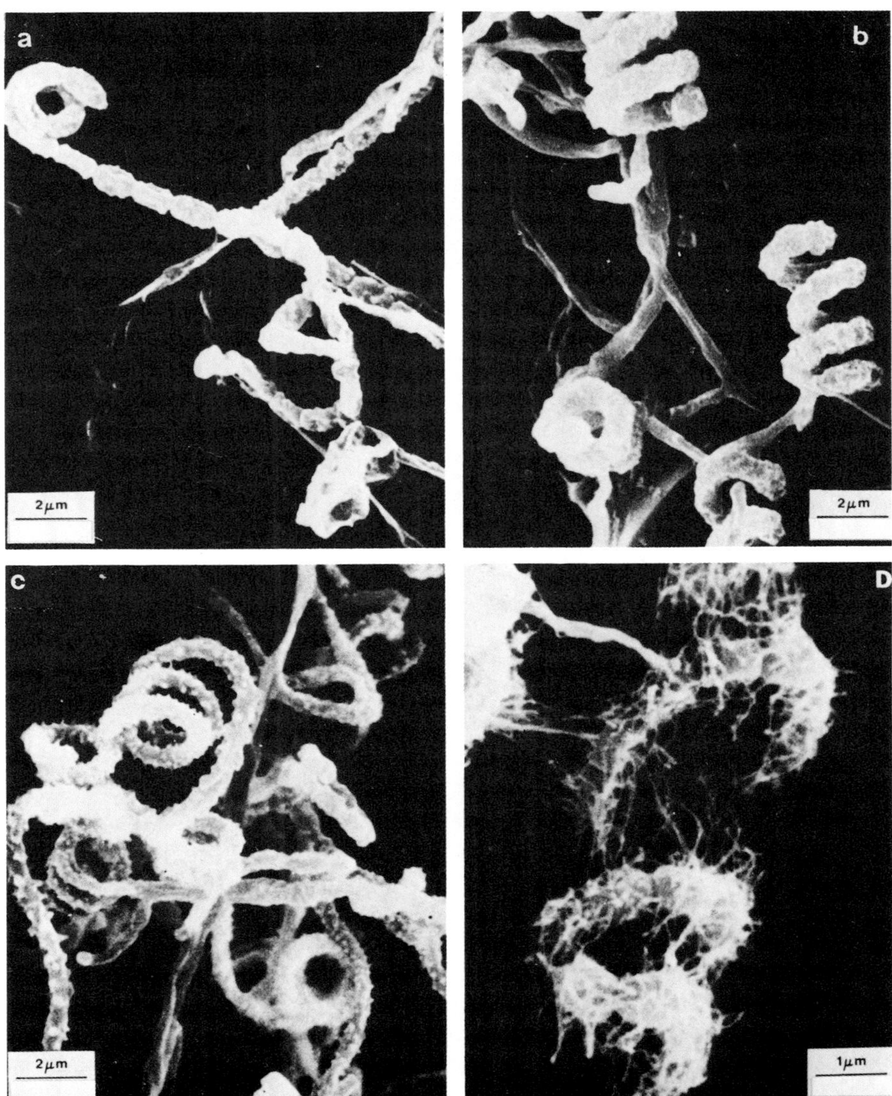

Fig. 160. Arthrospore chains of streptomycetes. a, *Streptomyces torulosus* (knobby); b, *S. antimycoticus* (rugose); c, *S. bluensis* (spiny); d, *S. karnatakensis* (hairy). (From KUTZNER 1981)

Fig. 158. *Sporomusa*, a Gram-negative anaerobic bacterium forming endospores. a, strain E, demonstrating a multilayered, typically Gram-negative cell wall; CM = cytoplasmic membrane; ML = murein layer; OM = outer membrane; inset: vegetative and spore-forming cells; b, strain E, showing the outer membrane. Bar: 0.5 μm in a, 0.1 μm in b, 1 μm in the inset in a. (From MOELLER et al. 1984)

Fig. 159. Germinating spore of *Bacillus polymyxa*. As shown by the 6 smaller arrows (in a), the spore coat has cracked; under the cracks, areas of the germ-cell wall show additional layers; b, higher magnification. (From GHUYSEN & SHOCKMAN 1973, after MURRAY et al.)

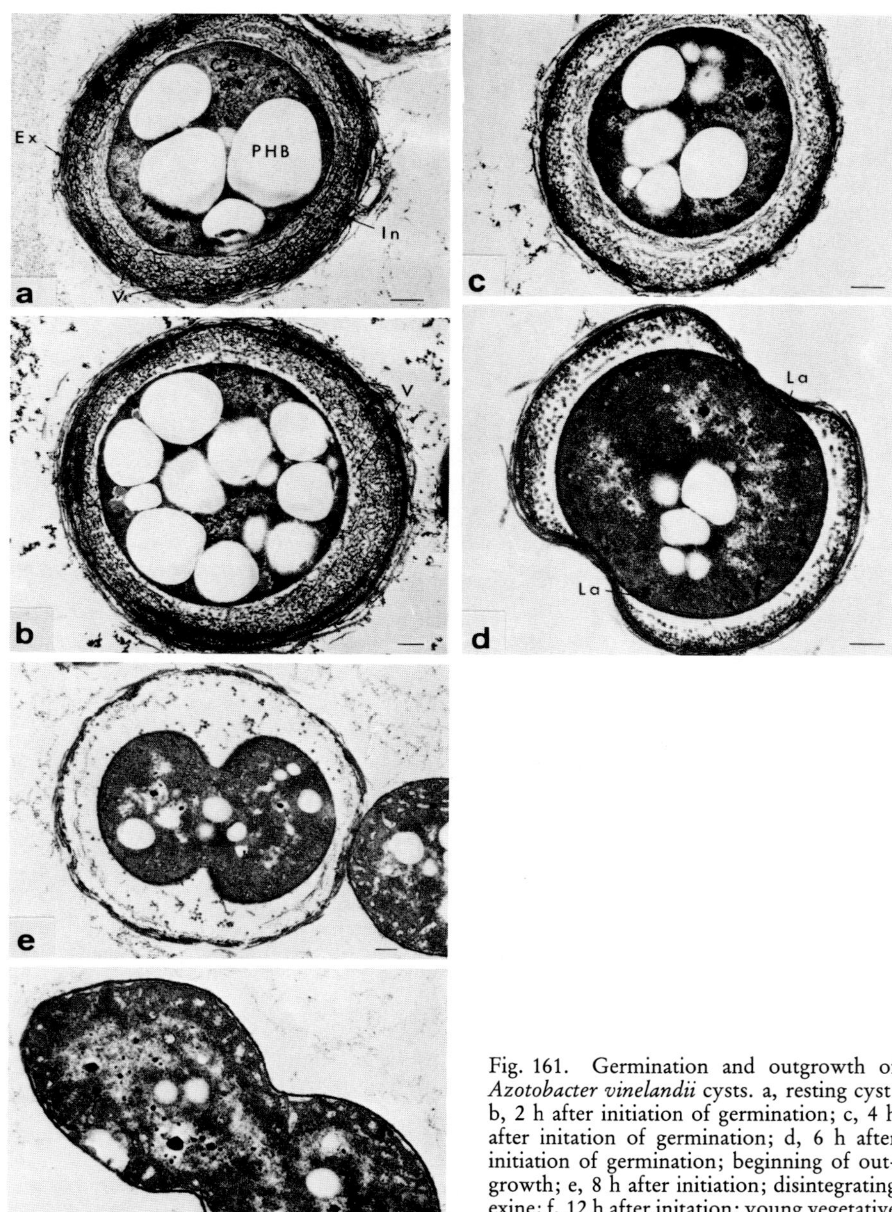

Fig. 161. Germination and outgrowth of *Azotobacter vinelandii* cysts. a, resting cyst; b, 2 h after initiation of germination; c, 4 h after initation of germination; d, 6 h after initiation of germination; beginning of outgrowth; e, 8 h after initiation; disintegrating exine; f, 12 h after initation; young vegetative cell. CB = central body; Ex = exine; In = intine; La = lamellar structures; PHB = polyhydroxybutyrate; V = vesicular intine. Bars: 0.2 μm. (From LIN et al. 1978)

iological properties are similar to those of *Bacillus* endospores. Some bacteria are capable of forming spherical cells with thickened walls; these differentiations are called cysts. During their formation, the whole cell contents are engulfed into the cysts, in contrast to endospore formation. *Azotobacter* and *Methylocystis* cysts are resistant against desiccation, mechanical stress and radiation; however, they are not heat-resistant. Ultrastructural and physiological changes occurring upon germination and outgrowth of *Azotobacter vinelandii* cysts (Fig. 161) have been investigated (CAGLE & VELA 1972, HITCHINS & SADOFF 1970, 1973, LIN et al. 1978, LIN & SADOFF 1968, 1969, LOPERFIDO & SADOFF 1973, SOCOLOFSKY & WYSS 1961, 1962, WYSS et al. 1961). Resting cysts have been found to be resistant against sonic oscillation, a property which was lost during their conversion to metabolically active vegetative cells. Samples examined by electron microscopy have revealed that, as germination progresses, vesicle-like and fibrillar structures became visible in the intine region. Lamellae associated with the cell membrane appeared in the central body at 6 hours post-initiation of germination. The PHB content of cysts decreased significantly after 4 hours of germination.

The persistent states formed by vegetative cells of *Myxococcus* and *Sporocytophaga* are called myxospores (KUPFER & ZUSMAN 1984, GHUYSEN & SHOCKMAN 1973). They have physiological and structural properties similar to cysts. Other persistent states have been described for *Arthrobacter globiformis*. Vegetative cells, under favourable nutritional conditions, are rod-like. Under starvation, coccoid cells are formed which are resistant, for a restricted period of time, against desiccation.

Akinetes (MILLER & LANG 1968) are persistent cells in cyanobacteria (Fig. 162). They are observed either in a subterminal position (next to a terminal heterocyst, *Cylindrospermum*), or within a chain of cells. They are characterized by a thick, multilayered envelope containing a major polysaccharide identical in composition and structure with that of the heterocyst in the same

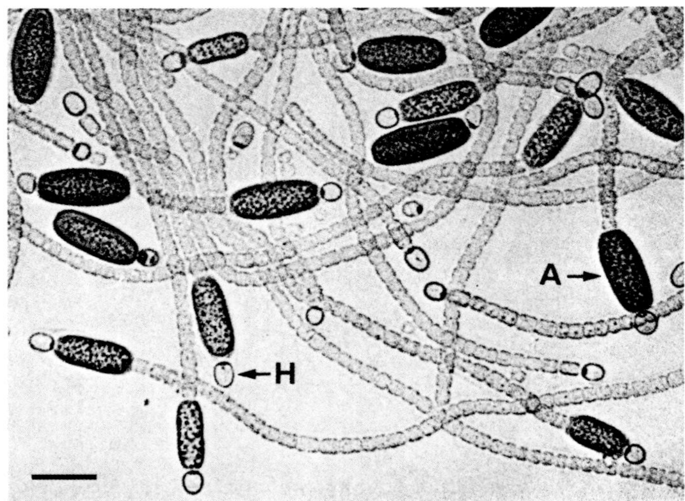

Fig. 162. A nostocacean cyanobacterium, *Cylindrospermum*, grown in the absence of a combined nitrogen source. Note the exclusively terminal heterocysts (H) and subterminal akinetes (A). Bar: 20 μm. (From STANIER et al. 1981)

species, and are often pigmented. The envelope is formed, during akinete development, around the pre-existing wall of a vegetative cell. This cell as a rule enlarges considerably and changes shape. At the time of maturation, the akinete is still unicellular. Its germination is accompanied by rapid transverse divisions. When it emerges from the akinete envelope, the germling often consists of a short chain of cells.

There are other persistent states such as motile kineospores (BARNARD & PARENTI 1983), and microcysts in *Spirillum itersonii* (CLARK-WALKER 1969).

Endospore morphogenesis is a special case of cell division (ANDREOLI et al. 1975, EATON & ELLAR 1974, ELLAR et al. 1975, HOLT et al. 1975, MACKEY & MORRIS 1971, MITRUKA et al. 1967, WALKER et al. 1966). It starts with axial accumulation of DNA followed by the separation of the chromosomes in the mother cell. One genome, the one which is later found in the spore, becomes separated from the vegetative cell cytoplasm by a septum. Formation of this septum is the first clearly visible step in sporogenesis. It is initiated by an invagination of the cytoplasmic membrane close to the cell pole, and is completed by centripetal ingrowth and membrane fusion. The future spore protoplast is engulfed by directed growth of the cytoplasmic membrane of the mother cell. As a consequence, the spore protoplast is enveloped by two cytoplasmic membranes, i. e. its own and one from the mother cell. Finally, the spore cell wall and the cortex, both of which contain peptidoglycan, are laid down, and the outer spore coat is differentiated. Labelling studies have demonstrated that murein from the vegetative cell is not incorporated into the spore coat. The enzyme systems responsible for synthesis of spore murein and the murein of the vegetative cell are different. Protoplasts have been found to produce complete spores supporting the view that the outer layers of the spore are laid down de novo onto the prespore membrane. Peptidoglycan isolated from the cortex of resting spores (BAILLIE & MURRELL 1974, FREHEL & RYTER 1980, GHUYSEN & SHOCKMAN 1973, IMAE et al. 1976, LINNETT & TIPPER 1976, PITEL & GILVARG 1970, TIPPER & LINNETT 1976, WARTH et al. 1963, WARTH & STROMINGER 1972) has been found to differ from the wall peptidoglycan of vegetative cells as follows: a high ratio of muramic acid residues in the spore peptidoglycan occurs as non-acetylated internal amides, the rest as N-acetylmuramic acid, of which about 50% are substituted by only a C-terminal L-alanine residue. The remainder is substituted by the usual peptides and exhibits a low extent of cross-linking. A consequence of these biochemical differences is a high flexibility of the spore peptidoglycan, indicating its specific functional aspects in the spore cortex.

The resting spore has a complex organization. Ultrathin sections and freeze-fracture preparations have revealed an outermost sculptured surface of the sporangium, various laminae making up the spore coats, the cortex, the wall primordium, the cytoplasmic membrane and the electron-dense spore body. It contains, as a specific constituent, up to 15% of the dry weight of the spore, dipicolinic acid.

Persistent states are not obligatory within the usual growth cycle of a bacterial cell able to form spores. Their formation is induced by lack of growth substrates or accumulation of certain products of the cell metabolism. Lack of water is no reason for the induction of spore formation. Manifestation of the induction of spore formation occurs within a period of several hours as determined under experimental conditions. Spore formation is repressed up to a few hours after the onset of the conditions appropriate for spore induction (as shown for *Bacillus cereus* var. *mycoides,* by sudden lack of growth substrate in the medium) when glucose is present in the medium. The "point of no return" is reached after about 6 hours of starvation. After this period of time, the induction (derepression) of spore formation remains and leads to spores about 10 to 13 hours after the onset of starvation.

Survival of bacteria as spores has been found to be up to two hundred or more years as observed in the case of spores isolated from soil taken from the Kew Gardens Herbarium. After a period of about 50 years, around 90% of all spores in samples containing *Bacillus subtilis*, *B. licheniformis* or *B. coagulans* had lost their viability. Spore germination (SCHLEGEL 1981) usually occurs under humid conditions and, occasionally, only when suitable growth substrates are available in addition to water. Stimulatory procedures such as treatment with elevated temperatures (60°C for *B. subtilis*, 100°C for some other species) improve the rate of germination.

Spore germination (Fig. 159) is preceded by water uptake and swelling. The physiological changes that occur here are: respiration and increase of enzyme activity; amino acids and dipicolinic acid are secreted; heat resistance is lost. The structural changes which have been observed include a disintegration of the exterior parts of the spore, i. e. the outermost surface of the sporangium and the laminae of the spore coats, paralleled by a thickening of the wall primordium and an increase in cell size. Patches of additional wall components composed of repeating units appear under the cracks of the coat (Fig. 159). The wall then enlarges and eventually covers the entire surface of the new cell. Finally, after rupture of the spore coat, a germination tube is formed and a typical vegetative cell is developed.

5.12 Bacteria lacking peptidoglycan in their walls

Bacteria lacking typical peptidoglycan (DOMINGUE 1983; MCCOY 1981) can be subdivided into two major groups: one, obviously, is that of wall-deficient forms (mycoplasmas, or, as originally called, *PPLO*, pleuropneumonia-like organisms) (BIBERFELD & BIBERFELD 1970, BOATMAN & KENNY 1971, BRUNNER et al. 1979, MENG & PFISTER 1980, PETERSON et al. 1973, SCHIEFER et al. 1974, 1975, SMITH 1971, TULLY et al. 1983, WILSON & COLLIER 1976), and bacterial L-forms (BROWN et al. 1971, MADOFF 1981, NISHIYAMA et al. 1984, PACHAS & CURRID 1974) which have the cytoplasmic membrane as the outermost layer. The other major group is that of bacteria having a rigid wall containing other polymers, or a non-rigid proteinaceous cell envelope.

5.12.1 Wall-deficient bacteria

These bacteria are small prokaryotes without a defined wall; they are bounded alone by their cytoplasmic membrane. Depending on the requirement of sterols for growth, the group can be devided in *Mycoplasma* (requiring sterols) and *Acholeplasma* (which does not require sterols). The cytoplasmic membrane of these organisms has been found to be stabilized by cholesterol, a lipid missing in other prokaryotes. This finding explains the observation that mycoplasmas are quite resistant to osmotic lysis or mechanical shock, such as sonication. Glycoproteins have not been found in *Mycoplasma* membranes. However, in some species, evidence has been obtained for the occurrence of carbohydrates, probably as glycolipids, and a study using plant lectins has shown specific sugar residues on the surface of various mycoplasmas.

Bacterial L-forms (or L-phase variants) are wall-less forms (Fig. 163). Since the first observation of this type of growth phase *(Streptobacillus moniliformis)* L-forms have been produced by suitable treatment, i. e. wall degradation, of normal cells of many Gram-positive and Gram-negative bacteria. Growing L-forms may revert to normal vegetative bacteria. Often, several antibiotic treatments were found to be necessary until L-forms could be considered stable. Even "stable" L-forms have been observed to be reverted to vegetative cells under certain modifications of growth conditions. It has been suggested that a protein, an "autolysine" of the vegetative cell which

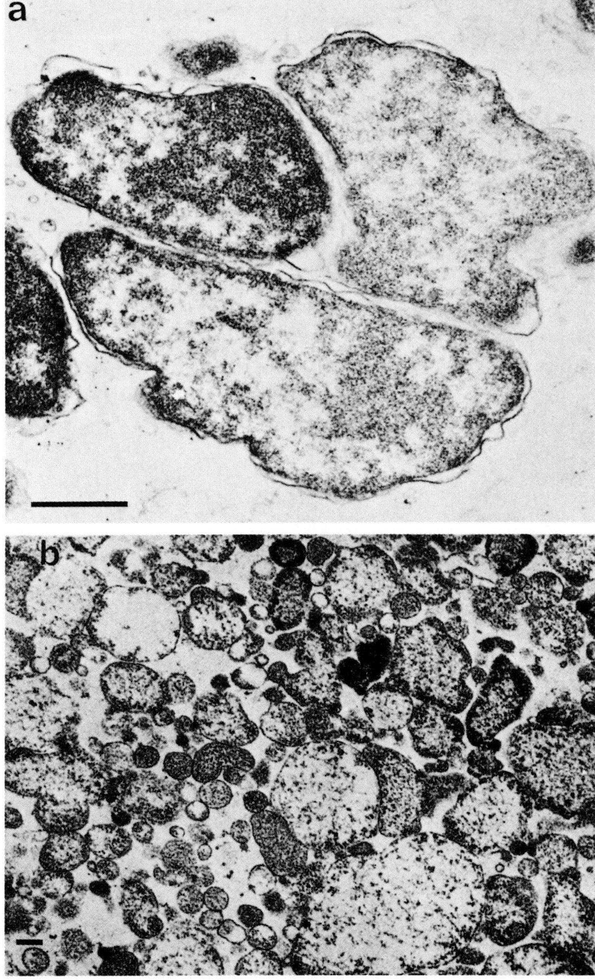

Fig. 163. L-forms of *Proteus*. a, the B-type shows cell wall constituents; b, the A-type shows only the cytoplasmic membrane. Bars: 0.5 μm. (From MADOFF 1981, after PACHAS & MADOFF)

continues to be produced by L-forms, might be responsible for the inhibition of the re-establishment of peptidoglycan in "stable" L-forms. Anything that destroys this autolysine or prevents its production favours reversion. Consequently, L-forms and states of reversion have been used for investigations of cell wall growth and assembly.

5.12.2 Bacteria with cell envelopes lacking peptidoglycan

In contrast to the first group, these bacteria contain defined cell envelopes. However, the peptidoglycan characteristic of most other bacteria is lacking. Bacteria with these properties have been detected both as members of the kingdom of eubacteria as well as within the archaebacteria.

5.12.2.1 Eubacteria

Two main sublines with these characteristics have been found within the eubacteria (SCHMIDT & STARR 1981, STARR et al. 1983): one is represented by the stalked *Planctomyces* strains (STALEY & BAULD 1981), the other by stalkless *Pirella* isolates. More than thirty strains belonging to these groups have been isolated from various aquatic habitats. All of these are characterized by a terminal holdfast substance. Budding has been observed, occurring opposite to the attachment pole. Some strains produce a terminal polymer rich in glutamine or glutamic acid. All strains have been found to be Gram-negative, aerobic heterotrophs, and they exhibit proteinaceous cell envelopes resistant to 10% SDS. Data on fatty acid content initially revealed a relationship between these organisms and the eubacteria. This finding has however been questioned after the discovery of the absence of peptidoglycan. In the meantime a detailed phylogenetic analysis involving 16S rRNA cataloguing has clearly indicated that the *Pirella-Planctomyces* group belongs to the eubacteria. Summing up all available data it can be postulated that these bacteria branched off very early during evolution and, thus, are of ancient origin.

5.12.2.2 Archaebacteria

A mosaic pattern of proteinaceous subunits, similar in appearance to the surface layers (S-layers) of many eubacteria, has been found to be characteristic of several archaebacteria (Figs. 164, 166 and 167). Other superficial structural aspects similar to those of eubacteria have been described. Subtler differences, not visible under the electron microscope, e. g. molecular sequences and details of function at the molecular level, are among the features which appeared to necessitate a separate taxonomic position for the archaebacteria (BALCH et al. 1979, BROCK 1967, GRIEF et al. 1984, KANDLER 1981, STETTER et al. 1981, WOESE 1981, ZILLIG et al. 1982).

The kingdom of archaebacteria was assumed to be a phylogenetically distinct and defined group of microorganisms on the basis of 16S rRNA cataloguing. It turned out that several (probably three) major groups make up the kingdom: the halobacteria (GIBBONS 1974, LARSEN 1962, 1967), the methanogens (DAVIS & WHITTENBURY 1970; EDWARDS & MCBRIDE 1975, KANDLER 1979, LANGENBERG et al. 1968, MAH & SMITH 1981, PROCTOR et al. 1969, ROMESSER et al. 1979, WHITTENBURY 1969, WHITTENBURY et al. 1970, ZEIKUS et al. 1975), and the thermoacidophiles (BROCK 1981).

The halobacteria may be divided into two subgroups: one consists of Gram-positive bacteria that can tolerate high NaCl concentrations (e. g., *Halococcus morrhuae*). The structural component of their thick walls is a sulphated heteroglycan containing glucose, mannose, galactose, glucosamine, galactosamine, glucuronic acid and galacturonic acid. A wall polymer of this composition can adequately fulfill the function of a cell envelope. It has a branched structure, acidic groups in form of sulphate and carboxylates that mimic the functions of carboxyl and phosphate groups of peptidoglycan and wall teichoic acids. Segregation of specific polymers from the heteroglycan has not been successful, indicating that a single large envelope molecule might possess all the required functions. The members of the second subgroup were found to be Gram-negative *(Halobacterium halobium, H. cutirubrum, H. salinarium)*; they have no peptidoglycan in their envelope and no other carbohydrate polymers in its place (BLAUROCK et al. 1976, CHO et al. 1967, MESCHER & STROMINGER 1976, MULLAKHANBHAI & LARSEN 1975, NICHOLSON & FOX 1983, STEENSLAND & LARSEN 1969, STOECKENIUS & ROWEN 1967). They are sensitive to osmotic damage. Both types of halophiles have high intracellular salt concentrations. Besides rod-shaped and coccoid forms, box-

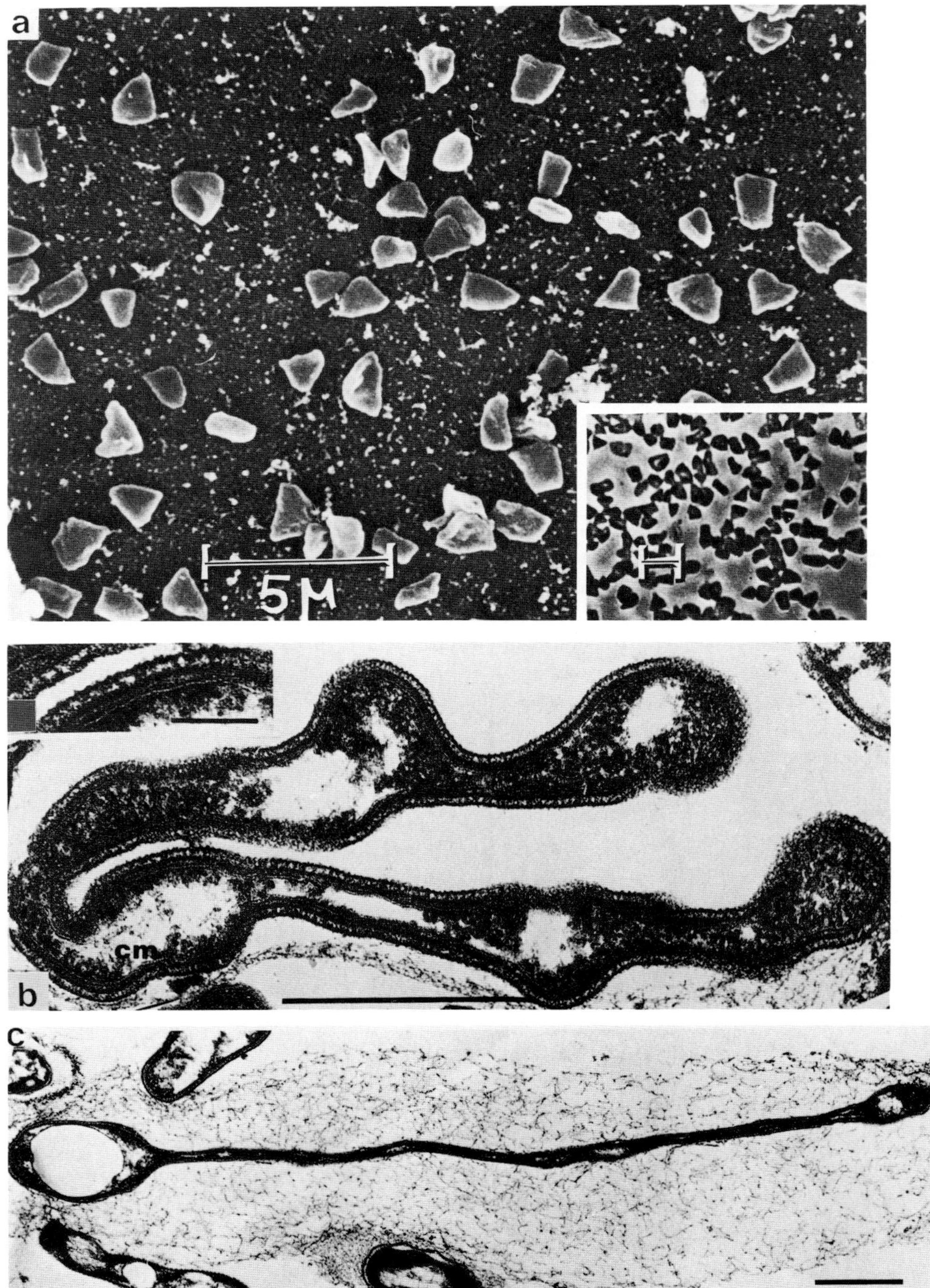
a
5M
b
cm
c

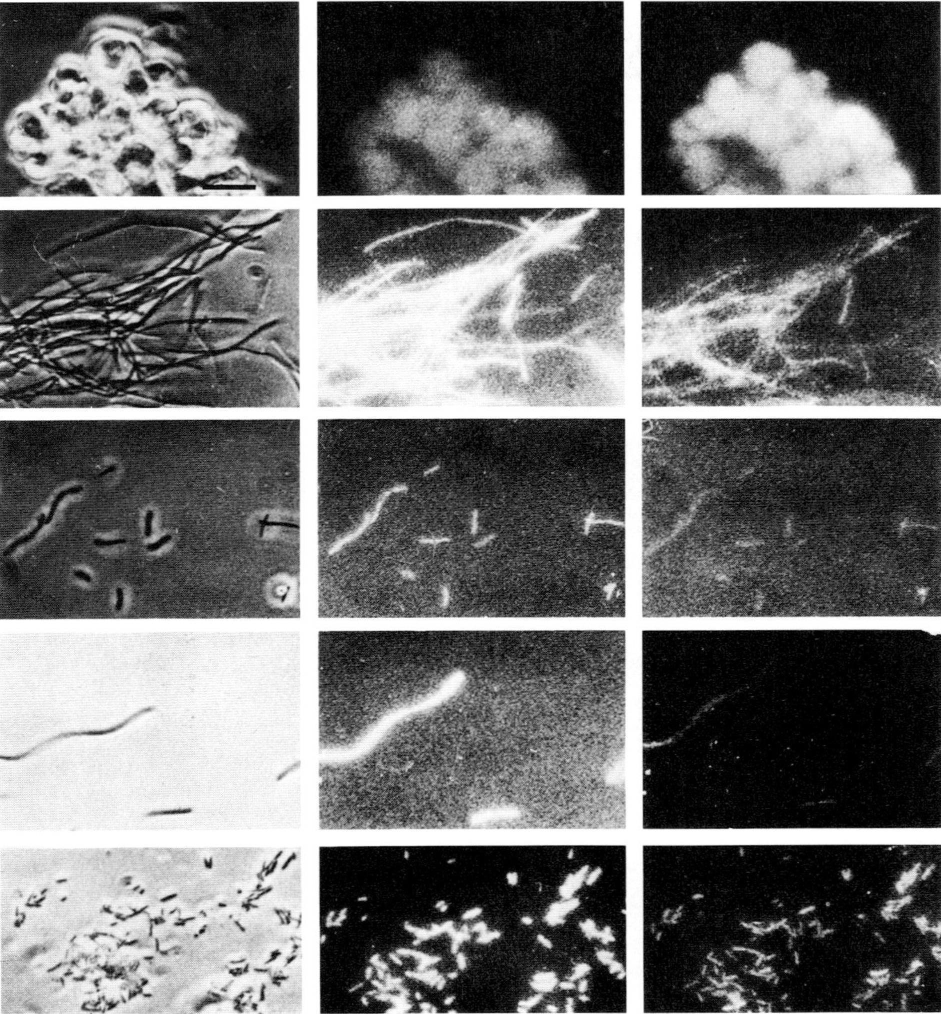

Fig. 165. Identification of methanogenic bacteria by fluorescence microscopy. Photomicrographs of different methanogenic bacteria. The columns show, from top to bottom: *Methanoscarcina barkeri, Methanobacterium formicicum, Methanobacterium* strain *M.o.H., Methanospirillum hungatii,* and *Methanobacterium* strain *TH*. The columns from left to right show identical fields of phasecontrast illumination, fluorescence when illuminated at 420 nm, and at 355 nm. Bar: 10 µm. (From DODDEMA & VOGELS 1978)

◀ Fig. 164. a, scanning electron micrograph (inset: light micrograph) of the box-shaped halophilic bacterium *GN*. Bar in the light micrograph also represents 5 µm. (From JAVOR et al. 1982); b, square bacterium bent back upon itself, showing lability of the cell wall. Inset: higher magnification to show cell wall details. cm = cell membrane. Bar: in b, 0.5 µm, in the inset in b, 0.1 µm; c, well-defined fibrillar sheath surrounding a square bacterium seen in cross section. Swellings are noticeable at the cell periphery and may be the sites of the storage granules. Bar: 0.5 µm. (b, c, from KESSEL & COHEN 1982)

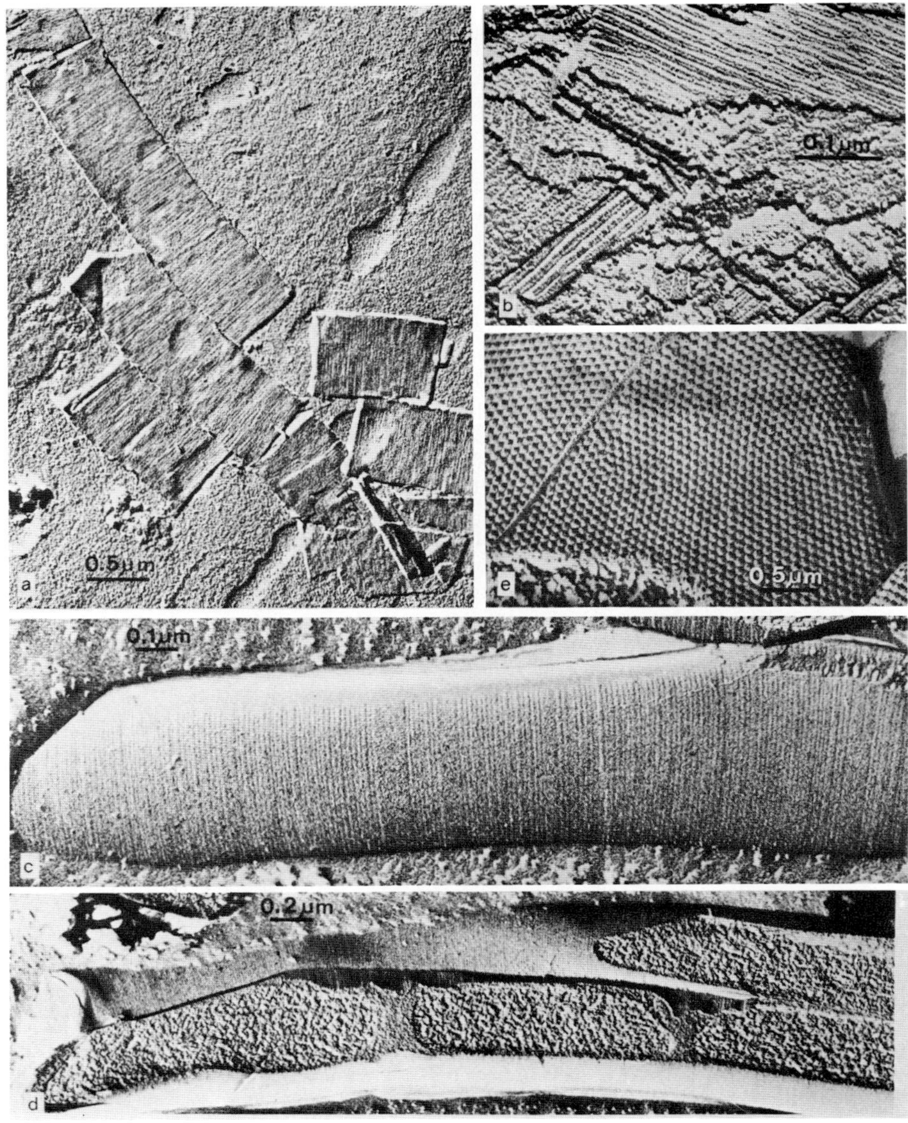

Fig. 166. Cell wall structures in methanogens. a, b, isolated protein sheath *(Methanospirillum hungatii);* c, surface view of a trichome of *Methanospirillum hungatii;* d, broken open trichome of *Methanospirillum hungatii,* e, surface layer of *Methanogenium marisnigri.* (From KANDLER 1979)

Fig. 167. Methanogenic bacteria. a, *Methanospirillum hungatii* (after ZEIKUS & BOWEN); arrow = gap between cells; b, *Methanosarcina barkeri* (after ZEIKUS & BOWEN); c, *Methanococcus vannielii* (after JONES et al.); Bars: in a, and b, 0.1 µm. (From MAH & SMITH 1981) ▶

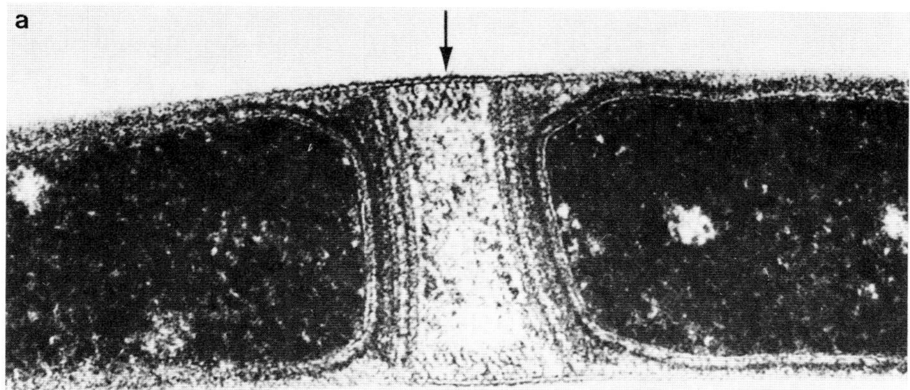
a

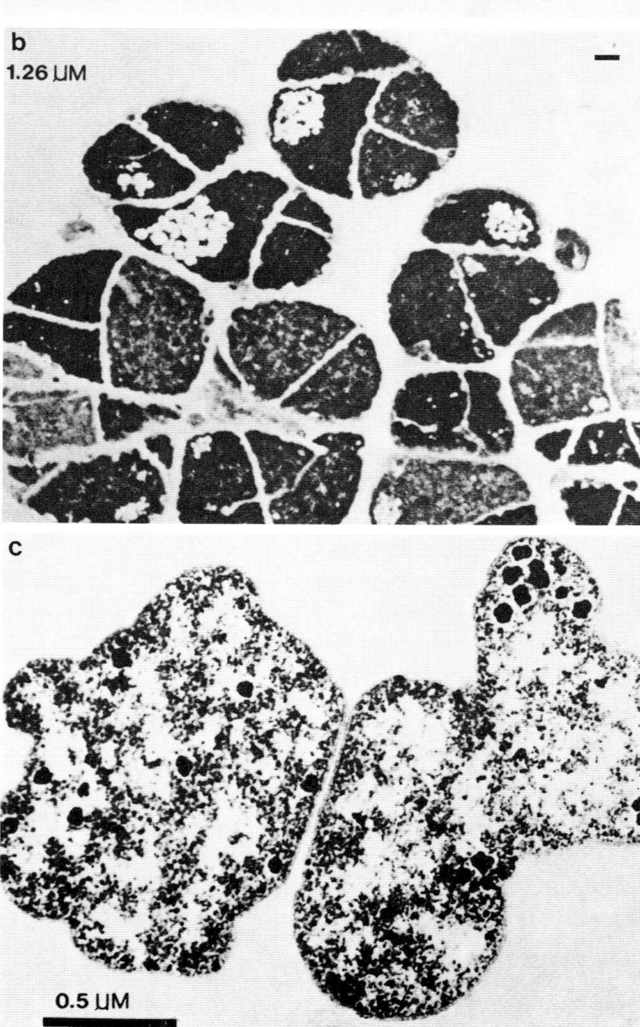
b
1.26 μM
c
0.5 μM

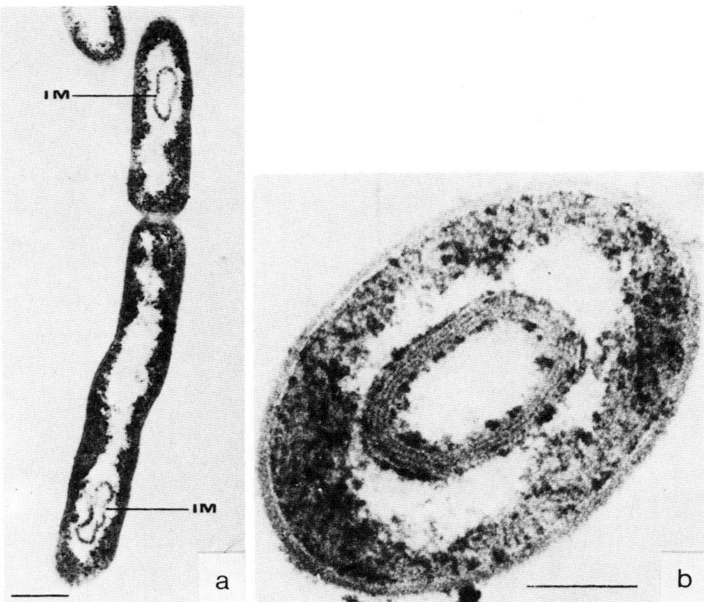

Fig. 168. Fine structure of *Methanobacterium thermoautotrophicum.* a, longitudinal section of a paired cell showing intracytoplasmic membranes (IM). Bar: 0.25 μm; b, cross section illustrating closely stacked membranes in concentric circles. Bar: 0.13 μm. (From ZEIKUS & WOLFE 1973)

shaped (or square) bacteria have also been isolated from hypersaline environments. As in a variety of non-halophiles, electron micrographs of the surface of several halobacteria show a regular hexagonal surface pattern. *H. salinarium* has, as a constituent of its envelope, a glycoprotein containing amino sugars. Glycoproteins may well be involved in the preservation of cell morphology. It has been shown that treatment with bacitracin, which interferes with the semi-dephosphorylation of polyprenol-pyrophosphate, or with proteolytic enzymes, results in the transformation of normal rod-shaped cells into spheres. The glycoprotein of *H. salinarium* has been found to be highly acidic, being localized on the outer surface of the cytoplasmic membrane. In addition, sulphated acidic polysaccharides have been described in envelopes of halophiles.

For light energy conversion under specific conditions, halobacteria may exhibit, as a constituent of their cytoplasmic membrane, a "purple membrane" (s. above) with bacteriorhodopsin as the only component (DANON & STOECKENIUS 1974, LINDLEY & MACDONALD 1979, MACDONALD et al. 1979, WEBER & BOGOMOLNI 1981, 1982). This membrane has been shown to be a rigid structure. Electron microscopic studies have led to the proposal that the bacteriorhodopsin molecule contains seven α-helices arranged almost perpendicular to the plane of the membrane.

A second major group of archaebacteria are the methanogenic bacteria (Figs. 165–169). They are strict anaerobes and generate methane by reduction of carbon dioxyde or certain other simple sources of carbon. Rod-shaped *(Methanobacterium)* (BARKER 1940, BRYANT et al. 1967, PANTSKHAVA 1977, PAYNTER & HUNGATE 1968, SMITH & HUNGATE 1958, SMITH & RIBBONS 1970, ZEHNDER et al. 1980, ZEIKUS & WOLFE 1972a, b, 1973), coccoid *(Methanococcus)* (JONES et al. 1983), sarcina-like *(Methanosarcina)*, spirillum-like *(Methanospirillum)*, plate-shaped *(Methano-*

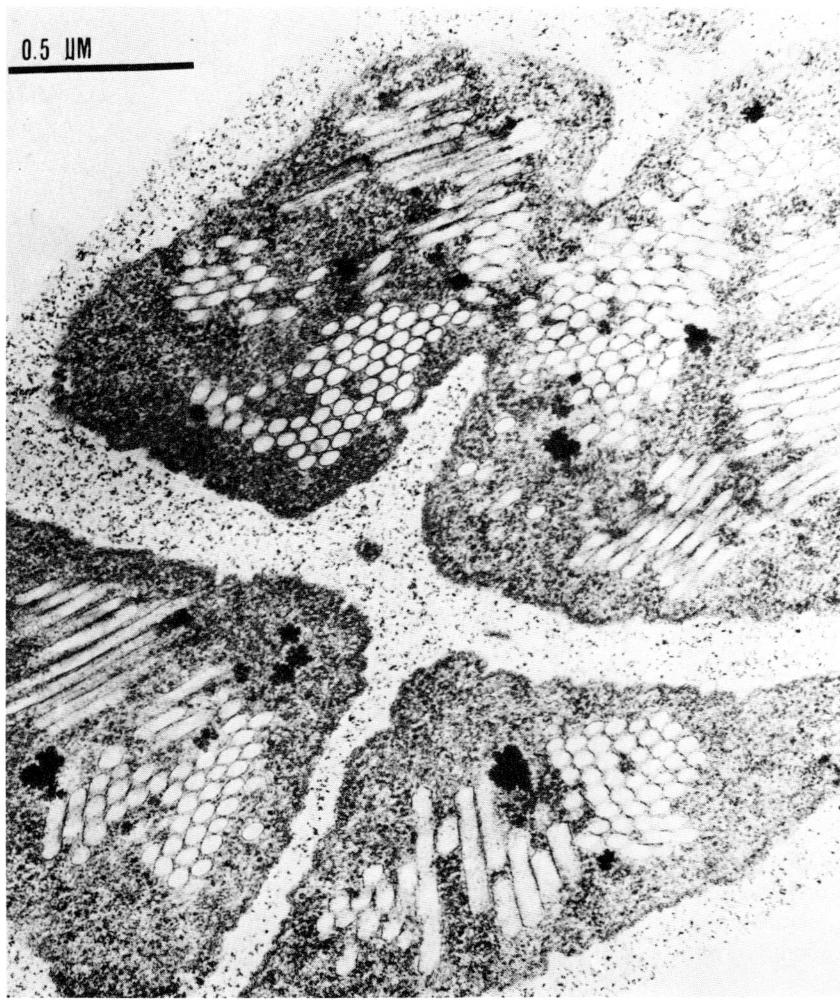

Fig. 169. The methanogenic bacterium *Methanosarcina* strain *W*, showing gas vesicles, compartmentalization, and unusual outer cell layer (after PANGBORN). (From MAH & SMITH 1981)

planus) (WILDGRUBER et al. 1982) and filamentous *(Methanothrix)* (HUSER et al. 1982) methanogens have been described so far. Instead of peptidoglycan, the walls of these bacteria (KANDLER 1979) consist of polypeptide envelopes *(Methanococcus)* or modified polypeptides *(Methanobacterium)*, polypeptides and polysaccharides in addition *(Methanobacterium)*, protein sheaths *(Methanospirillum)*, or heteropolysaccharides *(Methanosarcina)* containing uronic acid, neutral sugars and amino sugars. As with peptidoglycan, the polypeptides of methanobacteria probably comprise covalently linked units, each of which has L-glutamic acid, L-lysine and L-alanine in the ratios 2:1:1 and which may have a residue of either N-acetylglucosamine or N-acetylgalactosamine attached. The term "pseudomurein" has been introduced for the wall polymers of the order Methanobacteriales.

Cytological studies on *Methanobacterium thermoautotrophicum* have revealed the presence of intracytoplasmic membranes (Fig. 168) (s. above). They were described to be of the vesicle type, and were found irregularly distributed within the cytoplasm. Intracytoplasmic membranes have also been found, in different structural variations and types, in other methane bacteria (BALCH et al. 1979). ZEIKUS et al. (1975) have examined several methanogens in thin sections and concluded that these bacteria contain internal membranes formed by invaginations of the cytoplasmic membrane. However, in thin sections of *Methanolobus tindarius* (KÖNIG & STETTER 1982), of *Methanoplanus limicola* (WILDGRUBER et al. 1982), and of a thermophilic coccobacillus (FERGUSON & MAH 1983), internal membranes were neither emphasized nor evident. ZEIKUS & WOLFE (1973) suggested that intracytoplasmic membranes might be the site of ATP and methane synthesis. SAUER et al. (1980) have presented evidence indicating that CO_2 reduction to methane by cell extracts requires the participation of membrane vesicles. KELL et al. (1981) coined the term "methanochondrion" to describe, as an organelle, the invaginations of the cytoplasmic membrane assumed to be the site of methane synthesis. This idea could be used to explain some data obtained with *Methanobacterium thermoautotrophicum* such as uncoupler-insensitive ATP synthesis (DODDEMA et al. 1978), differential sensitivity of methane synthesis to various uncouplers (SAUER et al. 1981), and the presence in vesicles of an adenine nucleotide translocase activity (DODDEMA et al. 1980). Furthermore, cytochemical staining has suggested that hydrogenase activity in *M. thermoautotrophicum* may be localized exclusively at the site of internal membranes (DODDEMA et al. 1979). SPROTT et al. (1984), however, have provided indications for an alternative explanation of these results. They have found that the frequency of intracytoplasmic membranes in several methanogens grown on H_2–CO_2 varied with the conditions of growth, and from one strain to another. *Methanobacterium thermoautotrophicum* often generated large numbers of intracytoplasmic membranes, while *Methanospirillum hungatei* only rarely produced these membranes. Conditions allowing for rapid growth, including optimal temperature and high agitation rates, increased the production of intracytoplasmic membranes. These membranes consisted mainly of vesicles composed of one or several layers, often positioned in the central region of the cytoplasm. Several mesophilic methanogens could be grown such that intracytoplasmic membranes were rarely or never observed in thin sections or in replicas of cross-fractures from frozen cells. Since high rates of methane synthesis still occurred in these cultures, it follows that the intracytoplasmic membrane system is not a necessary organelle for methane formation in these strains. SPROTT et al. (1984) also found that negative staining for electron microscopy is not an accurate method to visualize intracytoplasmic membranes in these

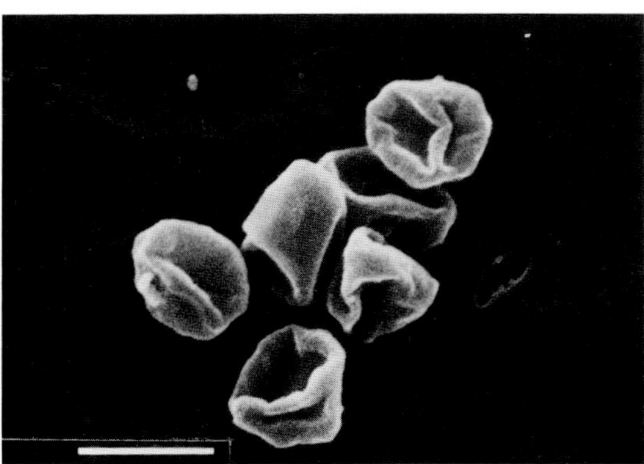

Fig. 170. The extreme thermophile bacterium *Sulfolobus* showing the lobed nature of the cells; scanning electron micrograph of a critical point-dried sample. Bar: 1 µm. (Original micrograph M. LÜBBEN & A. WEITH)

bacterial cells. They have suggested that during the drying process the cytoplasmic membrane might possibly be retracted away from the rigid cell wall material. Accordingly, the cytoplasmic membrane would buckle and the resulting depression would fill with the staining agent to become electron-dense.

Recognition of methanogenic bacteria in mixed cultures can be achieved by fluorescence light microscopy (DODDEMA & VOGELS 1978) using excitation wavelengths of 420 nm or 350 nm (Fig. 165). Fluorescence at 420 nm-excitation is greenish-yellow, at 350 nm-excitation blueish white, and a yellow fluorescence compound has also been described. Fading may occur within a few seconds.

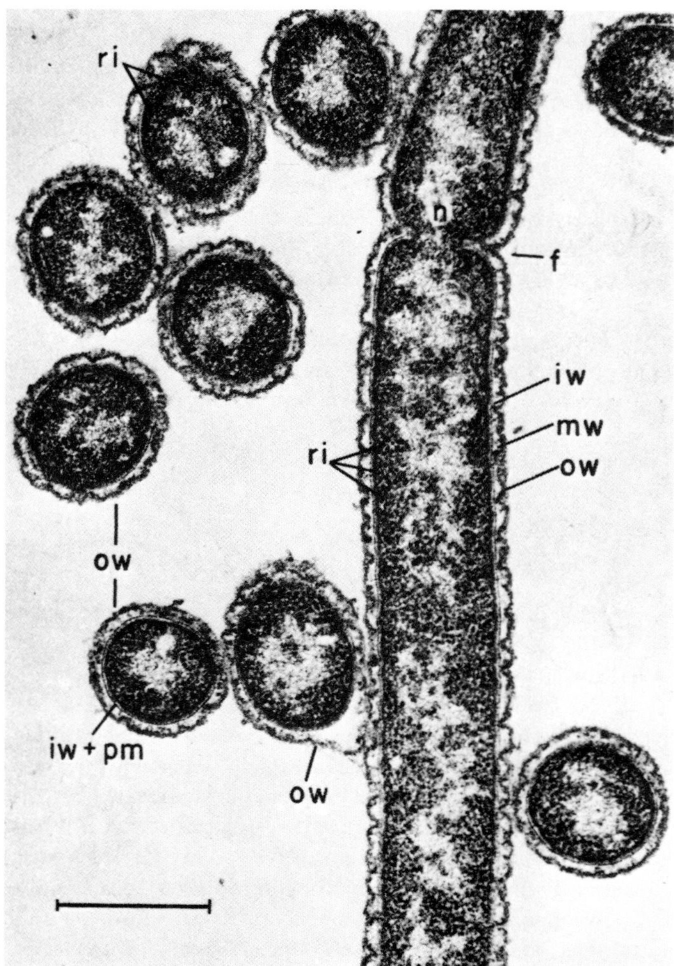

Fig. 171. The extreme thermophile bacterium *Thermus aquaticus;* longitudinal, transverse, and oblique views. – n = nucleoplasm; ri = ribosomes; pm = cytoplasmic membrane; ow = cell wall, outer dense layer; mw = cell wall, middle light zone; iw = cell wall, inner dense layer; f = cell division by furrowing. Bar: 0.5 µm. (From BROCK 1981, after BROCK & EDWARDS)

The fluorescence appears to be caused by several compounds (CHEESEMAN et al. 1972, GUNSALUS & WOLFE 1978). The fluorescence at excitation wavelength 420 nm is due to the presence of a 5-deazaflavin analogue as shown in isolates from *Methanobacterium* strain *M.o.H.* The fluorescence spectrum at 350 nm-excitation is probably caused by a substance similar to reduced nicotinamide adenine dinucleotide or pteridines. By an improved method, using optimized filter combinations, possible errors in the identification of methanogens can be reduced.

Among archaebacteria, the methanogens are the dominant form. Although widely distributed, they are not commonly encountered as they are killed by oxygen. Usually, methanogens are found in close association with other bacteria, e. g. *Clostridium*, that metabolize decaying organic material and give off hydrogen as a waste product. In fact, a thermophilic methanogenic consortium has been discovered (BOCHEM et al. 1982). It consists of methanogenic bacterial "tissue" colonized by other bacteria, and it grows at 60°C (s. below). An extremely thermophilic bacterium with pseudomurein as one layer of the cell wall and a surface layer made up of protein subunits is *Methanothermus fervidus* (STETTER et al. 1981), a member of the order Methanobacteriales.

Thermoplasma, of the order Thermoplasmales, has been detected in smoldering piles of coal tailings. As *Thermoplasma* has no cell wall but merely the limiting cytoplasmic membrane, it has also been called a mycoplasma-like organism. It has been placed in the group of the thermoacidophiles. The bacteria of this group are aerobic and very notable for their habitats. *Sulfolobus* (Fig. 170), one of the genera (order: Sulfolobales), has been isolated from hot sulfur springs. Growth above 90°C and at pH values even lower than 2 has been observed for several species of this genus.

Growth under similar extreme conditions (BROCK 1967, 1981, WOESE 1981) has also been found for bacteria of the order Thermoproteales. Known members of this group are *Thermoproteus, Thermofilum, Desulfurococcus* and probably *Thermococcus* and *Thermus aquaticus* (Fig. 171).

Thermodiscus, Pyrodictium and several other isolates have not yet been finally classified. All of these bacteria have in common that each of them exhibits very specific cytological features regarding cell shape, mode of division, wall structure and wall composition.

6. Cell growth

Bacterial cell growth involves a variety of structures and events (INGRAHAM et al. 1983, KLUG 1967, NANNINGA & WOLDRING 1985, STANIER et al. 1979, YOUNG et al. 1972). It encompasses the assembly of cellular structures from macromolecules into ordered planar and globular assemblages, indicating that this complex process is highly regulated (HILL et al. 1972). It comprises synthesis and assembly of components forming the cell envelope including the cytoplasmic membrane, peptidoglycan, components in the periplasmic space or "gel" (HOBOT et al. 1984, LAMPEN 1974), the outer membrane and the outermost surface layers such as ordered S-layers and capsules, intracytoplasmic membranes, flagella and other structures involved in locomotion, pili and fimbriae, the nucleoid, polysomes and all other components of the cytosol. The chemical reactions leading to the end product, i. e. the bacterial cell (Fig. 172), may be divided into several types (INGRAHAM et al. 1983): fueling reactions, biosynthetic reactions, and polymerization and assembly reactions, depending on their primary function in growth. Fueling reactions produce the precursors for biosynthetic pathways and provide ATP and reducing power for biosynthesis. Biosynthetic reactions produce the building blocks (amino acids, nucleotides, sugars, fatty acids)

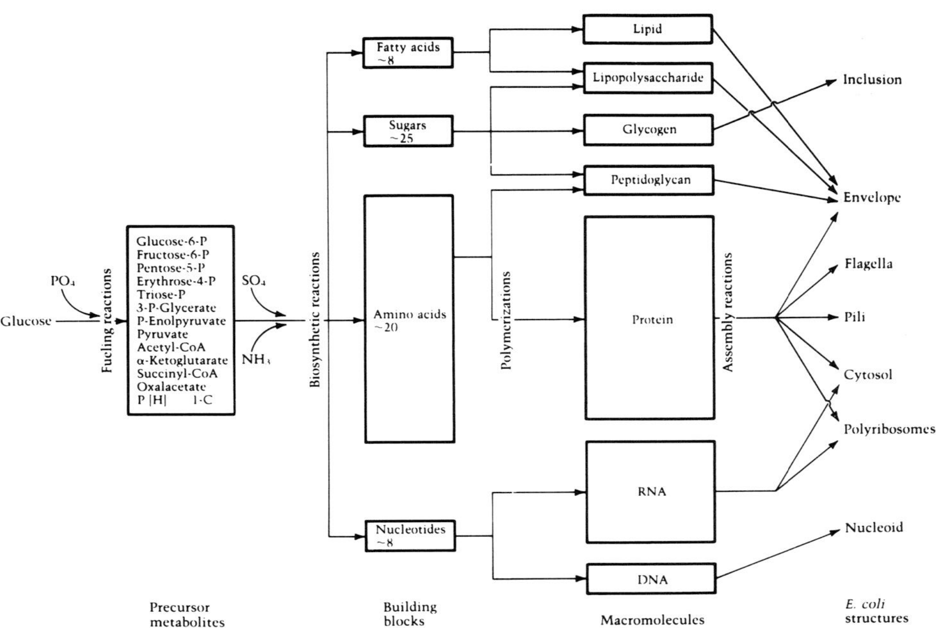

Fig. 172. Overall view of metabolism leading to chemical synthesis of *Escherichia coli* from glucose. Boxes are proportional to need in making *E. coli.* (From INGRAHAM et al. 1983)

needed for polymerization. Polymerization reactions consist of directed, sequential linking of these building blocks into large polymers. Assembly reactions involve the association of macromolecules to form cellular structures: envelope, membranes, appendages, nucleoid, polysomes, inclusions, and enzyme complexes.

INGRAHAM et al. (1983) have summarized the complex events leading to a typical bacterial cell structure, and to specific cell shapes as observed in nature, as follows:

(1) Assembly of cell structures occurs both by spontaneous aggregation (self-assembly) and by special, specific mechanisms. Some macromolecules are made at the sites of assembly and others must be transported to them.

(2) Assembly of the cytoplasmic membrane occurs by intercalation of protein and phospholipid in defined regions. Enzymes responsible for phospholipid synthesis are in the membrane. Proteins to be incorporated into the membrane may be made on membrane-attached ribosomes but are not, normally, made with a signal sequence; intercalation by a membrane-trigger mechanism (Fig. 173) based on the membrane-induced alteration of the conformation of the protein seems more likely. Lipid-lipid and lipid-protein interactions determine the specific composition of the two leaflets. Hydrophobicity is an important driving force in membrane assembly, but specific, directed forces must also be at work.

(3) Proteins secreted into the periplasmic space of Gram-negative bacteria appear to be made initially with a signal peptide sequence that facilitates passage through the cell membrane.

(4) Assembly of the peptidoglycan layer is accomplished by polymerization of the peptidoglycan chain after translocation of the precursors on a lipid carrier through the cell membrane. In

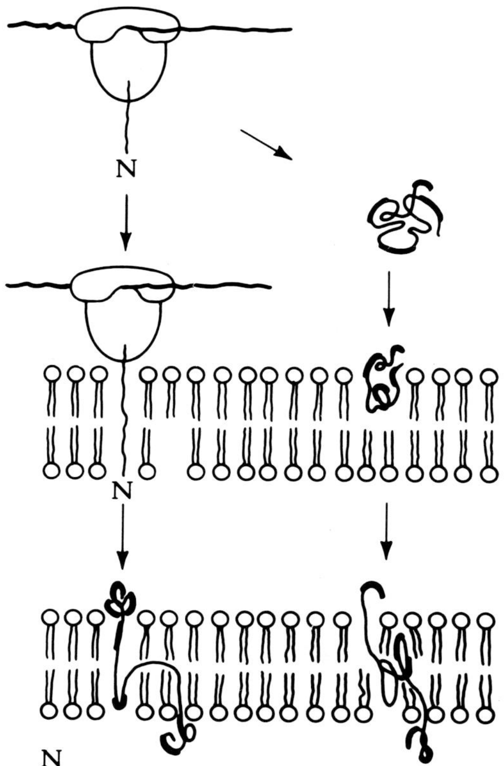

Fig. 173. Schematic presentation of the membrane trigger hypothesis. Thick lines represent polar regions, thin lines represent hydrophobic regions. – N = N terminus. (From INGRAHAM et al. 1983, after WICKNER)

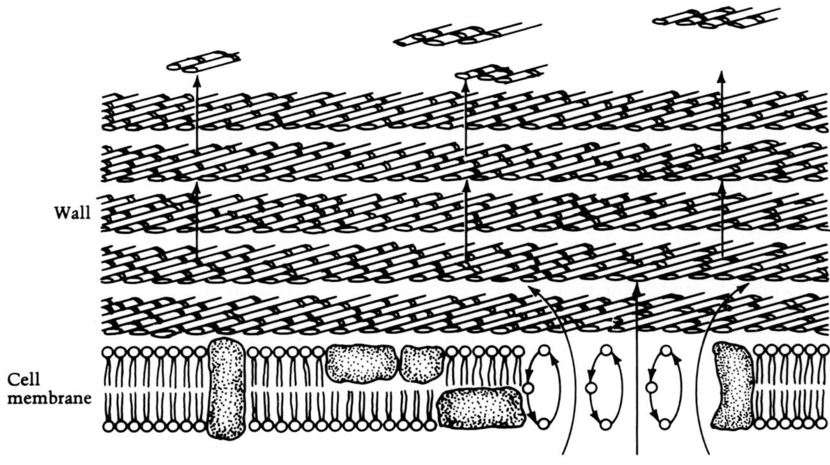

Fig. 174. Wall assembly and turnover in Gram-positive cells. The arrows show the progression of newly formed peptidoglycan from its site of synthesis at the cell membrane to the periphery of the cell where it is sloughed off. (From INGRAHAM et al. 1983)

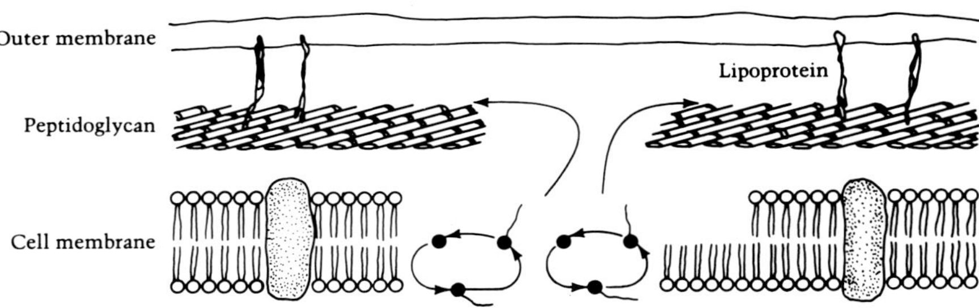

Fig. 175. Wall assembly in Gram-negative bacteria. The diagram depicts a region where peptidoglycan precursors are shuttled through the cell membrane on undecaprenol phosphate, polymerized into glycan chains, intercalated with covalent links to the existing peptidoglycan sac, and finally attached to the outer membrane by reaction with the lipoprotein. (From INGRAHAM et al. 1983)

Gram-positive cells, the thick multilayered peptidoglycan is in a dynamic state of synthesis at the membrane surface and sloughing off at the periphery (Fig. 174).

(5) Assembly of the outer membrane of Gram-negative cells (Fig. 175) occurs by independent assembly of each component (phospholipids, lipopolysaccharides, and proteins) rather than of multicomponent subunits. LPS subunits are synthesized in the cell membrane in two parallel processes: one produces the core polysaccharide on lipid A, the other the repeating polysaccharide side chain (Figs. 182 and 183). A transfer enzyme unites the two parts after translocation across the cytoplasmic membrane. Phospholipids are made in the cytoplasmic membrane and interact with lipid A and neighboring lipids to self-assemble in the inner leaflet of the outer membrane. Outer membrane proteins, probably made on ribosomes attached to the cytoplasmic membrane, are translocated by a process such as the bent loop model (Figs. 176–178) to the outer membrane, where LPS facilitates intercalation. Transport of proteins, LPS, and phospholipids to the outer membrane occurs at adhesive patches (called Bayer's junctions) where outer membrane and cytoplasmic membrane are in intimate association (Fig. 179). A special lipoprotein is important in anchoring the outer membrane to the peptidoglycan layer (Figs. 180 and 181). Some of the outer membrane proteins, called porins, self-aggregate to form diffusion pores.

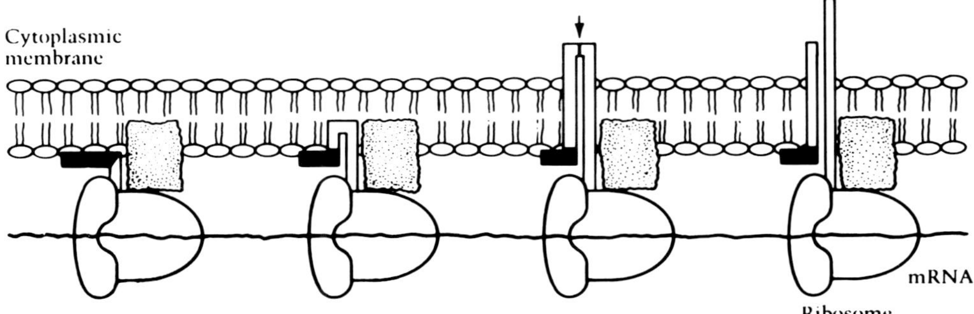

Fig. 176. Bent loop model for the translocation of secretory proteins across the membrane. An arrow designates the cleavage site in the precursor molecule. (From INGRAHAM et al., after INOUYE)

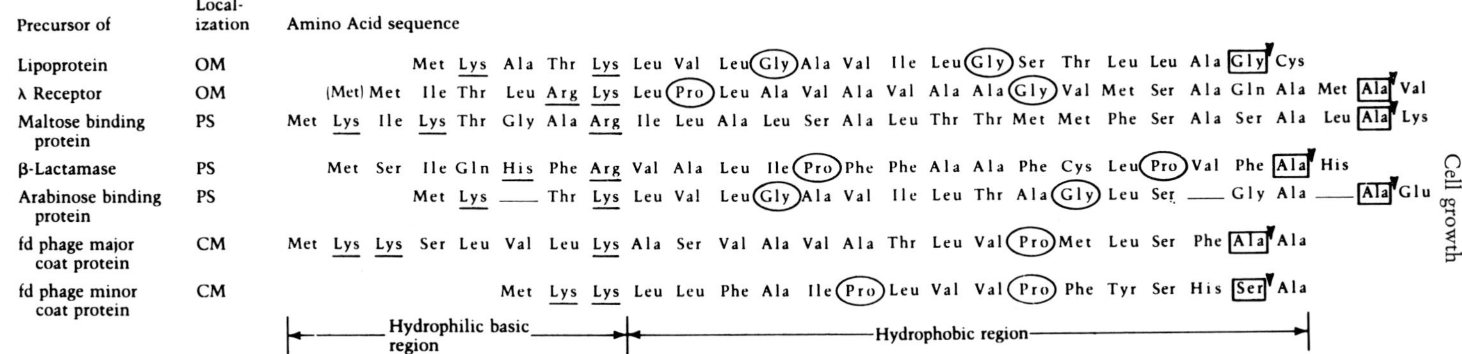

Precursor of	Local-ization	Amino Acid sequence
Lipoprotein	OM	Met Lys Ala Thr Lys Leu Val Leu Gly Ala Val Ile Leu Gly Ser Thr Leu Leu Ala Gly Cys
λ Receptor	OM	(Met) Met Ile Thr Leu Arg Lys Leu Pro Leu Ala Val Ala Val Ala Ala Gly Val Met Ser Ala Gln Ala Met Ala Val
Maltose binding protein	PS	Met Lys Ile Lys Thr Gly Ala Arg Ile Leu Ala Leu Ser Ala Leu Thr Thr Met Met Phe Ser Ala Ser Ala Leu Ala Lys
β-Lactamase	PS	Met Ser Ile Gln His Phe Arg Val Ala Leu Ile Pro Phe Phe Ala Ala Phe Cys Leu Pro Val Phe Ala His
Arabinose binding protein	PS	Met Lys — Thr Lys Leu Val Leu Gly Ala Val Ile Leu Thr Ala Gly Leu Ser — Gly Ala — Ala Glu
fd phage major coat protein	CM	Met Lys Lys Ser Leu Val Leu Lys Ala Ser Val Ala Val Ala Thr Leu Val Pro Met Leu Ser Phe Ala Ala
fd phage minor coat protein	CM	Met Lys Lys Leu Leu Phe Ala Ile Pro Leu Val Val Pro Phe Tyr Ser His Ser Ala

Fig. 177. Signal peptide sequences of proteins that enter the *Escherichia coli* envelope. – CM = cytoplasmic membrane; OM = outer membrane; PS = periplasmic space. (From INGRAHAM et al. 1983, after OSBORN)

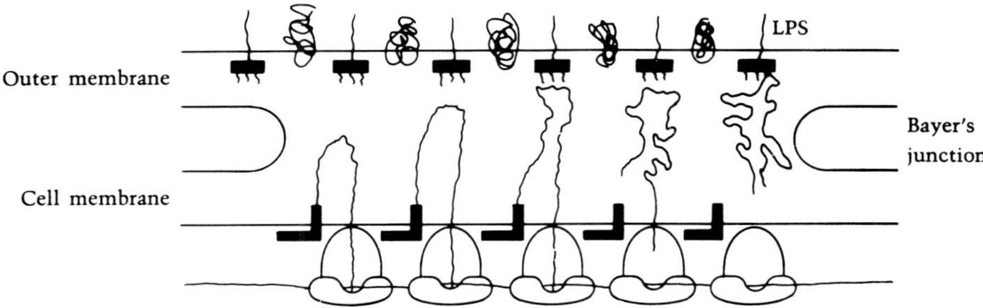

Fig. 178. Model for the assembly of proteins into the outer membrane. The interaction between a nascent protein chain and the lipopolysaccharide (LPS) of the outer membrane at one of the junctions of Bayer is depicted as guiding the final location of protein in the outer membrane. (From INGRAHAM et al. 1983, after OSBORN)

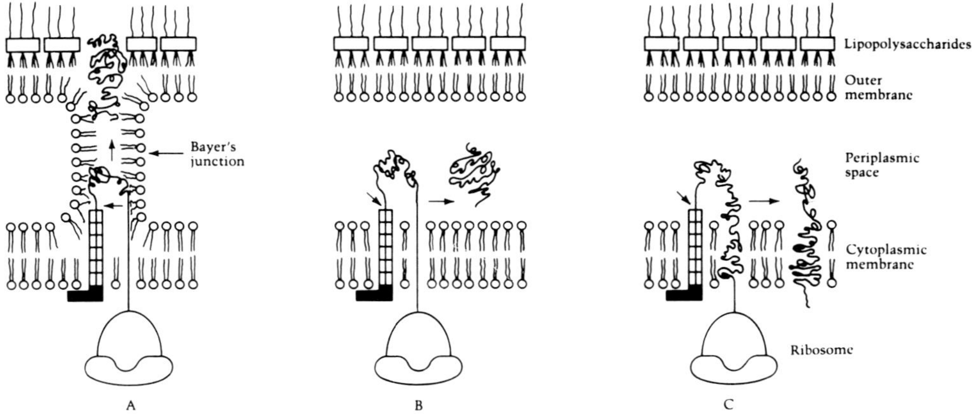

Fig. 179. Possible mechanism of secretion and final localization of envelope proteins in *Escherichia coli.* A, outer membrane proteins; B, periplasmic proteins; C, cytoplasmic membrane proteins. The peptidoglycan layer is not illustrated for simplicity. (From INGRAHAM et al. 1983, after INOUYE)

(6) Capsular material (Fig. 184) assembles outside the cell from subunits that are transported through the membrane(s) by a lipid carrier.

(7) Flagella are formed first by self-assembly of the proteins of the basal body and their appropriate intercalation into the envelope layers, followed by synthesis of flagellin and its transport (presumably through the central cavity of the growing flagellum) to the tip of its filament.

(8) The nucleoid forms by processes that involve methylation, supercoiling, and folding of newly replicated DNA. Some of these are enzymatic reactions that are well understood, but fascinating puzzles about nucleoid assembly (and function) remain.

(9) Polysomes form by the factor-promoted association of ribosomal subunits to special sequences on mRNA. The ribosomal particles themselves are the product of elaborate assembly reactions that involve the processing of 5 S, 16 S, and 23 S rRNA and the sequential attachment of

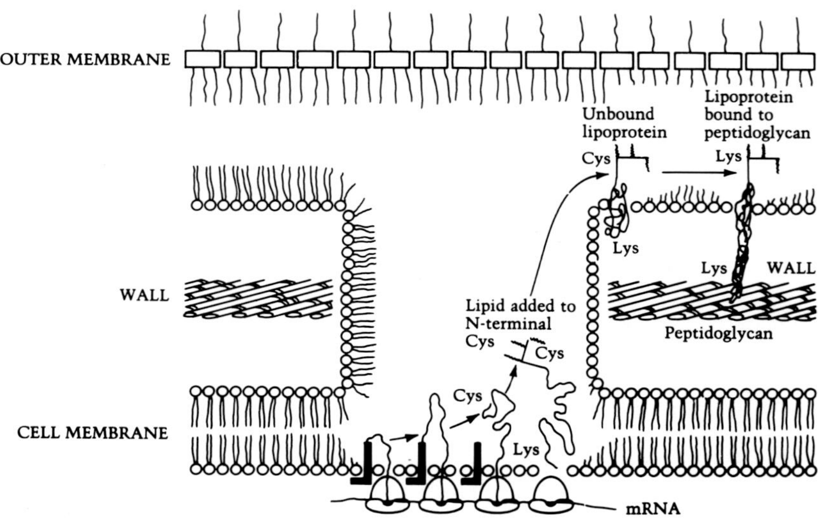

Fig. 180. Assembly of the lipoprotein. – Cys, Lys = amino acids. (From INGRAHAM et al. 1983, after OSBORN and after WU)

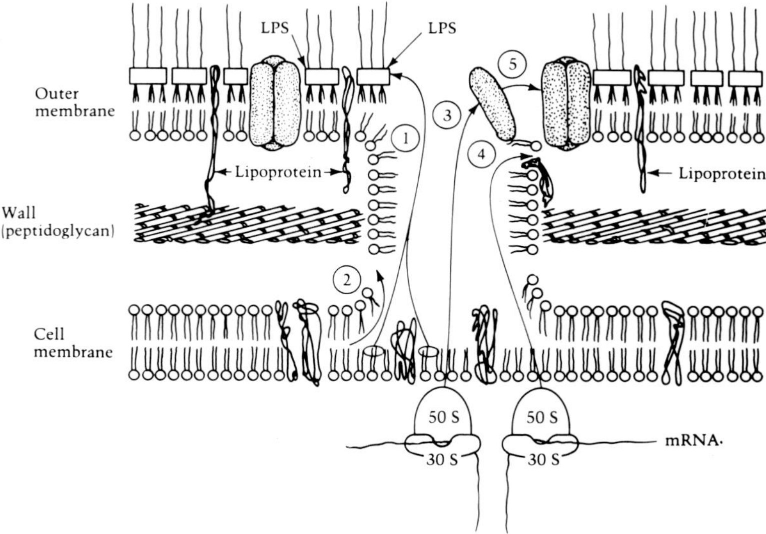

Fig. 181. Assembly reactions of the outer membrane of Gram-negative bacteria. – LPS = lipopolysaccharide. (From INGRAHAM et al. 1983)

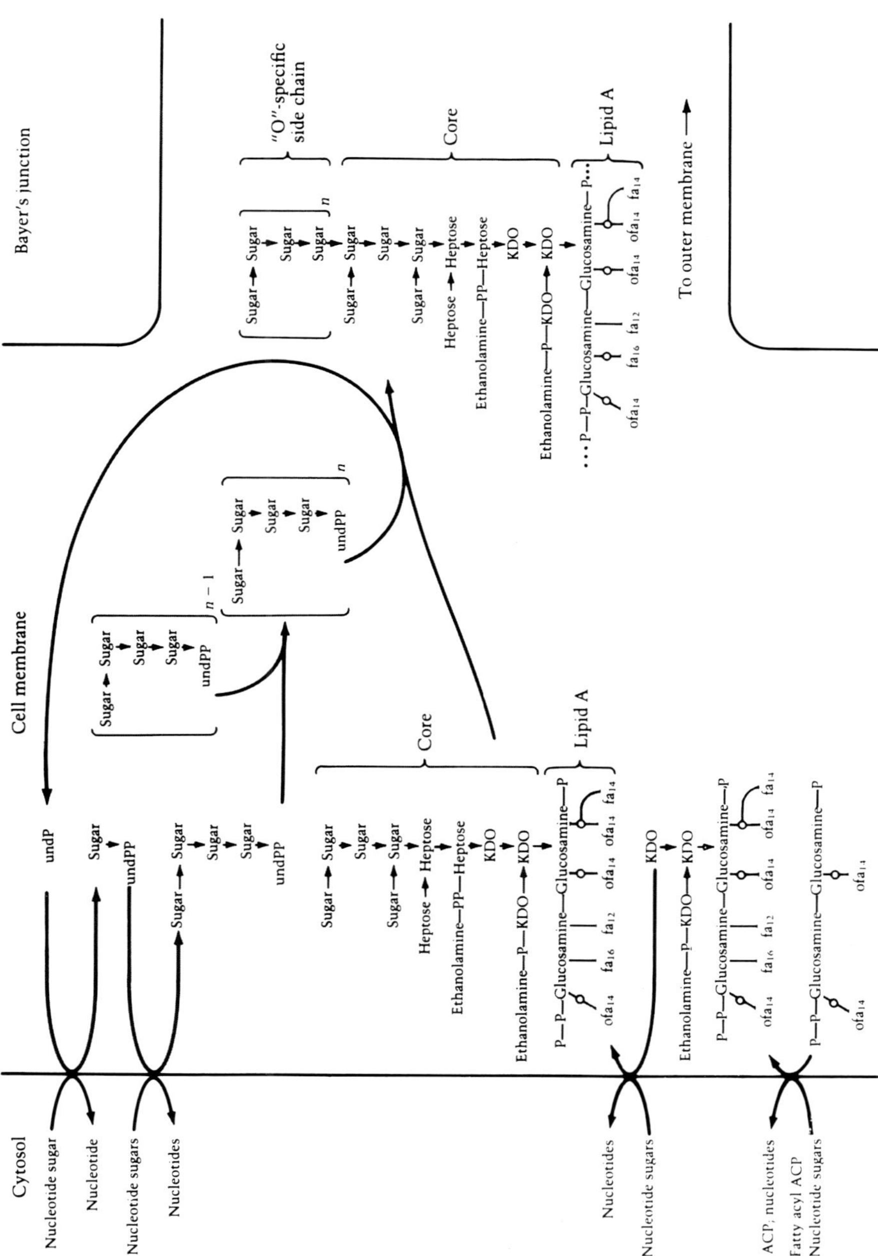

Fig. 182. Polymerization of lipopolysaccharide. – fa = fatty acids; KDO = 3-deoxy-D-manno-octulosonicacid. (From INGRAHAM et al. 1983)

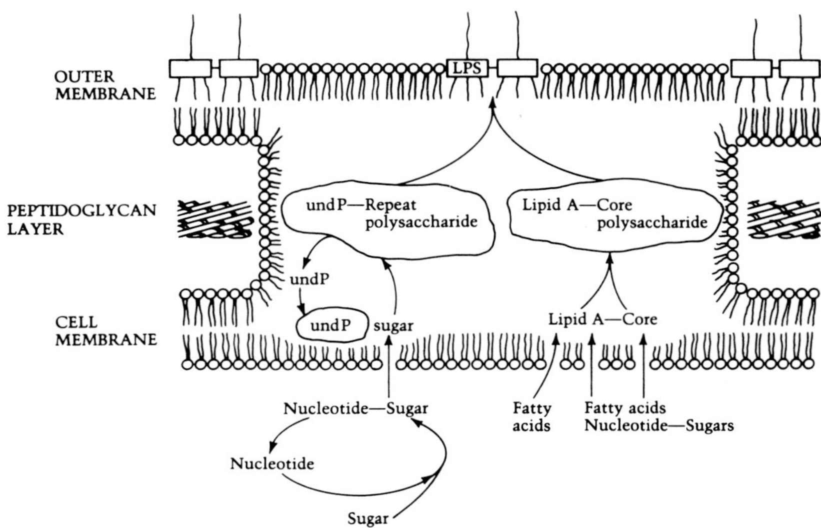

Fig. 183. Assembly of the lipopolysaccharide. – und P = undecaprenol-PO_4. (From INGRAHAM et al. 1983)

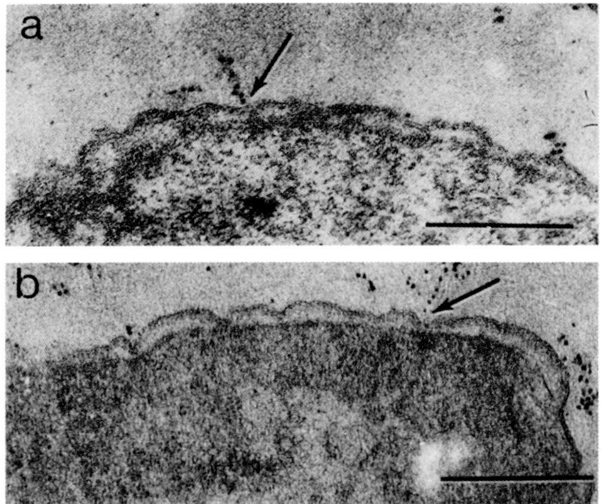

Fig. 184. Microscope study of the polysaccharide capsule of *Escherichia coli*. a, b, the immune-reactive filaments seem often to protrude from areas of wall-membrane adhesion (arrows). Bars: 0.25 μm. (From BAYER & THUROW 1977)

over 50 different proteins to those molecules to generate the 30 S and 50 S ribosomal subunits. The factors guiding this assembly in vivo are not yet known (Figs. 114 and 115).

(10) The cytosol is a fluid structure but probably has a loose organization brought about by affinities among proteins with cooperative or sequential functions. Some polypeptides must be processed in a variety of ways to become functional proteins. Some bacterial species contain organelles within the cytosol, including vesicles of various sorts bounded by protein membranes; their assembly has not been thoroughly studied.

7. General cytological aspects of cell division modes

The common bacterial cell division mode is the binary "fission"; its variations are "snapping" or "bending". These modes produce more or less equal daughter cells by a sequence of events (Fig. 185) finally leading to a division septum (ALLEN 1968, INGRAHAM et al. 1983, PARDEE et al. 1973).

Several other types of division, including inequal division, have been observed (Figs. 189, 191, 192). "Budding" is one of these modes occurring in a variety of bacteria (ARISTOVSKAYA & HIRSCH 1974, BAULD & TYLER 1971, BAULD et al. 1971, CONTI & HIRSCH 1965, DOW et al. 1976, HIRSCH 1968, HIRSCH & CONTI 1964, HIRSCH & RHEINHEIMER 1968, LEIFSON 1964, WHITTENBURY & MCLEE 1967). According to HIRSCH (1974), a process of new cell formation in a bacterium will be considered to be a true budding process (Figs. 187, 188, 190) if the following criteria apply:

a) Morphological: the new cell must be initially substantially smaller in size (i. e. narrower and shorter) than the mother cell. Even after separation, this usually holds true.

b) Developmental: all or most of the bud wall must be newly synthesized, although the age distribution pattern in the bud wall can vary in different organisms. Usually, buds are preformed

Fig. 185. Cell division; *Staphylococcus aureus* with growing cross wall. X = separating system; T = tubular structures. (From GIESBRECHT 1972)

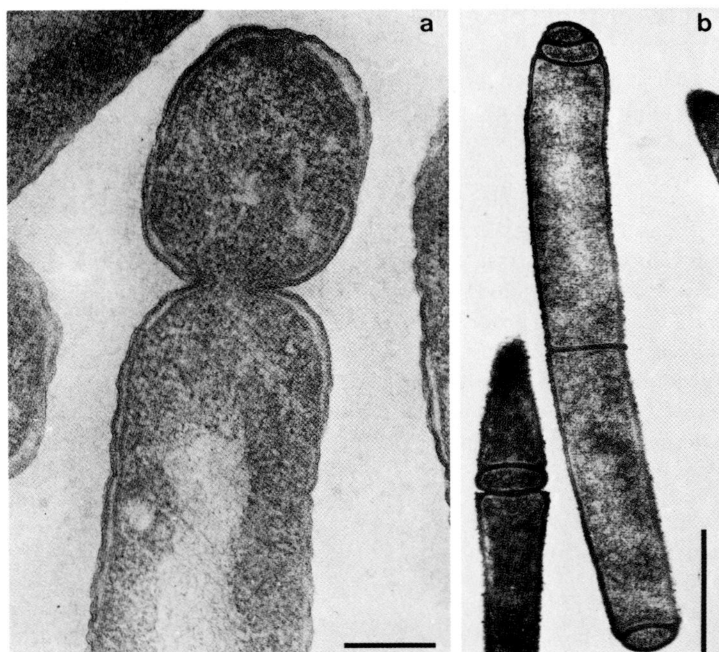

Fig. 186. a, cell of *E. coli* producing a minicell. Bar: 0.2 µm. (From PARDEE et al. 1973, after ADLER et al.); b, incorrect septa in a mutant of *Bacillus subtilis.* Bar: 1 µm. (From PARDEE et al. 1973, after van ALSTYNE & SIMON)

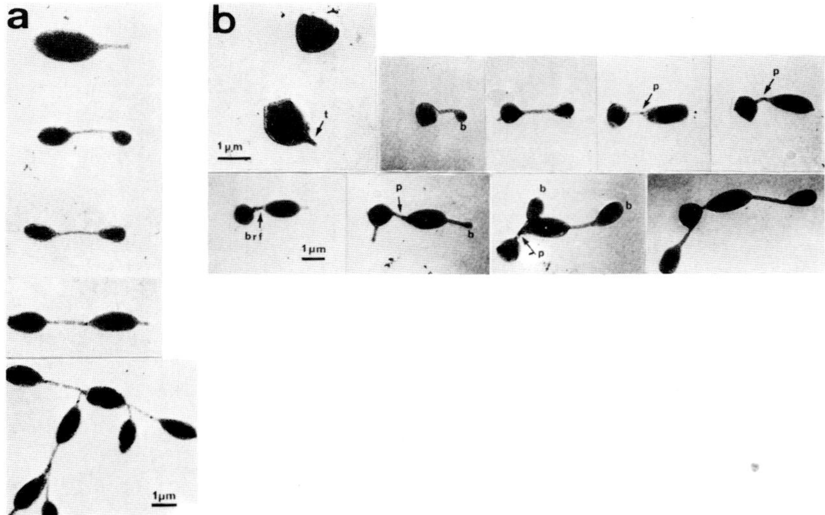

Fig. 187. a, developmental cycle of *Rhodomicrobium vannielii;* b, same bacterium as a, but exospore germinating sequence. – b = bud formation; brf = branch formation; p = plug formation; t = tube formation. (From DOW et al. 1976)

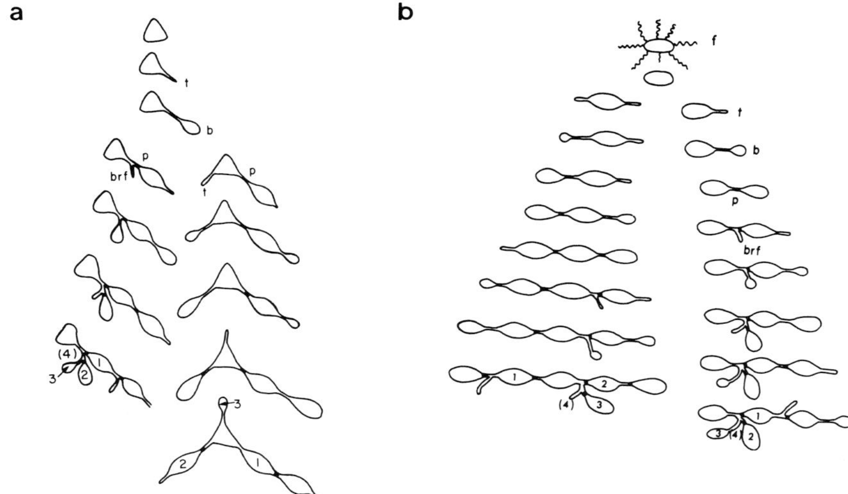

Fig. 188. a, germination sequence of *Rhodomicrobium vannielii* exospores; b, diagrammatic representation of progressive sequence of events during the vegetative cell cycle of *Rm. vannielii*. – b = bud formation; brf = branch formation; f = flagella; p = plug formation or plug; t = tube synthesis. (From Dow et al. 1976)

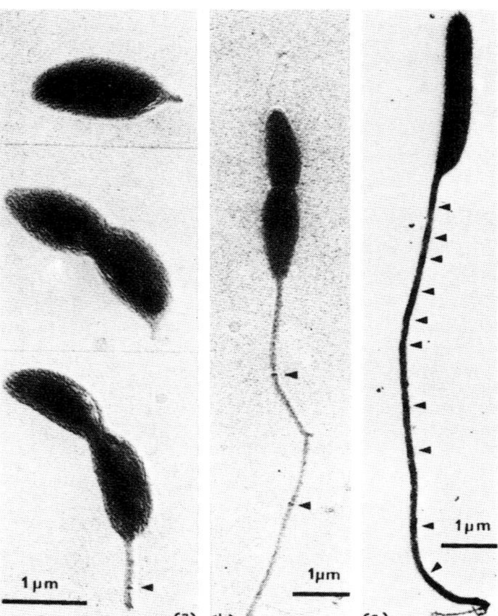

Fig. 189. a, *Caulobacter* life cycle; b, mother and flagellated daughter on the point of separation; c, mother cell which has undergone twelve generations as determined from the number of crossbands in the stalk. (From Dow et al. 1976)

before they are nucleated. Also, spherical and small buds increase in cell diameter during cell wall growth, contrary to elongation of rod-shaped non-budding bacteria.

c) Functional: budding must be the only mode of new cell formation. A bud is a creation, by the mother cell, of a new space outside, into which the new cell constituents either have to migrate or

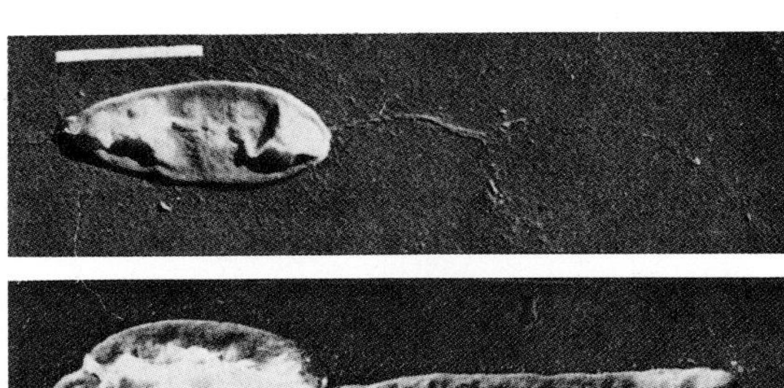

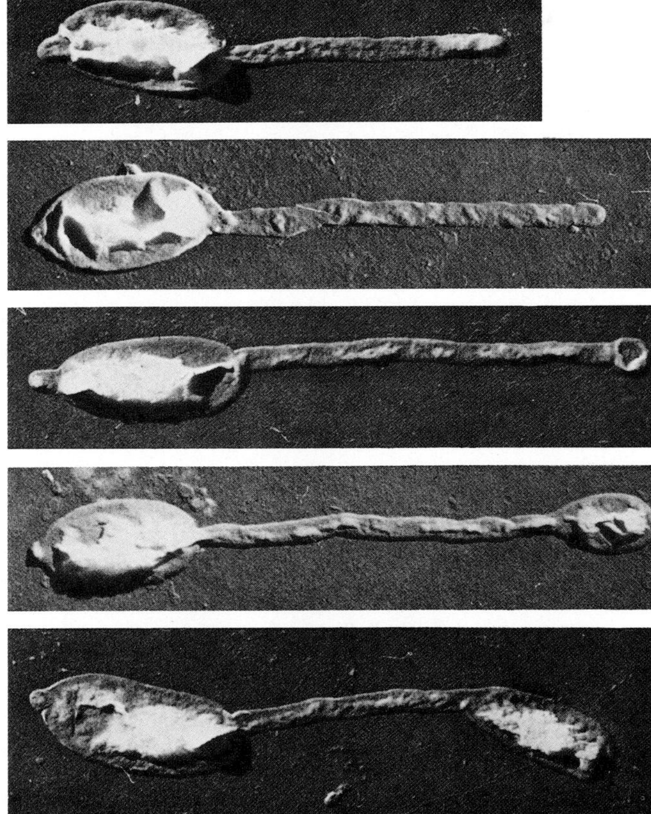

Fig. 190. *Hyphomicrobium* strain *B-522* at various stages of the cell cycle. Bar: 1 μm. (From MOORE 1981, after de COURAY CLARKE)

where they have to be formed de novo. Thus, budding does not necessarily describe a mode of cell separation (multiplication) but rather one of new cell wall formation.

In the case of a typical inequal cell division, septa are formed incorrectly (PARDEE et al. 1973). Examples are the generation of "minicells" (Fig. 186), that of abnormal septa, or situations where septation is lacking (filamentous growth without a sheath). Division mutants have been extensively studied in order to understand their unusual division and growth modes and the interactions of DNA replication, regulation of activity of enzymes involved in cell septation, and cytological appearance of these cells.

Besides the common division modes, true branching has been observed in filamentous cyanobacteria. The morphology of this type of cells has been described thoroughly at the light microscopic

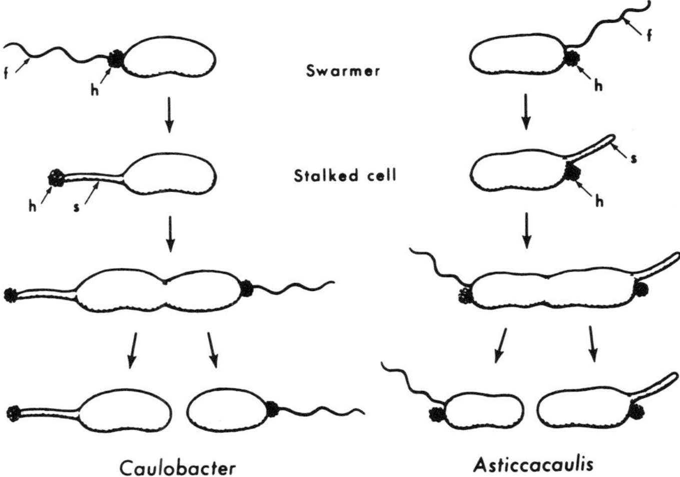

Fig. 191. Life cycle diagrams of a typical *Caulobacter* sp. and of *Asticcacaulis excentricus;* h = holdfast; f = flagellum; s = stalk. (From SCHMIDT 1981, after SCHMIDT & STANIER)

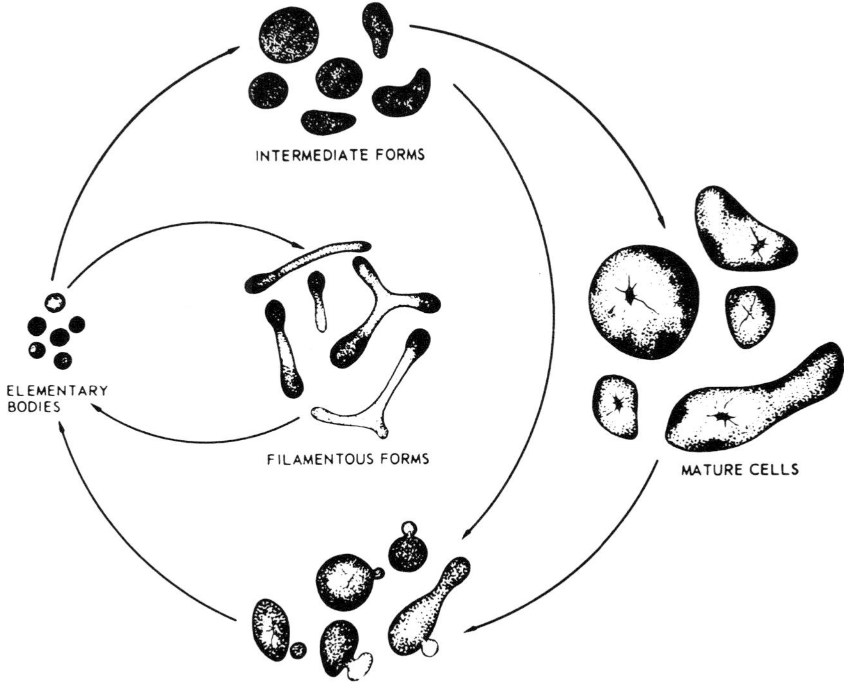

Fig. 192. Diagram of the proposed development of mycoplasma-like organisms in the aster plant. (From GRUNEWALDT-STOECKER & NIENHAUS 1977, after SINHA & PALIWAL)

level. THURSTON & INGRAM (1971) have carried out an ultrastructural study of *Fischerella ambigua.* They showed that this organism possesses a mechanism of cell division quite unlike that seen in unbranched, filamentous bacteria. NIERZWICKI et al. (1982) have provided further evidence that, in *Mastigocladus laminosus,* septum formation in older cells may occur parallel to the long axis of the filament, thereby confirming that true branching takes place.

Cell conglomerates and other complex cell aggregates have occasionally been observed. SKERMAN et al. (1983) have described a Gram-negative rod-shaped organism which exhibits unicellular and multicellular phases of growth. Unicellular-phase cells were motile, with two different types of flagella per cell. Multicellular conglomerates arose from single cells which lost motility, became optically refractile, and reproduced by multiplanar centripetal septation. This process was enhanced by sodium ions. Under suitable conditions, conglomerates dissociated into single cells which produced clear colonies with, initially, only a few unicellular-phase cells.

8. Interrelationships between cell growth and division

Cell division in its typical sequence of events and in its various modifications has been investigated in different bacterial species (CHAI & LARK 1967, MURRAY & DOUGLAS 1950, NANNINGA & WOLDRINGH 1985, ROGERS et al. 1980, ZAVARZIN 1960). When analyzing bacterial growth and cell division (ROGERS et al. 1980; further references therein), the most evident events are observed to occur periodically; this gives the appearance of a cycle. The most obvious event, cell division, is actually the terminal one. However, a description of the events could begin with less obvious phenomena, e. g. the biochemical events that occur between cell divisions. The sequence of steps in the total process comprises the initiation of bi-directional DNA synthesis, chromosome replication, septation, and finally cell separation. At the end of the process the rest of the cell content – ribosomes, enzymes, wall and membranes – appear to be partitioned more or less equally between the daughter cells. The entire round of replication, from formation of the DNA initiation apparatus to cell separation, takes some time, at least one hour for *E. coli.* Division times of *E. coli* can be as long as several hours or as short as 20 min, depending on growth conditions. Thus, the cycle can be longer or shorter than the interval between cell divisions. As a consequence, several cycle sequences can go on in the same cell, each at a different stage, and new septum sites have been observed before the first had been completed. Several attempts have been made to describe the major events of the cell cycle in detail. In 1968, HELMSTETTER & COOPER devised a method for obtaining synchrony in cell growth by selectively eluting newly divided cells from a randomly growing culture adsorbed to an inverted membrane filter. Using this system, they obtained accurate timing of the key events of the cycle including DNA initiation, DNA completion, and cell division. They developed a model which separates the division cycle into three parts that occur in sequence, without, however, giving further details regarding the progression from one stage to the other. Some of the results obtained by various workers on detailed aspects of interrelationships between cell growth and division are described below, especially those with relevance to cytological problems.

A special situation appears to exist in the assembly of *Rickettsia.* These bacteria are obligatory intracellular parasites. They are generally believed to maintain their morphological integrity throughout their life cycle. However, HASE (1983) has come to other conclusions as a result of his investigations on the assembly process of *Rickettsia tsutsugamushi* in irradiated L cells. In his studies, nascent forms, three days after infection, lacked limiting membranes. During "maturation", they assumed a round shape. Faint membranous components became visible in fuzzy zones covering the surface of these round structures. The rickettsial membranes were assembled along the fuzzy zones. This assembly was paralleled by formation of a filamentous network in one zone (f area) of the parasite-host cell system and manufacture of rickettsia ribosomes in another zone (g area).

8.1 Special aspects in Gram-positive bacteria

A comprehension of the events taking place during bacterial growth and division (Figs. 193–199) has been facilitated by the application of different electron microscopic techniques in a variety of bacterial species and by various experimental approaches aimed at altering natural growth conditions (e. g., by inhibition of DNA synthesis). Cell wall assembly and growth have been followed by high resolution autoradiography. With this technique, De Chastellier et al. (1975 a, b) found that the total radioactivity incorporated per bacterium *(Bacillus megaterium)* doubled during the life cycle. This doubling occurred in the cylindrical part of the cell wall but not in the polar caps. It was deduced that elongation of *B. megaterium* cells is not constant during the life cycle but increases with the length of the cell. This was taken to confirm conclusions drawn by Collins & Richmond (1962) from their analysis of the cell size frequency in a non-synchronized culture. This mode of growth implies that at the end of a division cycle the two daughter bacteria already possess the number of growth sites they will need after their separation. In addition, it was concluded from autoradiographic studies that, in *B. megaterium,* cell wall elongation actually occurs by diffuse intercalation of newly synthesized murein precursors along the cylindrical part of the cell wall, the precise growth zone observed in the middle of the cell being responsible for cross wall formation. Until recently, the majority of studies favored a diffuse intercalation of new material in the case of the wall of Gram-positive rods. Schlaeppi et al. (1982), however, demonstrated zonal insertions by establishing that the cell wall of *B. subtilis* consists of a limited number of large "subunits" which segregate during cell growth.

A further aspect of studies on the growth of bacterial cells has been the examination of cosegregation of cell wall and DNA (Fig. 196). Schlaeppi & Karamata (1982) investigated this phenomenon in a lysis-negative mutant of *B. subtilis* by continuously labelling cell wall, or DNA, or both cell wall and DNA. After four or five generations of chase in liquid media it was found by light microscope autoradiography that the numbers of wall segregation units per cell were 29 and 9 in rich and minimal medium, respectively. Under the same conditions, the numbers of segregation units of DNA were almost 50% lower: 15 and 5, respectively. Simultaneous labelling of cell wall and DNA provided figures almost identical to those obtained for the cell wall alone, implying cosegregation of the two components. Statistical analysis ruled out their random distribution into daughter cells. Measurements of the positions of grain clusters at the end of the chase period along chains of cells, each derived from a single cell at the beginning of the chase, showed that cell wall units were localized according to a symmetrical pattern whereas those of DNA were distributed in an asymmetrical but highly regular way. It appears that of the two cell wall units of the same age in *B. subtilis* only one has a strand of DNA attached to it. For *Streptococcus faecalis* (Edelstein et al. 1980) evidence has been obtained that wall growth takes place in the region of the cross-wall eventually forming two cells out of one. Fig. 195 illustrates the sequence of events. The diagram is based on observations from experiments where protein synthesis had been shut off and subsequently was re-initiated. In part, the findings were made possible by the special behaviour of the autolytic system of *S. faecalis* (Figs. 198 and 199). The autolysine present in this bacterium is an N-acetylmuramidase (like lysozyme) and it liberates free reducing groups of N-acetylmuramic acid

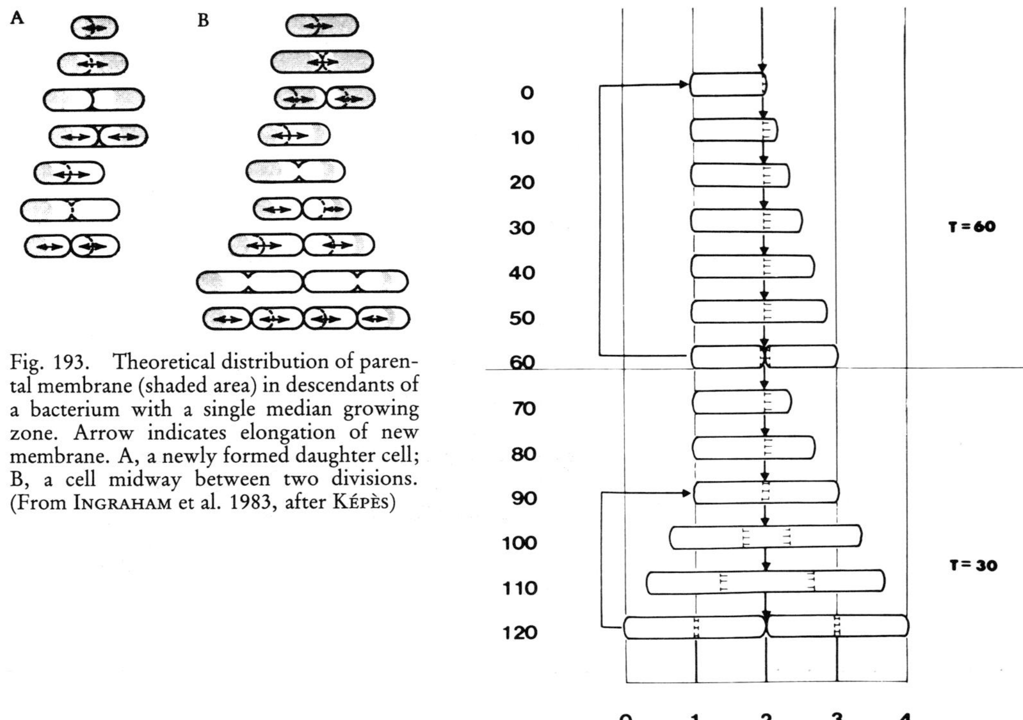

Fig. 193. Theoretical distribution of parental membrane (shaded area) in descendants of a bacterium with a single median growing zone. Arrow indicates elongation of new membrane. A, a newly formed daughter cell; B, a cell midway between two divisions. (From INGRAHAM et al. 1983, after KÉPÈS)

Fig. 194. Model of bacterial cell growth; a unit cell is assumed to be grown for one cell cycle in a medium where the mass doubling time is 60 min (T = 60); at 60 min the daughter cell on the left is transferred to a richer medium where the mass doubling time is 30 min (T = 30). The growth sites are shown as dashed vertical lines, and the direction of growth at each site is shown by horizontal lines attached to them. Line 2 corresponds to the spatial location of cell divisions. The triangles show the penicillin-sensitive sites. (From PARDEE et al. 1973, after DONACHIE & BEGG)

from the glycan chains. The autolysine had been shown to be located in the region of the cross-wall, in contrast to bacilli which are much more lytic and have an active wall turnover, with total cell lysis often occurring when protein synthesis is shut off. With *S. faecalis* an experimental approach based on these specific properties of the autolysine system could be used for the determination of the location of wall synthesis prior to cell division. In order to test the behaviour of newly added wall material compared to that of the old wall, exponentially growing cultures were either pulse-labelled with ^{14}C-lysine, or labelled continuously with this amino acid for at least 10 generations followed by a pulse of ^{12}C-lysine for 0.8 or 0.1 of a generation time. The walls were isolated and allowed to autolyse. From the amounts of liberation of radioactivity relative to the loss in turbidity, the conclusion was drawn that the newly synthesized wall was always the first to be solubilized by the autolysine, indicating wall synthesis in the region of cross walls.

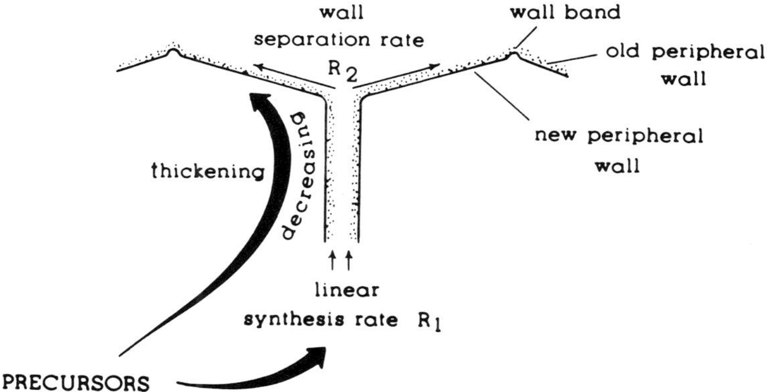

Fig. 195. Model of wall growth for *Streptococcus faecalis*. (From GHUYSEN & SHOCKMAN 1973, after HIGGINS & SHOCKMAN)

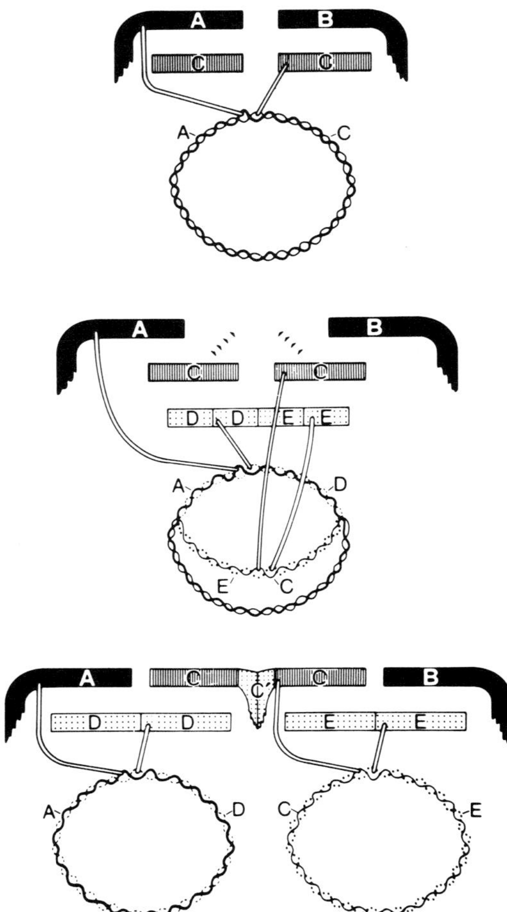

Fig. 196. Cosegregation of cell wall and DNA in *Bacillus subtilis*. Model of cell wall organization and DNA-cell wall association. – A, B, C, D, E = segregation "units"; C' = new septum. (From SCHLAEPPI & KARAMATA 1982)

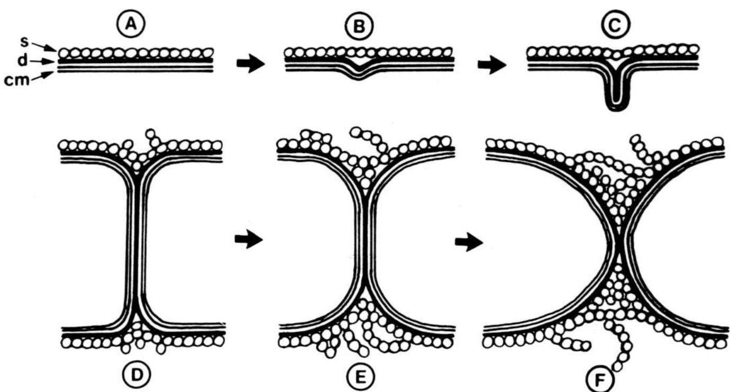

Fig. 197. Diagram illustrating the cell division process in *Clostridium thermohydrosulfuricum* and *C. thermohydrosaccharolyticum.* A, non-dividing cell; B, C, early stages of septum formation; D, completed septum; E, F, cell separation. – cm = cytoplasmic membrane; d = dense layer (mucopeptide); s = S layer. (From SLEYTR 1978)

For *Streptococcus faecium* (GIBSON et al. 1983) it has been shown that the cell grows at discrete growth sites in which septa separate and expand into pairs of polar caps. The initial event in the formation of a new growth site is the synthesis of an annular ridge of material on the internal surface of the cell wall. This ridge forms a nascent septum which is usually assembled directly under a raised band found on the external surface of the cell wall.

These wall bands have been used as markers to study the number and geometry of growth sites. The results of these studies indicated that the volume of the two completed poles produced by a single site was constant and thus independent of growth rate and that the initiation of new sites could not be related to the timing of rounds of chromosome synthesis but instead occurred at a constant cell volume. A model was proposed in which a new site of surface growth would be produced when the volume within the two segments of wall joined by a single wall band reached a critical size. Depending on the growth rate, these two segments could be two poles or a pole and a nascent pole. Cell wall assembly during inhibition of DNA synthesis by mitomycin C has revealed additional information. When mitomycin C was added to an exponential-phase culture doubling in mass every 64 min, DNA synthesis was inhibited, and eventually cell division stopped. The growth sites formed before and after inhibition of DNA synthesis enlarged until they contained about 0.25 μm^3 of cell volume, at which point they ceased to increase in size. When these sites approached the limit, new sites were initiated. This result had also been observed in untreated cells undergoing a large range of exponential-phase mass doubling times. Thus, regardless of whether chromosome replication is inhibited or uninhibited, sites have the same finite capacity to enlarge to about 0.25 μm^3, and when this capacity is reached, new sites are initiated. Although initiation of new growth sites seems to be independent of normal chromosome replication, these results confirm studies showing that chromosome replication is necessary for the terminal events of growth site development which result in the division of a site into two separate poles. Different models for the regulation of growth site initiation have been discussed.

Special image evaluation procedures and cell reconstruction techniques (ROGERS et al. 1980; further references therein) ("section-rotation method") have been devised for the detailed analyses

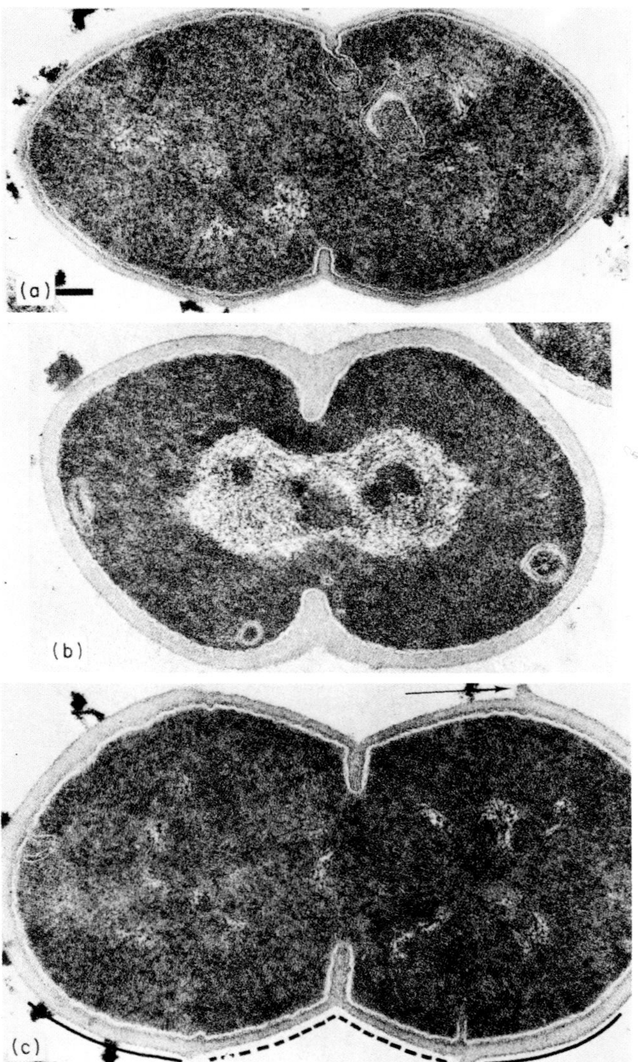

Fig. 198. Cell division and peptidoglycan synthesis. Sections of *Streptococcus faecalis*. a, a cell from an exponentially growing culture; b, a cell from a culture deprived of threonine for 20 hr; extensive thickening of the wall; c, a cell recovering from 10 hr of threonine starvation. Note the conservation of polar wall thickened during starvation. Bar represents 0.1 µm and applies to all three micrographs. (From GHUYSEN & SHOCKMAN 1973, after HIGGINS & SHOCKMAN, and HIGGINS et al.)

of growth zones. They have been used for comparative studies of the events taking place in growth and cell division in *Streptococcus faecalis* and *Bacillus subtilis* as well as other bacilli. One of the results is the following: whereas the peripheral wall of the streptococci is fed out by the layers pealing apart from the open septum, the walls of the poles of *B. subtilis* are formed during and after

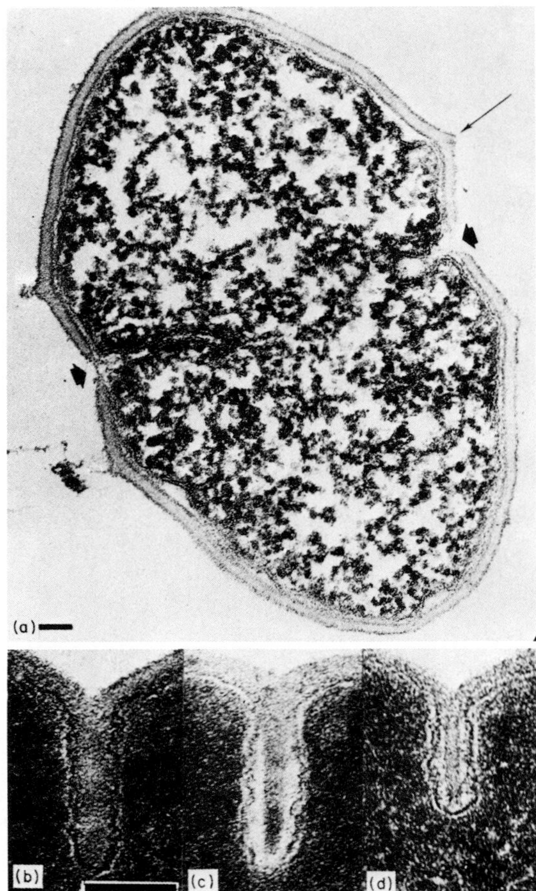

Fig. 199. Autolysis of *Streptococcus faecalis.* a, 60 min after the beginning of autolysis. The cross wall is no longer visible (large arrows). The wall bands (small arrow) mark the separation of "new" equatorial, peripheral wall from "old" polar wall; b, control; non-autolyzing cross wall; c, d, primary autolytic attack at the leading edges of the cross wall. Bars: 0.1 µm. (From GHUYSEN & SHOCKMAN 1973, after HIGGINS et al.)

septal closure. As in streptococci, the peripheral wall of the bacilli grows thicker as it is fed out, and the total volume of the polar wall is much greater than that present in a flat disc. Thus, the formation of the poles of *B. subtilis* not only has some characteristic features in common with peripheral wall formation by *S. faecalis* but also demonstrates clear differences.

Besides the application of radioactive material which eventually gets incorporated into the growing cell wall, other approaches to the study of the topology of growth of Gram-positive rods have been used (Ingraham et al. 1983, Rogers et al. 1980; further references therein); examples are the binding of bacteriophages which involves measuring the change of the distribution of teichoic acids along the length of the walls as a discriminating factor in cells grown in continuous culture, and manipulation of the cell wall composition by certain nutrient limitations for certain periods. In many Gram-positive bacteria grown under phosphate limitation, teichoic acids are replaced by uronic acid containing phosphate-free polymers which do not fix bacteriophages. Monitoring the distribution of wall teichoic acids on the growing cell wall surface by specific reaction with concanavalin A linked to a fluorescent dye or ferritin has also been used in these growth studies. Labelling of bacilli with fluorescent antibodies is yet another technique for the analysis of wall growth. Possible errors in the interpretation of the results obtained by this method might be caused by incorrect labelling, especially that to capsular or other extracellular material instead of wall components.

Detailed investigations of cell division in staphylococci have yielded further insight into the sequence of events and the structural components involved (Giesbrecht 1985). Punching of holes into the peripheral wall in untreated cells is a normal process which follows cross-wall completion and represents the first visible step of cell separation. With penicillin, however, analogous holes appear to be punched by the content of "murosomes", i. e. extraplasmatic vesicular structures located exclusively within the bacterial wall material. These holes have been found to be places at sites of presumptive cross-wall separation even if no sufficient cross-wall material had been assembled before at this site. Consequently, because of the internal pressure of the protoplast, lytic death is the inevitable result of this perforation of the protective peripheral cell wall. Hence, the very mechanism of penicillin-induced bacteriolysis in staphylococci can be considered to be mainly the result of a special morphogenetic wall defect. The data indicate that all hypotheses on the mechanism of bacteriolysis which are based purely on biochemical events such as simple changes in total wall enzyme activities or total cell wall synthesis cannot be sufficient to explain penicillin-induced lytic death.

8.2 Special aspects in Gram-negative bacteria

The structure of the envelope in Gram-negative bacteria is more complex than that of Gram-positive strains (Burdett & Murray 1974 a, b). Therefore, the problems involved in looking at the topology of the growth in Gram-negative bacteria (Figs. 200–205) are somewhat different from those in Gram-positive cells (Begg & Donachie 1977, Ingraham et al. 1983, Ito et al. 1977, MacAlister et al. 1977 a, b; Nanninga et al. 1982, Rogers et al. 1980, Schwarz et al. 1969, Verwer & Nannniga 1980). One distinct feature is that peptidoglycan turnover in Gram-negative bacteria hardly exists or, at least, is very low. Again, as with Gram-positive bacteria, different approaches have been used in studying cell growth and division, such as labelling with fluorescent or ferritin-conjugated antibodies, or bacteriophages, or autoradiography after radioactive labelling. The analysis includes the outer membrane, the peptidoglycan layer and the cytoplasmic membrane. As to the outer membrane, the general impression is that it may well be built up by diffuse intercalation over its surface with material being exported via adhesion sites as first recognized by Bayer (1979). Experiments to follow growth of the outer membrane by the use of bacteriophages (Ryter et al. 1975) have been interpreted in favour of the hypothesis that cells of *E. coli* grow from their poles. However, this work has been strongly criticized by Smit & Nikaido

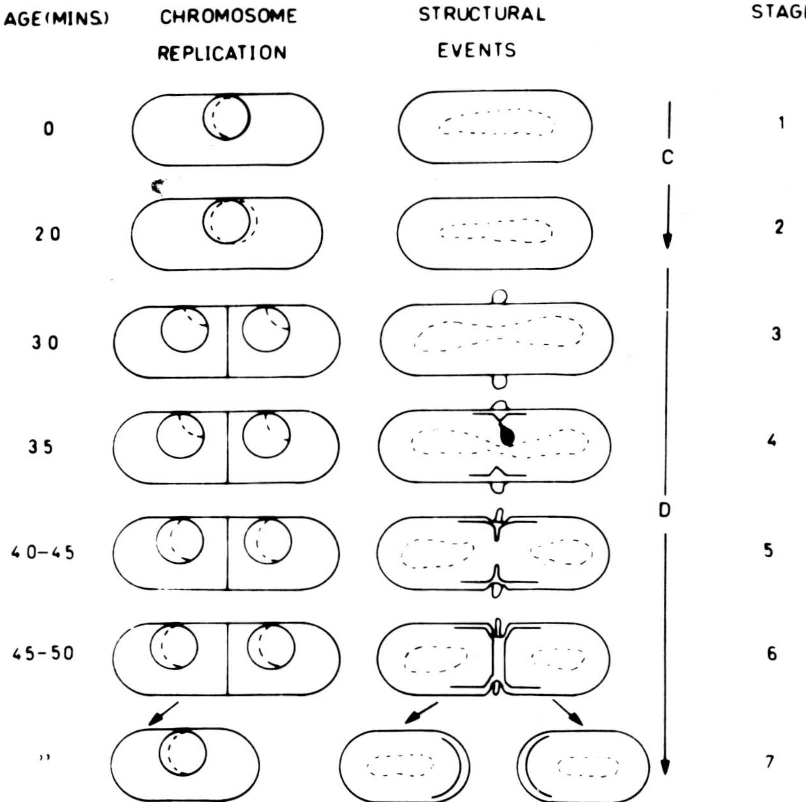

Fig. 200. Diagram of basic events during the cell cycle of *E. coli* in glucose minimal medium. DNA replication is completed at 20 min and a new round of replication is then initiated; separation of nucleoids also occurs at about this time. (From BURDETT & MURRAY 1974)

(1978) who have examined the insertion of one of the major outer membrane proteins, a porin, in *Salmonella typhimurium.* They found no evidence for polar growth in these bacteria.

With respect to the growth of the peptidoglycan layer, earlier studies and more recent results are somewhat contradictory (BURMAN et al. 1983, RYTER et al. 1973, SCHWARZ et al. 1975). The older findings, derived from application of labelled diaminopimelic acid, indicated that growth of the peptidoglycan layer occurred by insertion of new material at a large number of sites. More recent investigations lead to the conclusion that genome doubling and the new spatial orientation of the genomes with respect to the cell envelope could be the trigger for new growth areas (beside the central one) (HELMSTETTER & COOPER 1968, KOPPES & NANNINGA 1980, KOPPES et al. 1978a, b, 1980, NEWMAN & KUBITSCHEK 1978, WOLDRINGH 1976). This interpretation is similar to the general ideas developed by SARGENT (1979) working with a Gram-positive bacterium *(B. subtilis,* s. above).

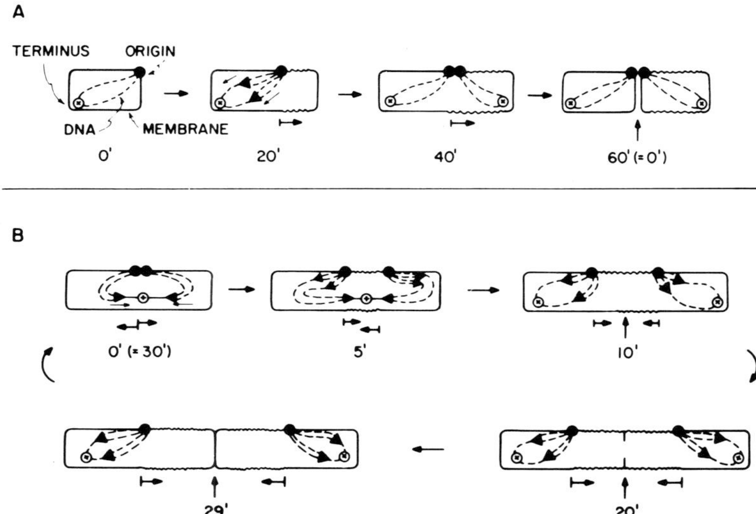

Fig. 201. Schematic representation of chromosome and membrane growth of *Escherichia coli B/r*. A, division time of 60 min; B, division time of 30 min; these cells show dichotomous replication. (From PARDEE et al. 1973)

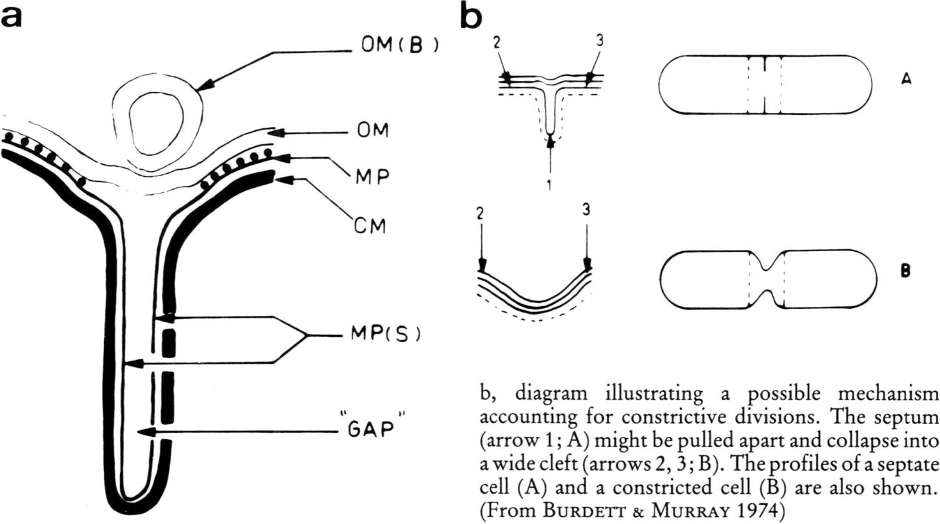

b, diagram illustrating a possible mechanism accounting for constrictive divisions. The septum (arrow 1; A) might be pulled apart and collapse into a wide cleft (arrows 2, 3; B). The profiles of a septate cell (A) and a constricted cell (B) are also shown. (From BURDETT & MURRAY 1974)

Fig. 202. a, diagram of septum structure in *Escherichia coli*. – CM = cytoplasmic membrane; MP = mucopeptide; MP (S) = mucopeptide lamellae in septum; OM = outer membrane; OM (B) = bleb formed by outer membrane.

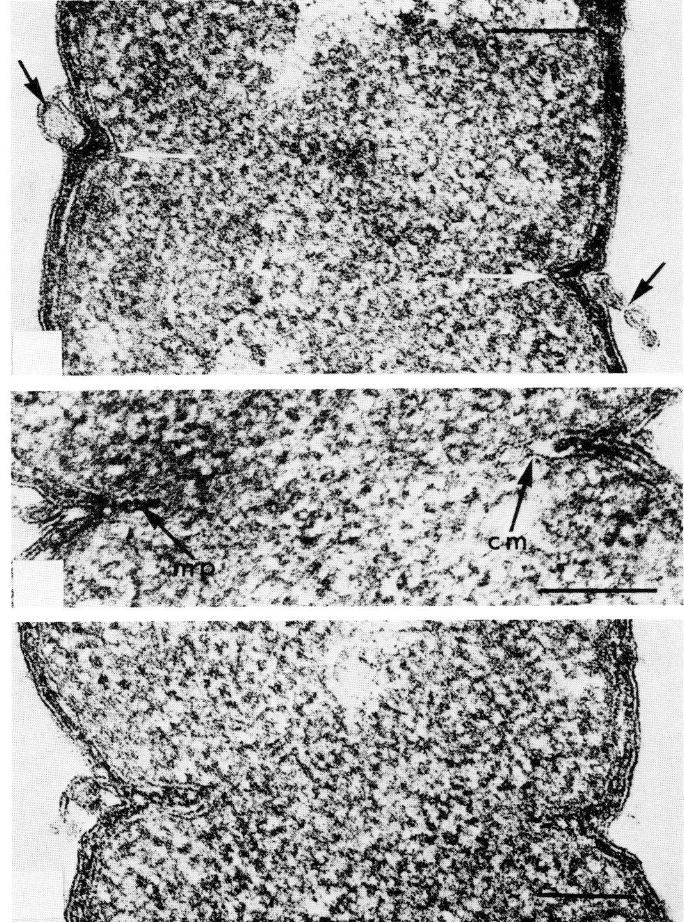

Fig. 203. Septum formation in *Escherichia coli.* Note vesicles of outer membrane (black arrows) and ingrowth of mucopeptide (mp) and cytoplasmic membrane (cm) (arrows). Bars: 0.1 µm. (From BURDETT & MURRAY 1974)

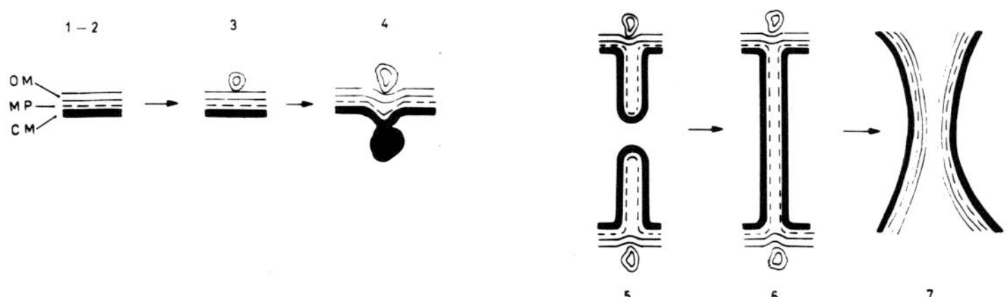

Fig. 204. Sequence of septation (stages 1–7) in synchronous cultures of *E. coli.* The outer membrane (OM) forms a bleb (stages 3–6) but only enters the septum at cell separation (stage 7); an intermediate layer (not shown) between OM and mucopeptide (MP) may also be excluded from the septum. – CM = cytoplasmic membrane. (From BURDETT & MURRAY 1974)

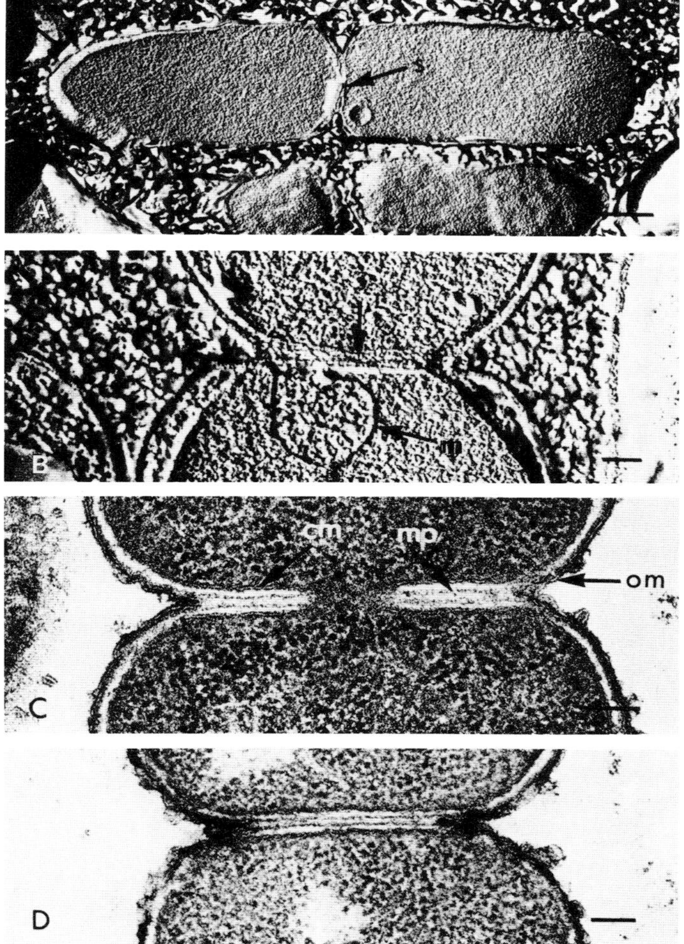

Fig. 205. Cell division in *Escherichia coli.* A, B, septa (s) and mesosome-like membranes (m); C, septation showing partial septa; D, complete septa. mp = murein; om = outer membrane; Bars: 0.5 μm in A, and 0.1 μm in B–D. (From BURDETT & MURRAY 1974)

It is important to realize that the growth of the wall layers could be thought to be simply the consequence of growth of the cytoplasmic membrane (INGRAHAM et al. 1983). Enzymes involved in important stages of peptidoglycan synthesis have been found to be firmly bound to the cytoplasmic membrane. One could imagine their lateral distribution and localized regulation as dictating the growth of the wall layers and thus being decisive for length extension or division. Unfortunately, defining the topologies of the expansion of the cytoplasmic membrane has so far been impossible since it is well known that the components of the cytoplasmic membrane are capable of rapid lateral movements.

9. Homologous and heterologous associations involving bacteria

In nature, bacteria can be found as single, assembled (BITTON & MARSHALL 1980, COSTERTON et al. 1978, JANNASCH & WIRSEN 1981), or "glued-together" (by external layers, s. above) cells being members of a colony, as primary cell aggregates of various shapes and complexity due to incomplete cell division, as secondary homologous associations brought about by specific properties of exposed surface structures or excreted compounds, and as heterologous associations, either with other bacteria, or with eukaryotic cells or tissues. The last group includes both symbiotic (exo- and endosymbiotic) and parasitic systems. As many of the typical properties of these associations or of their constituent cells are correlated with specific cytological aspects, some examples for the above mentioned types of aggregates will be described. However, the situation is by far too complex for being treated in detail. Therefore, the reader should consult appropriate sources such as the book edited by SCHWEMMLER & SCHENK (1980) on Endocytobiology, Endosymbiosis and Cell Biology, and the book edited by SCHENK & SCHWEMMLER (1983), Endocytobiology II, Intracellular Space as Oligogenetic Ecosystem.

9.1 Homologous associations

Primary and secondary homologous associations, i. e. associations between bacteria of a given

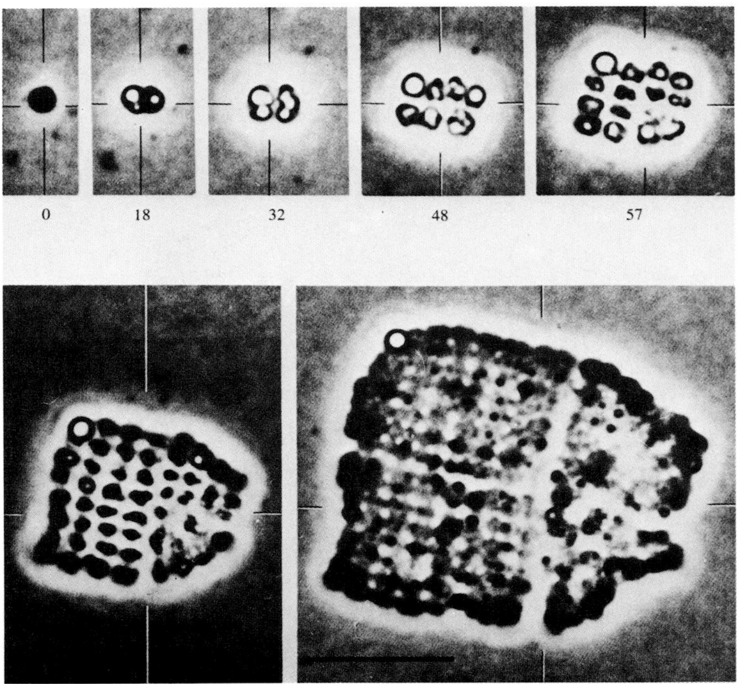

Fig. 206. Clone formation of *Lampropedia hyalina*. Cell division and formation of poly-β-hydroxybutyrate along two axes alternating synchronously by 90 degrees. Times given in hours; last two micrographs: left, 66 h; right: 86 h. Bar: 10 μm. (From STARR 1981, after KUHN & STARR)

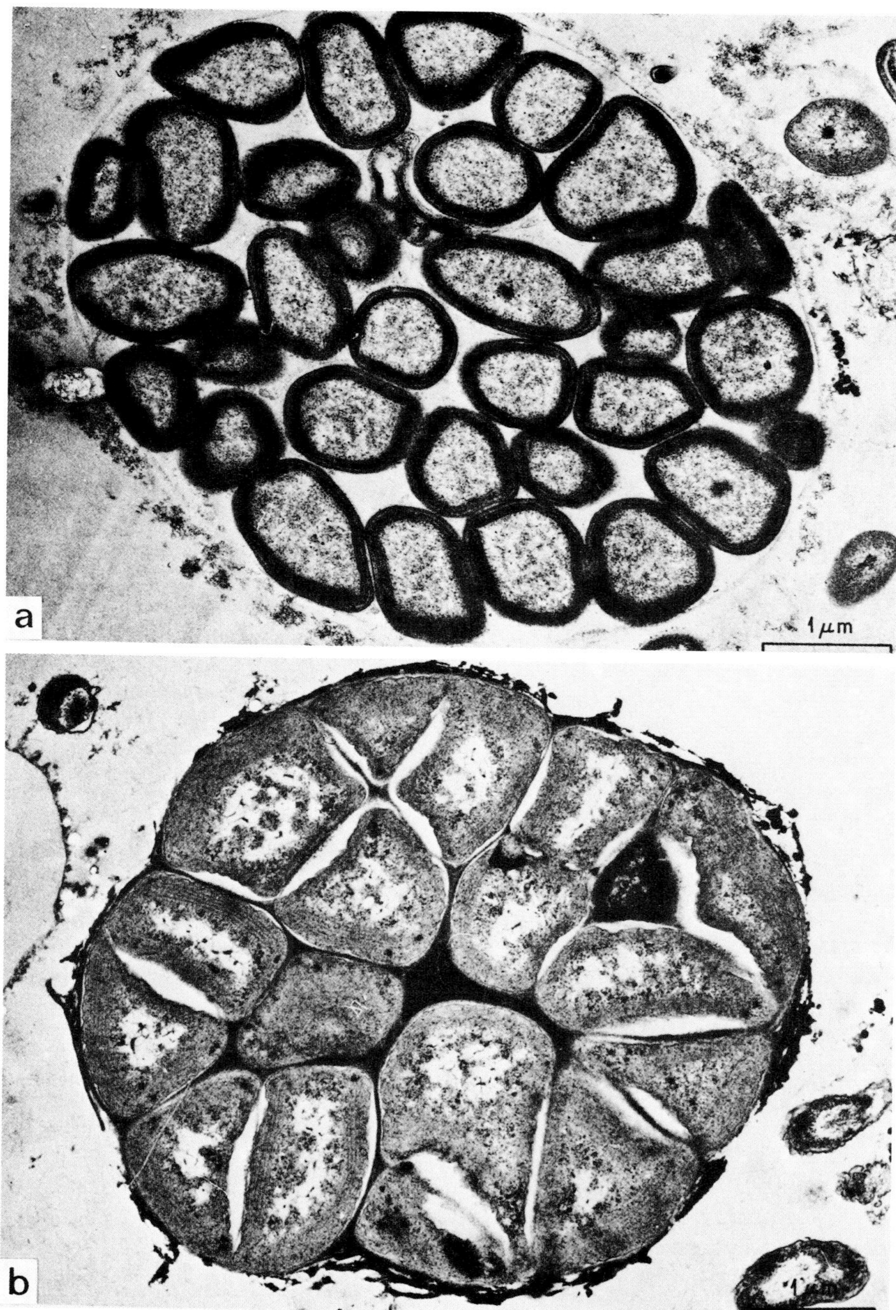

Fig. 207. Bacterial cell aggregates. a, loose aggregate of *Nitrosomonas europaea* referred to as a zoogloea; b, compact aggregate of *N. europaea* referred to as a cyst. (From WATSON et al. 1981)

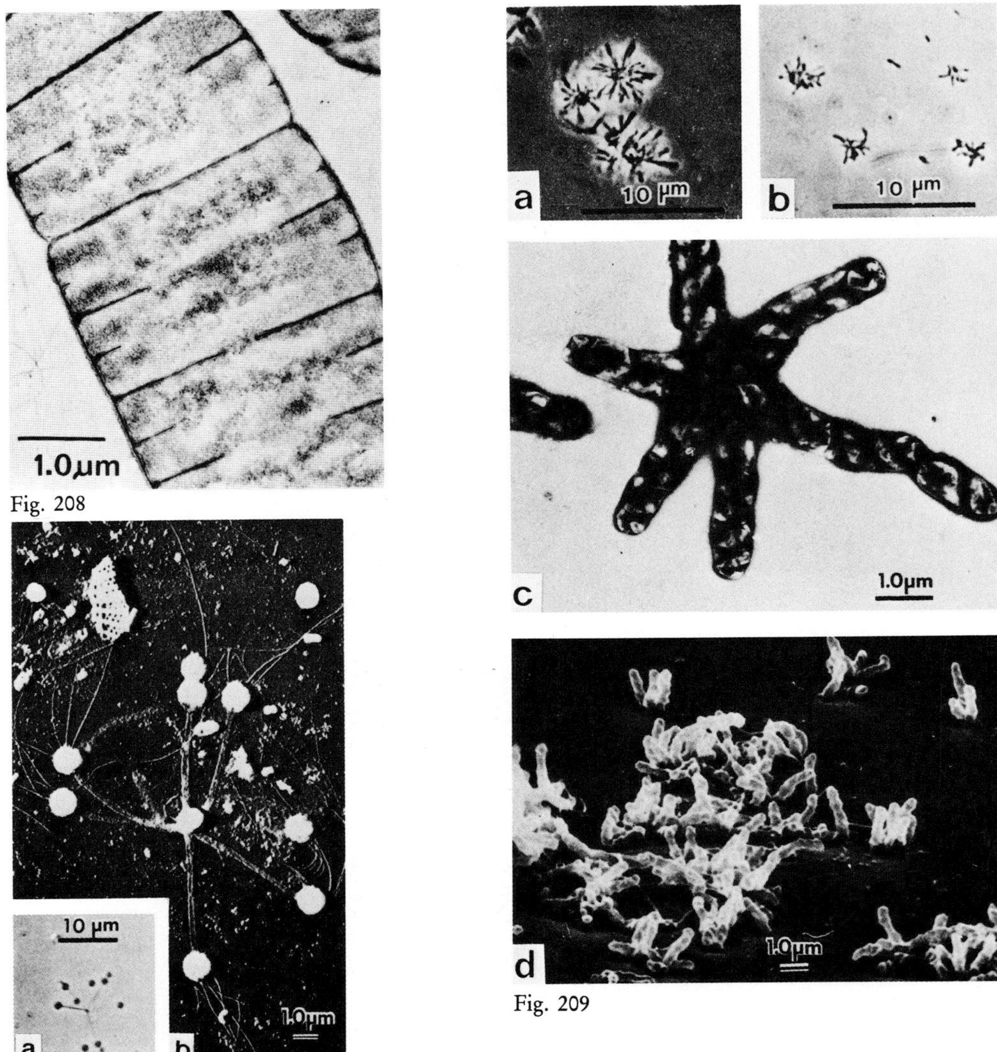

Fig. 208

Fig. 209

Fig. 210

Fig. 208. Arrangement of cells in chains; *Caryophanon latum;* isolate *J 1;* many septa at various stages of completion. (From TRENTINI 1981)

Fig. 209. Clusters of *Seliberia stellata,* a Russian soil strain, and *Seliberia*-like aquatic isolates. a, *S. stellata;* b, aquatic isolate; c, *S. stellata;* d, *ICPB* strain *4133,* grown on glass cover slip. (From SCHMIDT & SWAFFORD 1981)

Fig. 210. *Blastocaulis-Planctomyces* morphotypes. a, morphotype I rosette with buds; b, morphotype I rosette; shadow cast (after KRISTIANSEN). (From SCHMIDT & STARR 1981)

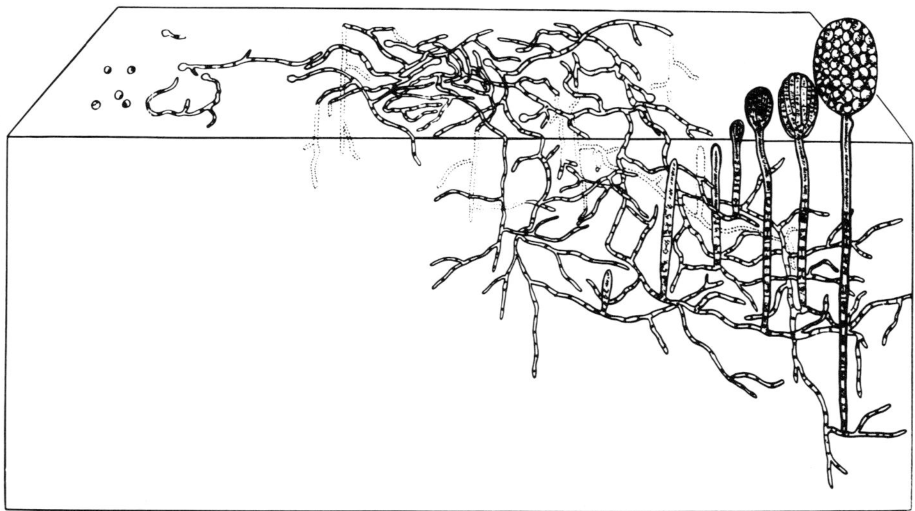

Fig. 211. Growth forms and life cycle of a typical actinoplanate. (From BLAND & COUCH 1981)

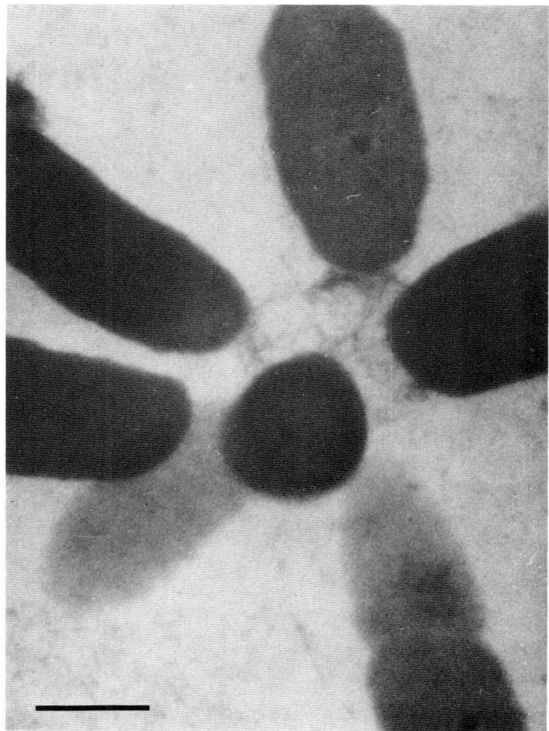

Fig. 212. "Star"-formation mediated by polar fimbriae in *Rhizobium* spec. Bar: 0.5 µm. (Original micrograph F. MAYER)

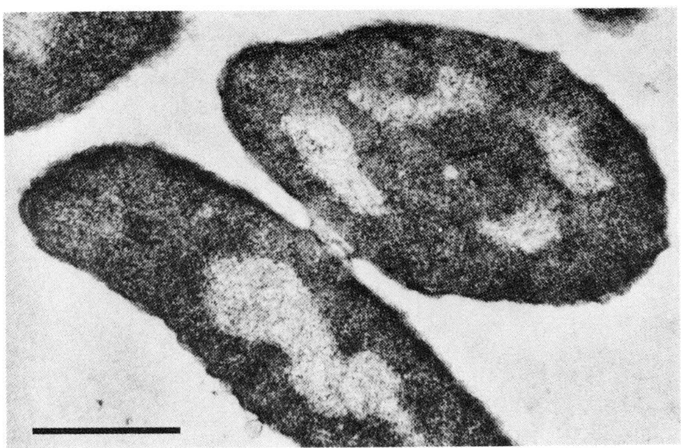

Fig. 213. *Escherichia coli* in contact during conjugation. Bar: 0.5 µm. (From STRYER 1981, after CARO)

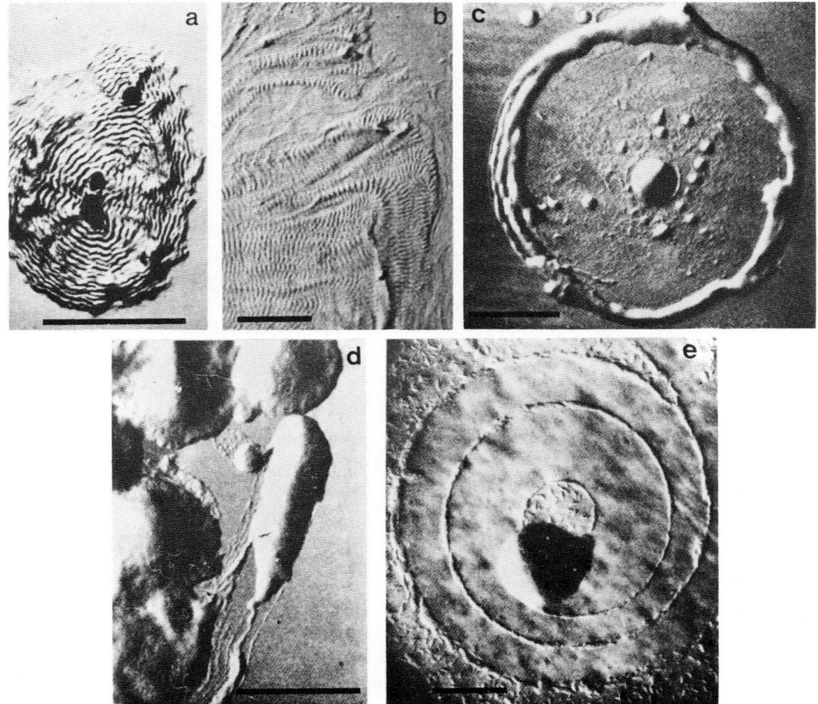

Fig. 214. Gliding bacteria. a, b, migrating waves; a, *Archangium* (Myxobacterales); b, *Stigmatella erecta* (Myxobacterales). Bars: 1 mm in a, 0.1 mm in b. (From REICHENBACH & DWORKIN 1981); c, d, e, migrating *Bacillus circulans;* bars: 1 mm c, ring colony; d, elongated colony; e, final part of migration has occurred in a spiral. (From GILLERT, and INST. WISS. FILM 1982)

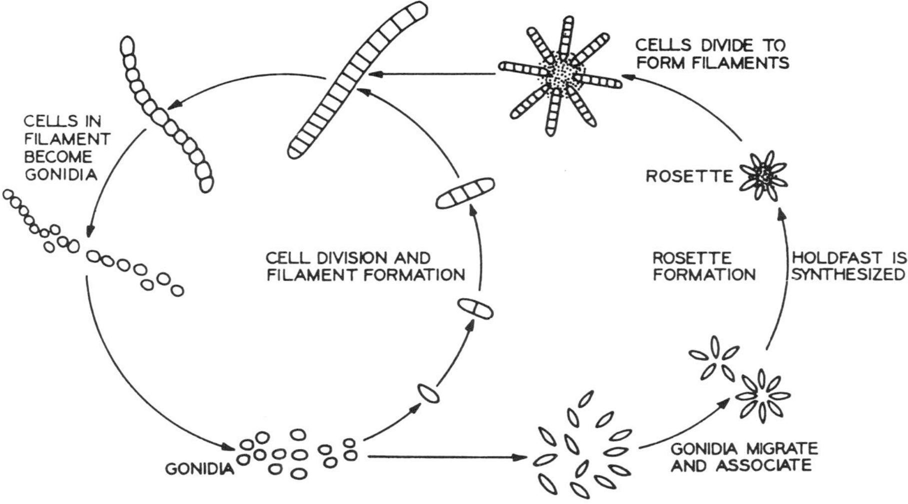

Fig. 215. Life cycle of *Leucothrix mucor*, a marine gliding bacterium. (From Brock 1981)

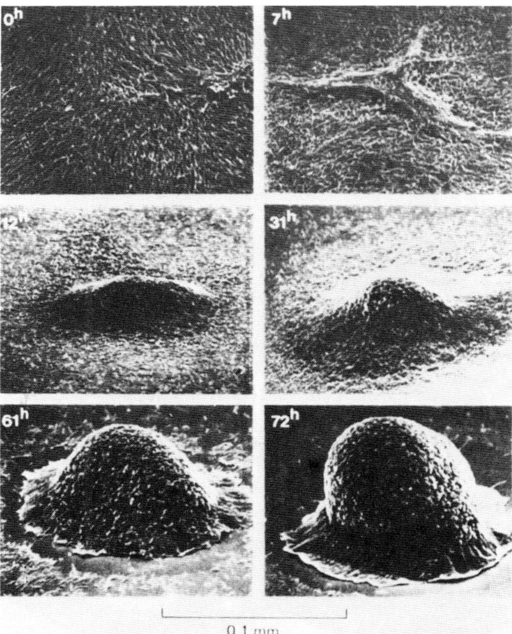

Fig. 216. Swarm of myxobacteria aggregating and forming a fruiting body in response to starvation. (From Alberts et al. 1983, after Kuner)

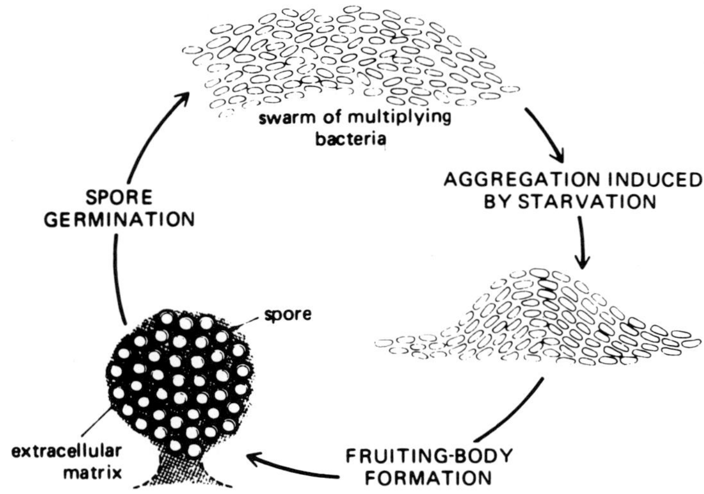

Fig. 217. The life cycle of myxobacteria. Starvation causes the cells to aggregate and differentiate to form a fruiting body. The spores in the fruiting body germinate when conditions become favorable again. (From ALBERTS et al. 1983)

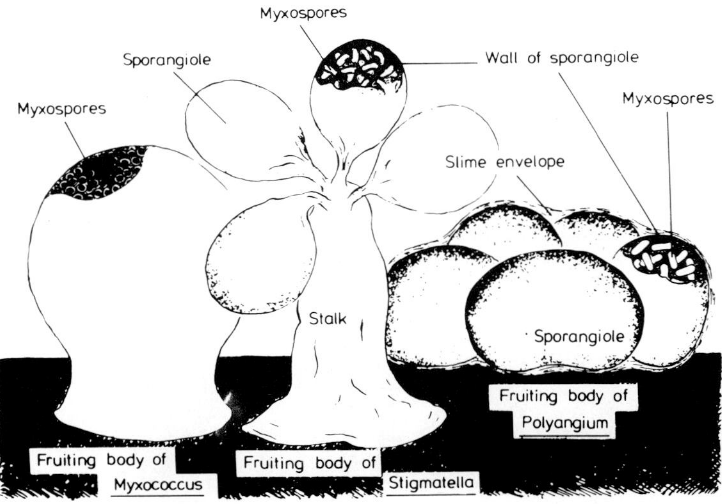

Fig. 218. The proposed terminology for myxobacterial fruiting bodies and their structural components. (From REICHENBACH & DWORKIN 1981)

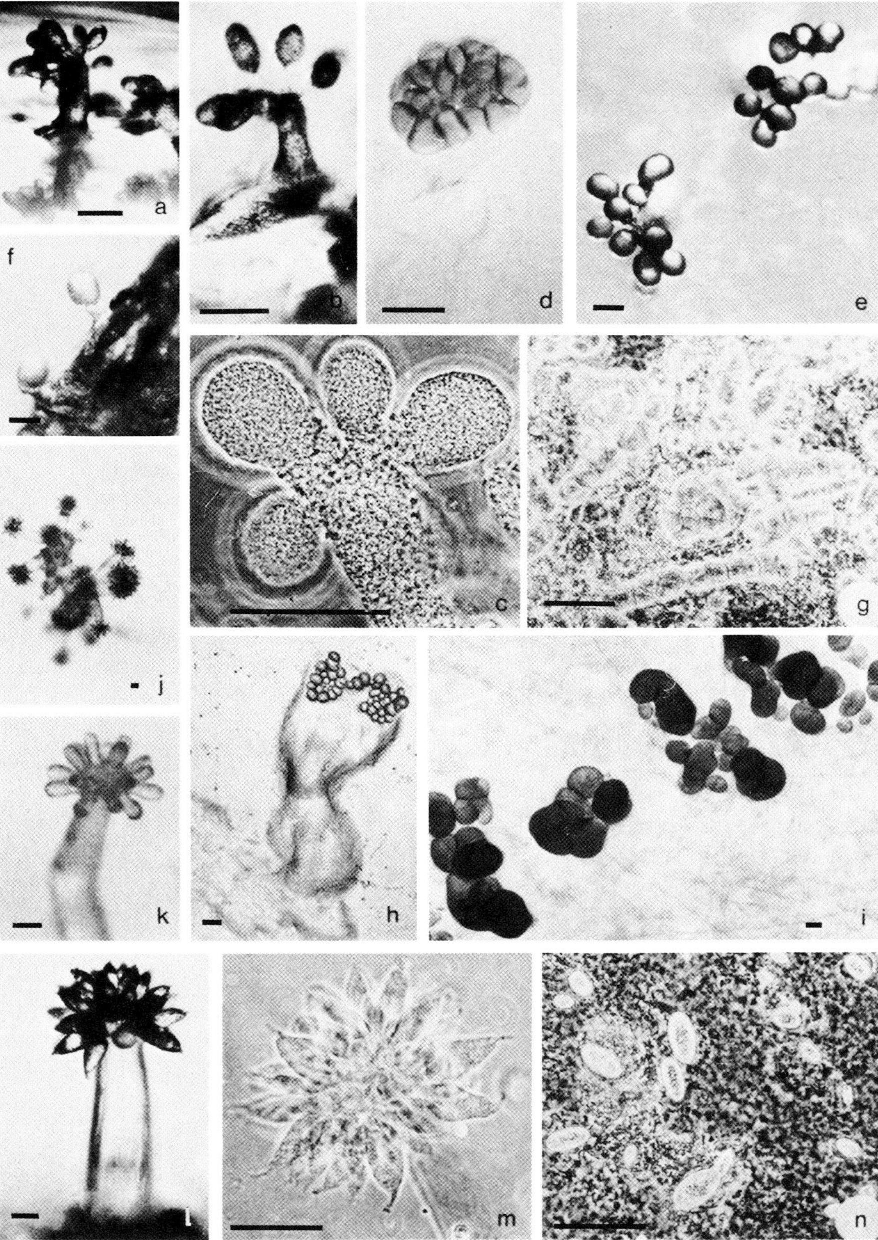

Fig. 219. Myxobacterial fruiting bodies. a–c, *Stigmatella aurantiaca;* d, *Cystobacter;* e, *Stigmatella erecta;* f, *Melittangium lichenicola;* g, *Sporangium cellulosum;* h, *Polyangium* sp.; i, *P. thaxteri;* j–m, *Chondromyces apiculatus;* n, *Nannocystis exedens.* Bars: 50 μm. (From REICHENBACH & DWORKIN 1981)

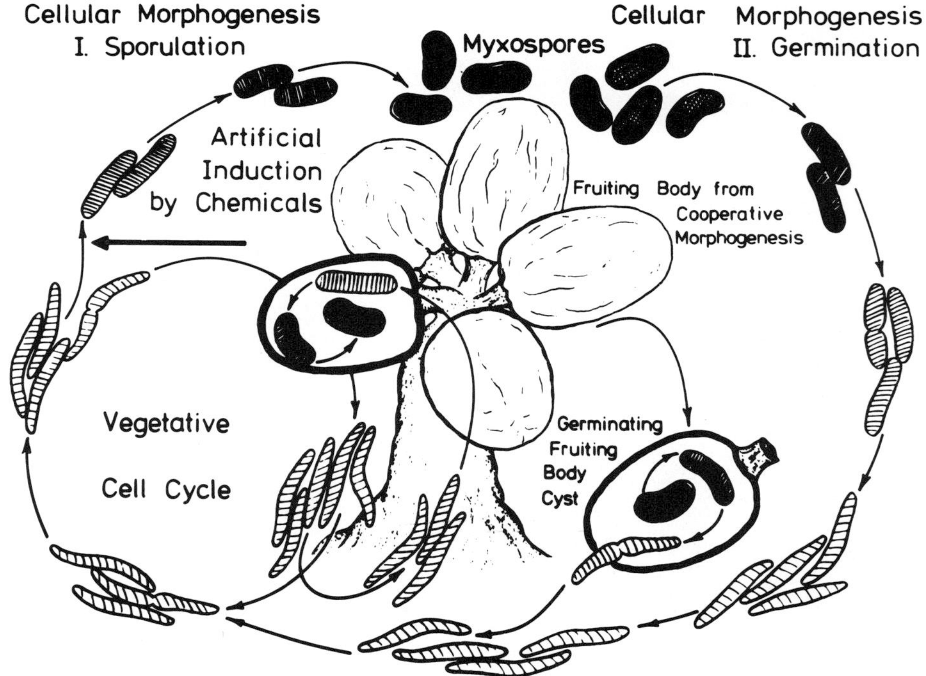

Fig. 220. Cellular morphogenesis in myxobacteria. Fruiting body and cells are not drawn on scale. (From Reichenbach & Dworkin 1981, after Gerth & Reichenbach)

strain or species, have been observed in a wide variety of cases (e. g. Gillert and Inst. Wiss. Film 1982; Murray & Elder 1949, von Muto 1904, Shinn 1940, Shukla & Iyer 1961). Examples are the formation of large cell aggregations made up of single cells (Figs. 206–211), sometimes mediated by exopolysaccharides or other extracellular compounds or structures, the "star" formation (Fig. 212) observed in soil bacteria (Heumann 1968, Mayer et al. 1974), brought about by the action of fimbriae, cell conjugation (Fig. 213) initiated by the action of F-pili (Achtman 1975, Achtman et al. 1977, Achtman & Skurray 1977, Lederberg 1956, Manning & Achtman 1979, Manning et al. 1980, Ou & Anderson 1970, Stallions & Curtiss 1972, Willetts & Skurray 1980), and the formation of fruiting bodies found in myxobacteria (Figs. 214–221) (Brockmann & Todd 1974, Kühlwein 1969, MacRae & McCurdy 1975, Peterson 1969; Rosenberg 1984, Varon et al. 1984, Voelz & Reichenbach 1969).

9.2 Heterologous associations (symbiotic bacteria, parasites)

Heterologous associations, i. e. associations between different species, occur either betwen bacteria of different species, or between bacteria and higher organisms. Symbiotic and parasitic associations involving bacteria are included in this type of interaction.

In all these cases, the initial step finally leads to a stable of at least temporarily stable association and is mediated by properties involving the surfaces of both partners (Bryant et al. 1967; Cheng et

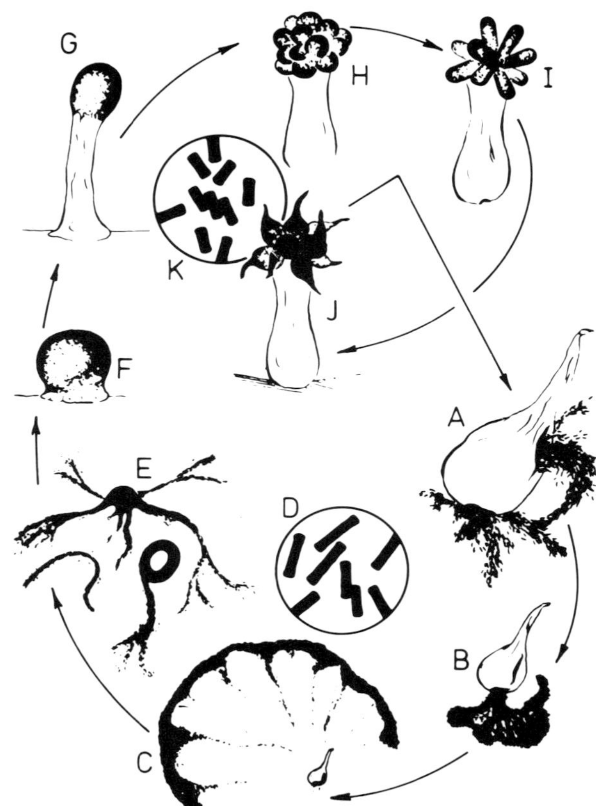

Fig. 221. Cooperative morphogenesis in myxobacteria (*Chondromyces apiculatus*). A, germinating sporangiole; B, C, development of swarm colony; D, vegetative cells; E, aggregation; F–J, fruiting body development; K, myxospores (From Reichenbach & Dworkin 1981)

al. 1981 a, b, Jones 1977, Jones & Isaacson 1983, Lennette et al. 1974, McCowan et al. 1980, Patterson & Hafeez 1976, Rosenberg et al. 1983, Scotland et al. 1983, Skerman et al. 1983, Smith 1977). In some circumstances the initial process of association is no longer evident as the associations remain stable or even "divide" (s. below, the "Tibi" grain). Some associations of this kind are less complex than others. An example for a relatively simple system is a granular consortium described by Bochem et al. (1982) (s. also above). It consists of a methanogenic bacterial "tissue" colonized by other bacteria, which grows at 60°C. The system converts acetic acid to methane, a process which appears to be very important for methanogenesis both in nature and in digesters. Thin sections of the consortium showed three morphologically distinct layers. The outer zone had a resemblance to a pseudoparenchyma and consisted of "macrocysts" and coccoid cells, similar to *Methanosarcina* cells. The inner zone was built up of loosely packed ovoid cells. The third zone showed internal cavities containing thermophilic rod-shaped bacteria. The cavities were the exclusive site for gas formation. However, the non-methanogenic nature of the rods was substantiated by pure culture isolations.

Another example is the well-known ectosymbiotic association of regularly arranged *Chlorobium* cells around a motile, colorless central bacterium. *Chlorochromatium consortium* is the green form of the association; in a brown variation of this type of association, *Pelochromatium consortium*

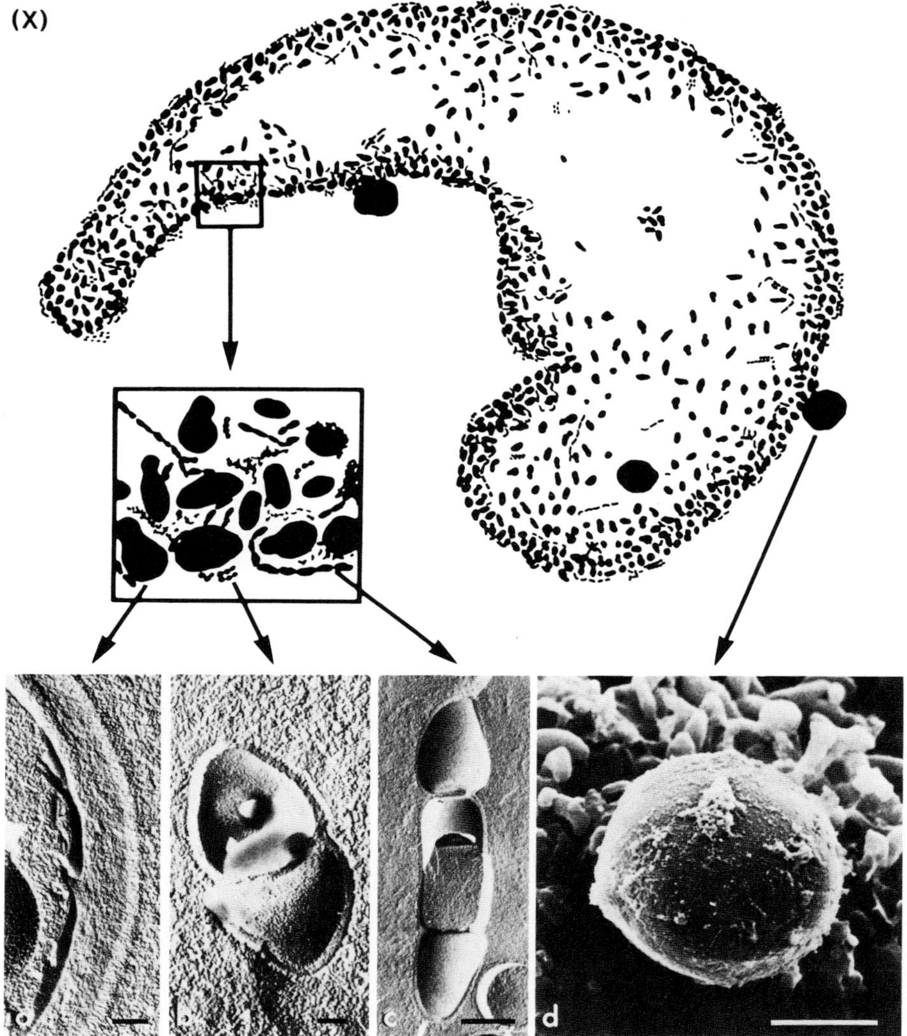

Fig. 222. General structure of the "Tibi" grain (X). Micrographs of freeze-etched *Saccharomyces cerevisiae* (a), *Streptococcus lactis* (b), and *Lactobacillus brevis* (c); d, scanning electron micrograph of a rough element of unknown origin. Bars: 0.2 μm in a, 0.1 μm in b, 0.5 μm in c, and 10 μm in d. (From Moinas et al. 1980)

replaces *C. consortium* (Lauterborn 1915, Pfennig & Trueper 1974). The members of the consortium were actually described as using their specific association in order to mutually benefit from specific physiological properties of the partner as is defined to be the case in symbiosis. Such a situation is the usual one. However, strong evidence does not always exist for it. An example where the mutual benefit does not appear to be obvious for the host (a plant) is the "Tibi" grain (Fig. 222). These grains originate from Mexico where they occur in the form of hard granules on the disc-

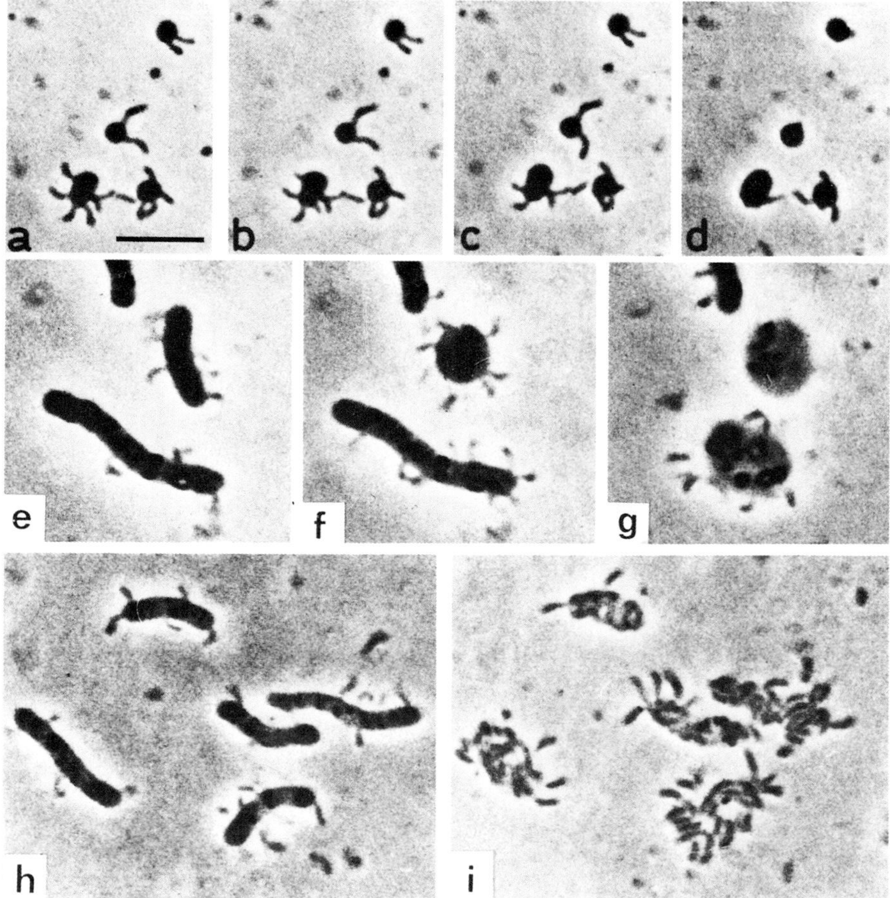

Fig. 223. *Bdellovibrio bacteriovorus.* a–d, penetration into *Salmonella typhimurium;* sequence of events within 22 min; e–g, spheroplast formation; sequence of events within about 60 min; h, i, release of *Bdellovibrio* from the lysed *Salmonella* cells; h, immediately after attachment of the parasite to the *Salmonella* cells; i, final stage 150 min later. Bar: 5 µm (same magnification in a–i). (From STOLP 1968)

shaped leaves of the *Opuntia* plant (DAKER & STACEY 1938, LUTZ 1899, WARD 1892). "Tibi" grains are used for the preparation of a drink by fermentation of sucrose solutions. The grains have been described as symbiotic associations of *Lactobacillus brevis, Streptococcus lactis* and *Saccharomyces cerevisiae* (HORISBERGER 1969). Fermentation and transfer of the grains can be carried out under non-sterile conditions without contamination. The grains are translucent and were found to contain a matrix of dextran consisting of a backbone of α-D (1 → 6)-linked glucopyranosyl residues with (1 → 3)-linked side chains (HORISBERGER 1969). MOINAS et al. (1980) presented some aspects of the structural organization of this association as revealed by light, scanning, and transmission electron microscopy. They observed that, in the presence of the growth medium, the grains increased in size and then divided by fission. The fission was determined by the inner pressure of

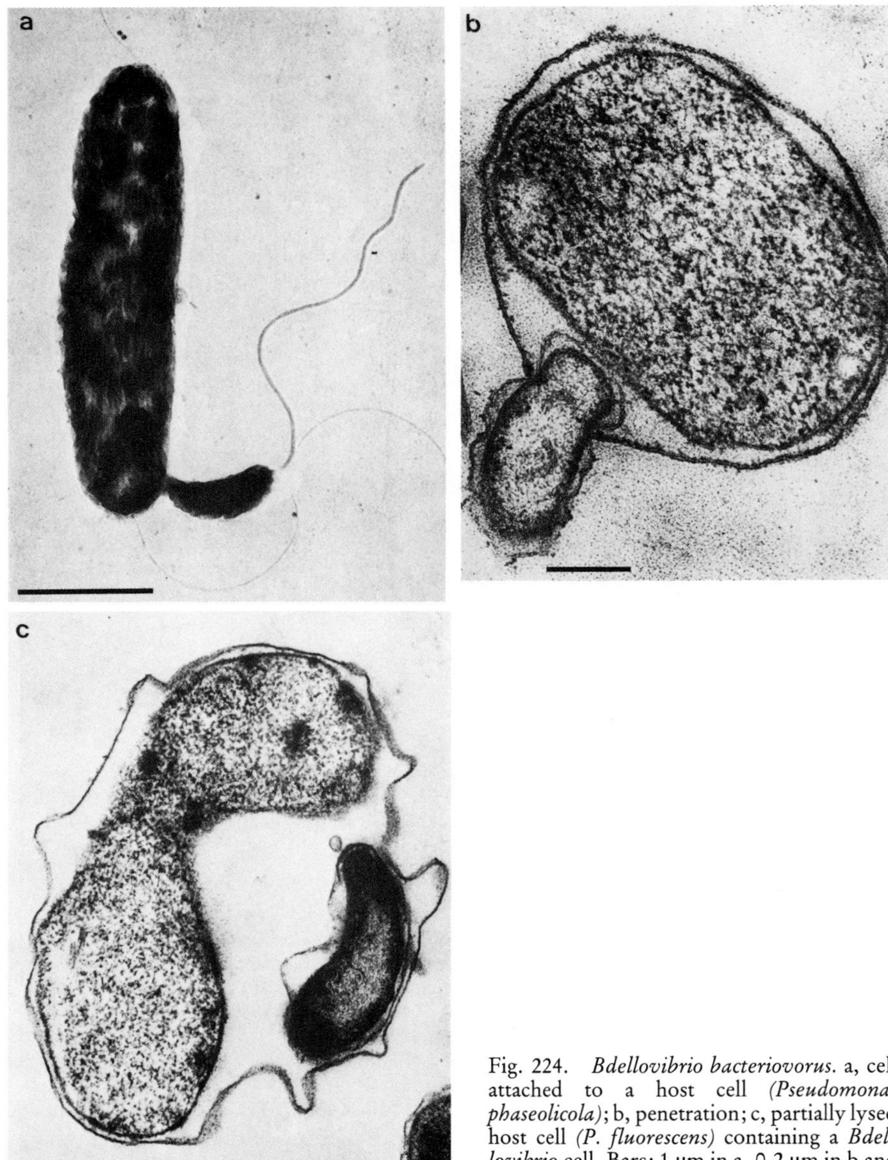

Fig. 224. *Bdellovibrio bacteriovorus.* a, cell attached to a host cell *(Pseudomonas phaseolicola)*; b, penetration; c, partially lysed host cell *(P. fluorescens)* containing a *Bdellovibrio* cell. Bars: 1 µm in a, 0.2 µm in b and c. (From STOLP 1968)

CO_2. Due to CO_2 formation during fermentation the grain was hollow. At high magnification the outer layer of the grain was shown to contain streptococci, polymorph lactobacilli, yeasts, and the dextran generated by the lactobacilli. The streptococci were embedded in the dextran. In the inner layer the lactobacilli were much shorter, and the outer layer was much more densely populated by microorganisms than the inner layer. The outer layer contained less dextran than the inner layer as

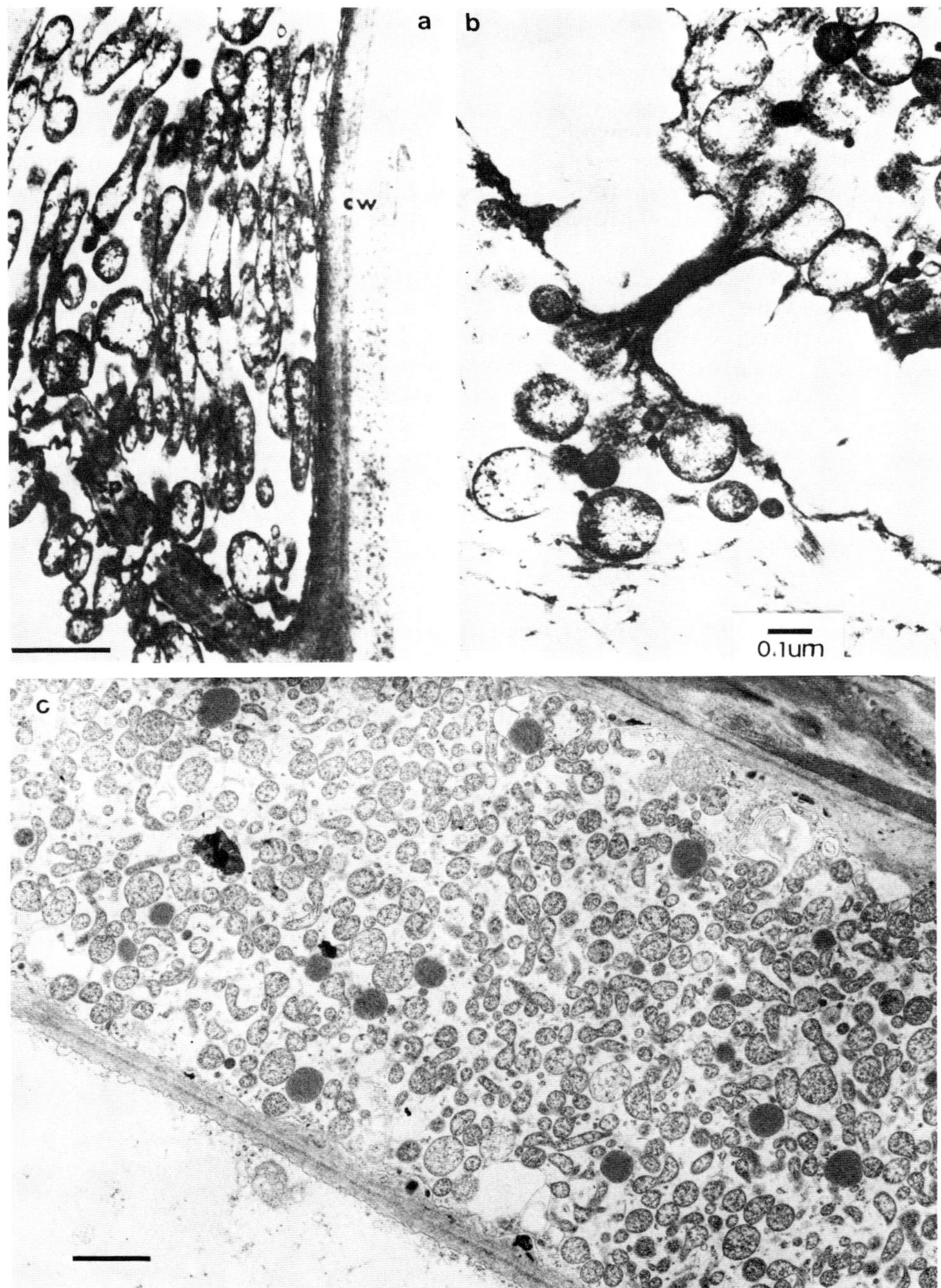

Fig. 225. a, mycoplasma-like organisms in *Zea mais*. cw = cell wall; Bar: 1 µm; b, passage of mycoplasma-like organisms from one cell of clover to the neighboring cell. (From GRUNEWALDT-STOECKER & NIENHAUS 1977; a, after GRANADOS; b, after HORNE); c, phloem-limited bacteria (mycoplasma-like) in *Catharanthus roseus*, infected with aster yellows (isolate F. NIENHAUS, original electron micrograph F. EBRAHIM-NESBAT). Bar: 1 µm.

revealed by the application of fluorescein-conjugated con A, a lectin which specifically combines with a variety of branched polysaccharides.

In contrast to the associations described so far, *Bdellovibrio bacteriovorus* is a parasitic bacterium (Figs. 223 and 224). The cell is relatively small. It possesses a flagellum enabling it to reach a high speed as compared to other flagellated bacteria. STOLP (1968) described a cycle of events finally leading to the death of the host bacterium. Various strains of *Bdellovibrio* were found to attack pseudomonads, enterobacteria and other Gram-negative bacteria. In contrast to bacteriophages which are only propagated in growing bacteria, *Bdellovibrio* also utilizes non-growing bacterial cells. The wild types are obligate parasites. However, some mutants have been found to be saprophytes, growing on complex media outside of the host.

Effects of the action of parasitic bacteria on other microorganisms should not be mixed up with killing due to lysis from outside as brought about by certain bacteria. STEWART & BROWN (1969), in seeking algal viruses, found a non-filtrable agent which killed or lysed living green algae and cyanobacteria. It was distinguished from viruses by its wide host range and its ability to digest autoclaved algae. It turned out to be a Gram-negative, unicellular, rod-like bacterium without flagella, forming slimy colonies. Motility was by gliding in a flexous manner, and it did not produce fruiting bodies. The bacterium was identified as a member of the order Myxobacterales and it was named *Cytophaga N-5*. Lysis of algae and cyanobacteria was caused by digestion of the cell walls from outside.

When associations of parasitic bacteria with higher organisms are described (NIENHAUS 1976, TULLY & WHITCOMB 1981), mycoplasmas (KAHANE et al. 1982, NIENHAUS & SIKORA 1979), rickettsias (HASE 1983, WEISS 1981), mycoplasma-like organisms *(MLO)* and rickettsia-like organisms *(RLO)* (NIENHAUS 1979, NIENHAUS et al. 1976, PARAMESWARAN 1983, PETZOLD et al. 1973) have to be mentioned besides others (Figs. 225–232). Mycoplasmas, belonging to the family Mycoplasmataceae or Acholeplasmataceae of the order Mycoplasmatales, and rickettsias are the causative agents for a variety of animal diseases. Rickettsias (Fig. 226) are of the Gram-negative cell wall type, which they share with chlamydias, the latter being known to cause severe diseases in man. In contrast, *MLO*s (Fig. 225) and *RLO*s, which are plant disease agents, were found not to be

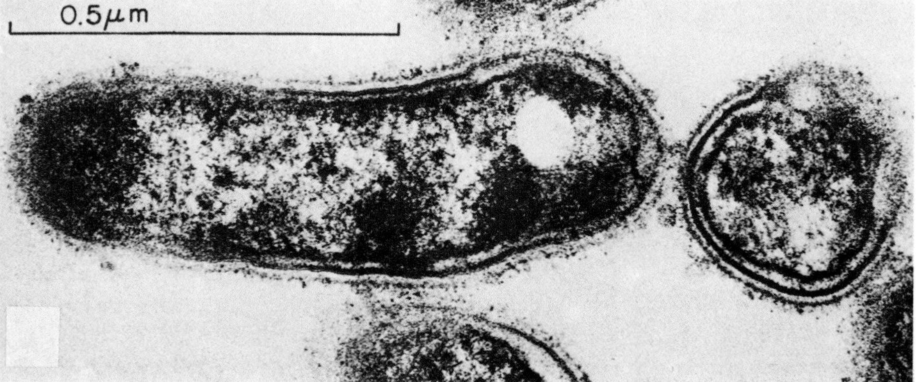

Fig. 226. Rickettsiae released from cultured chicken fibroblast cells in the presence of specific anti-rickettsial immune serum; *Wilmington* strain of *Rickettsia typhi.* (From WEISS 1981, after MCCAUL et al.)

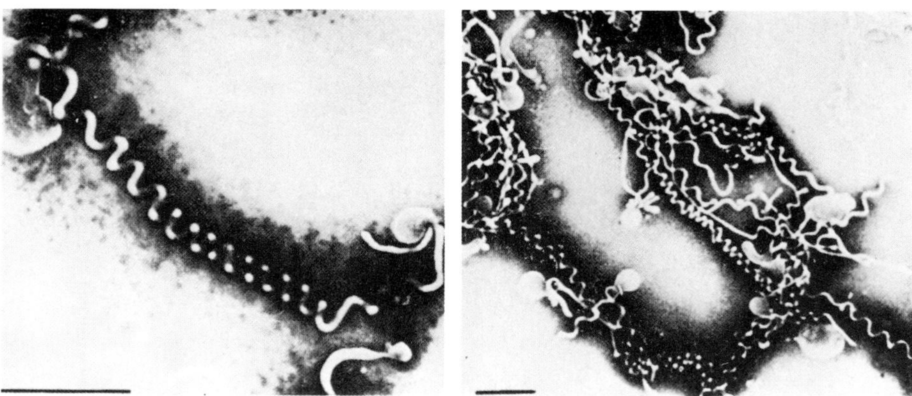

Fig. 227. Filaments and "blebs" formed by *Spiroplasma citri*. Bars: 1 µm. (From GRUNEWALDT-STOECKER & NIENHAUS 1977, after COLE et al.)

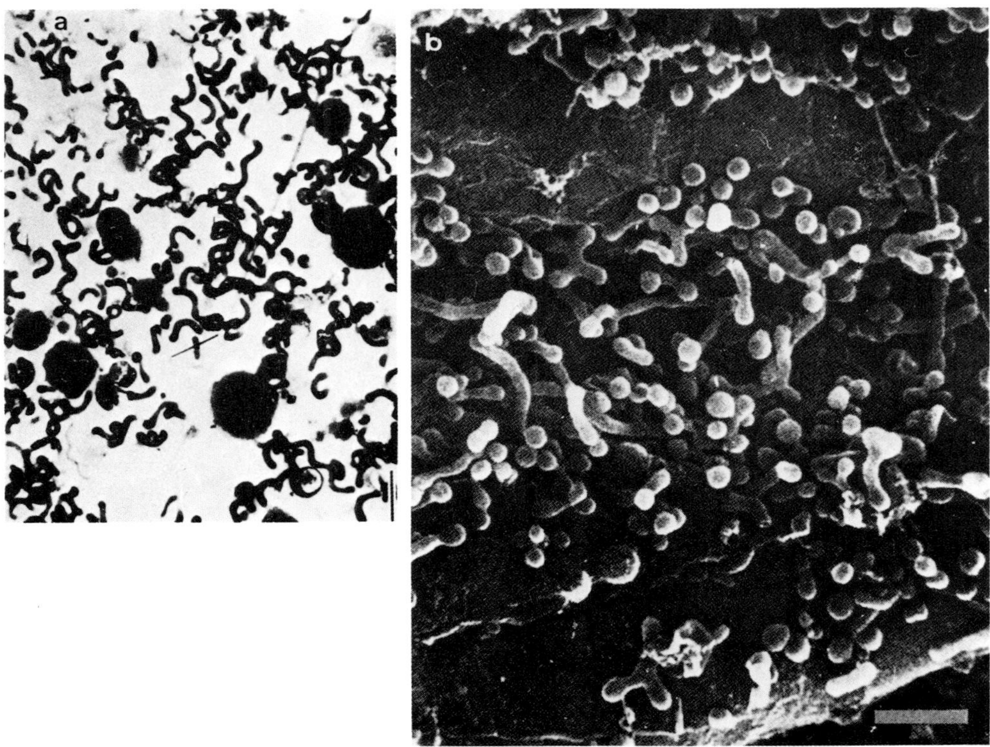

Fig. 228. a, *Spiroplasma citri*, two days old. Thin section 1 µm thick. Bar: 1 µm. (From GRUNEWALDT-STOECKER & NIENHAUS 1977, after COLE et al.); b, mycoplasma-like bodies in sieve element of clover phyllody-diseased aster. Filamentous and branching forms. Bar: 1 µm. (From MCCOY, after HAGGIS & SINHA)

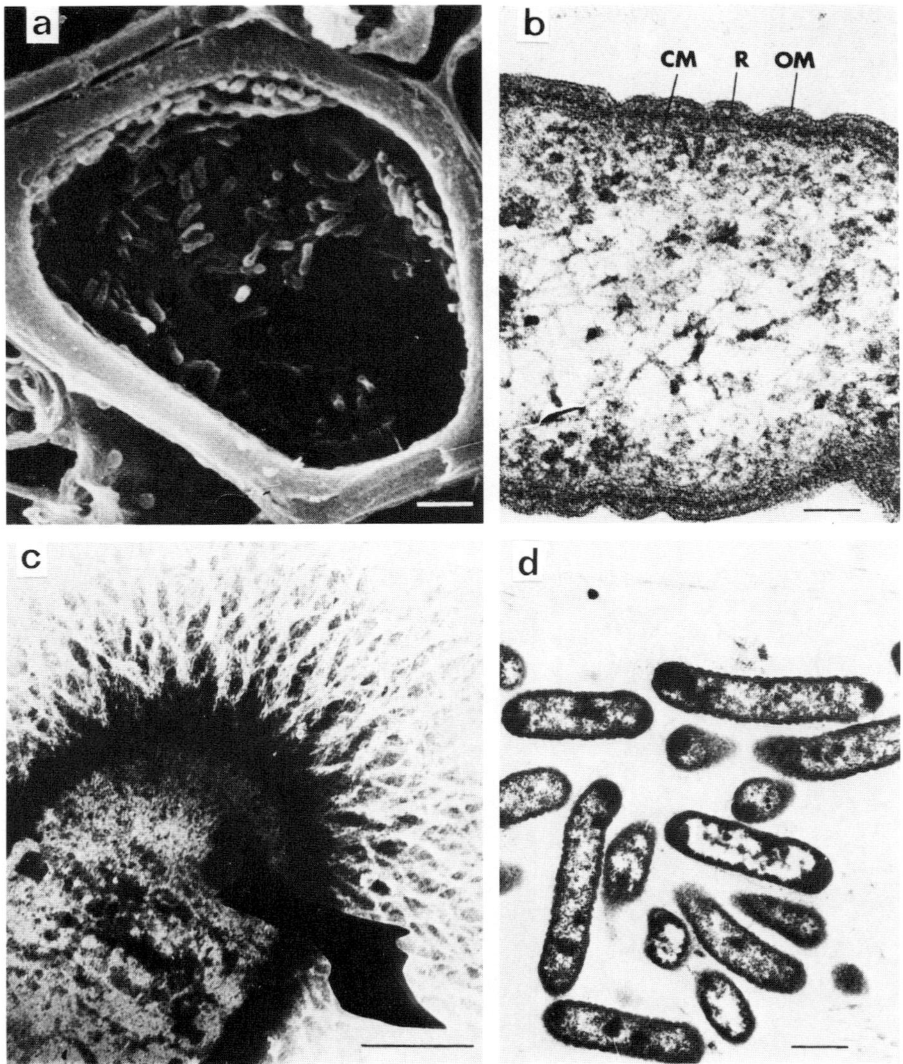

Fig. 229. Xylem-limited bacteria. a, tracheary element of grapevine with Pierce's disease, showing the causal bacterium in the lumen (bar: 2.5 µm; after GARROTT & DAVIS); b, Pierce's disease bacterium: – CM = cytoplasmic membrane; R = R-layer; OM = outer membrane (bar: 50 nm; after MOLLENHAUER & HOPKINS); c, bacterium associated with plum leaf scald disease showing fibrillar structures (bar: 0.2 µm; after FRENCH & KITAJIMA); d, bacteria associated with elm leaf scorch disease (bar: 0.5 µm; after SHERALD & HEARON). (From DAVIS et al. 1981); e, xylem-limited bacteria (rickettsia-like) in the tracheides of Norway spruce (*Picea abies* Karst.). Staining with methylen blue (light micrograph). Bar: 10 µm. (Original micrograph F. EBRAHIM-NESBAT); f, as e, but electron micrograph from an ultrathin section. Bar: 1 µm. (From EBRAHIM-NESBAT & HEITEFUSS 1985)

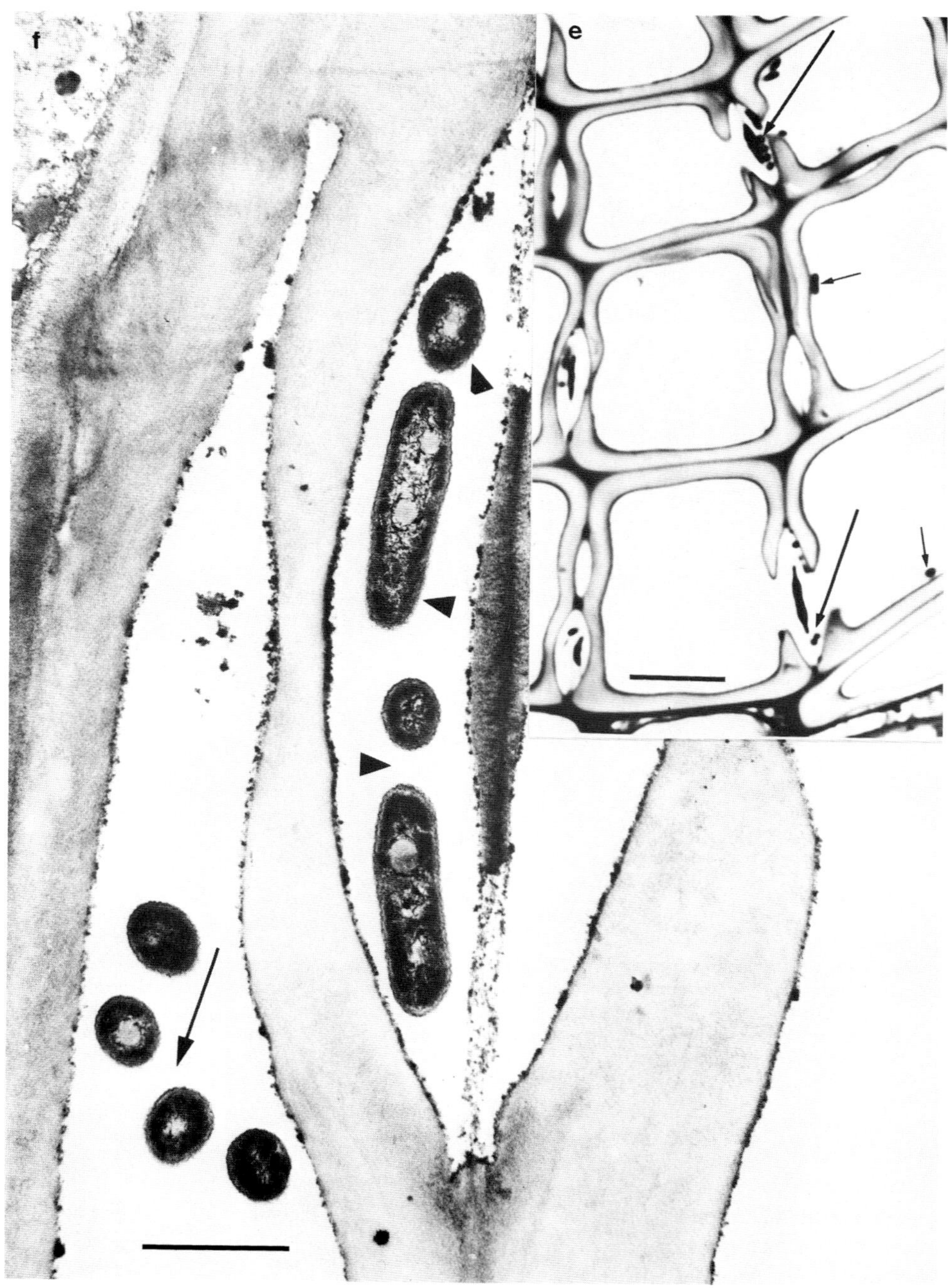
f
e

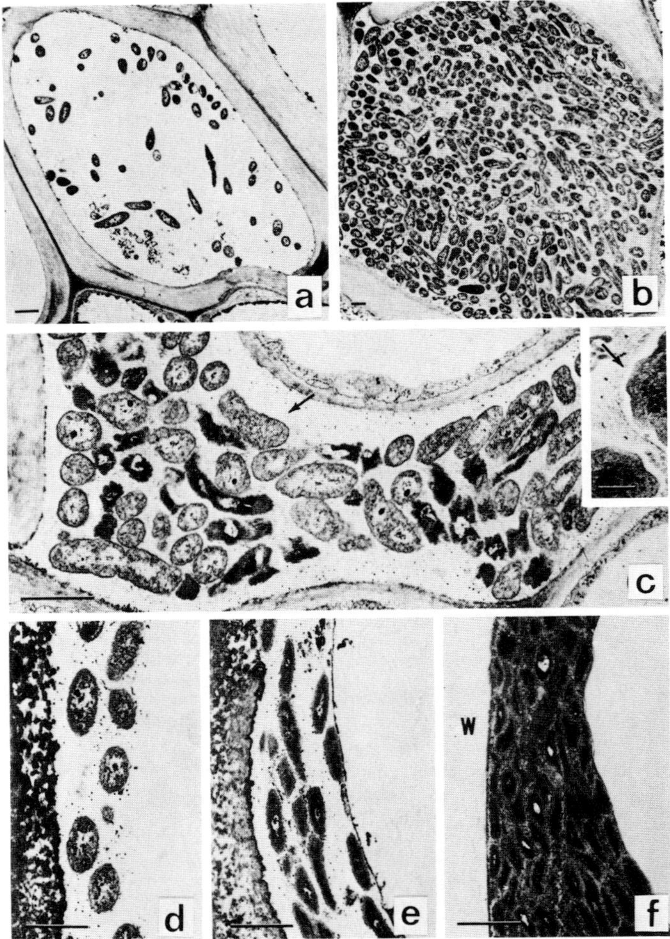

Fig. 230. a–f, ultrastructural study of Pierce's disease bacterium in grape xylem tissue, transverse sections through leaf veins showing common distribution patterns of intraxylem bacteria. Secreted products act as encapsulating substances. A large range of bacterial development is present only when the bacteria are distributed as shown in a, b, and c, i. e. when they are not attached to the xylem walls. Bars: 0.1 µm. (From Mollenhauer & Hopkins 1974)

members of the two families mentioned above (Razin 1973, 1978). However, they are very similar to them. The terms *MLO* and *RLO* are not satisfactory as they are reminiscent of the long-abandoned term "pleuropneumonia-like" organisms *(PPLO)*, which are now classified as mycoplasmas (Maramorosch 1979) and found to possibly have evolved from ancestral clostridia. *MLO*s and *RLO*s, together with plant pathogenic spiroplasmas (Saglio et al. 1973), though differing taxonomically (*MLO*s are wall-deficient, *RLO*s are walled), have certain features in common. Most seem to require insect vectors for their transmission to plants (Maramorosch 1963, 1969). The same species of vector can sometimes act as a transmitter of *MLO*s and virus, and

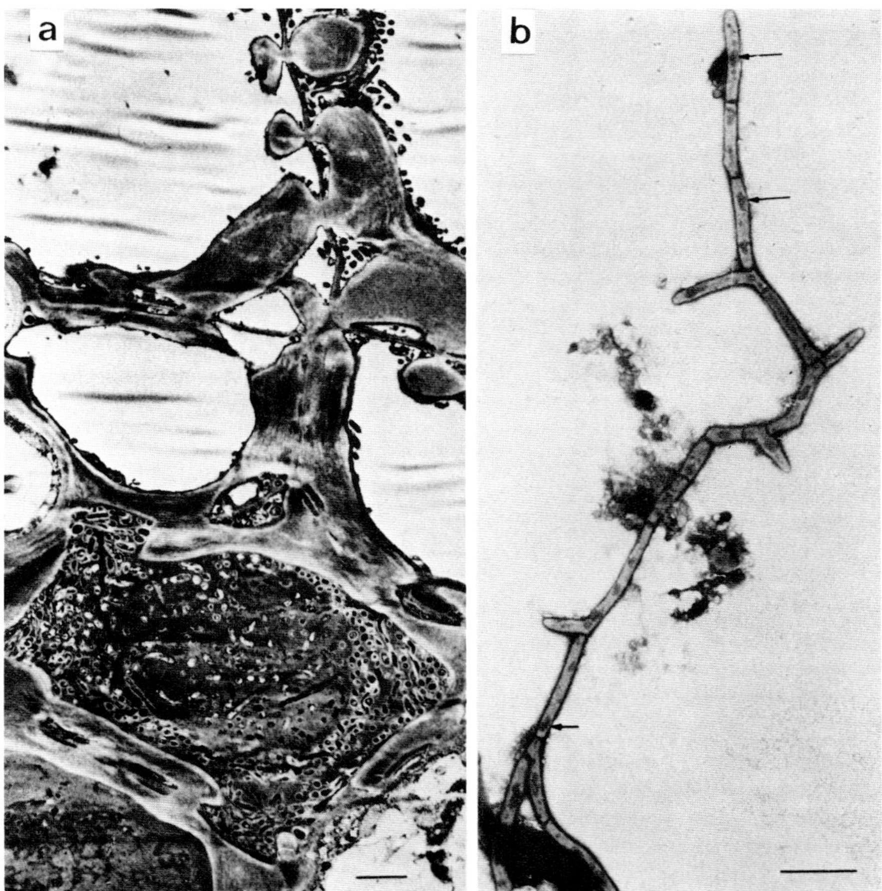

Fig. 231. Fastidious bacteria in plant vascular tissue. a, ratoon stunt-associated bacterium in thin section of Sudan grass xylem cells. Bacteria occur within the pit fields, near residual primary walls and, in some cells, within a matrix material (bar: 5 μm; after WORLEY); b, chain of 16 ratoon stunt-associated bacterial cells showing pseudobranching and "mesosomes" (arrows); septa between the cells (bar: 1 μm; after TEAKLE). (From DAVIS et al. 1981)

up to 1967 a large group of plant diseases was believed to be caused by viruses (MARAMOROSCH 1979). Typical diseases caused by *MLO*s and *RLO*s are yellows diseases. The knowledge of the mechanism of action of these agents is limited. DANIELS & MEDDINS (1974) reported that one, or perhaps two, toxins are produced by cultured *Spiroplasma citri* (Figs. 227 and 228), and that they cause cell injury in leaves including severe disturbance in the photosynthetic functions, growth, and hormonal changes. This might be due to upsetting the balance of the plant's metabolism by the fact that the invading spiroplasmas use cholesterol for their membranes. Cholesterol is also incorporated, as a normal constituent, into the plant membranes. The spiroplasmas also interfere with the disposition of lipids (RAZIN 1978).

Well known and studied in detail is the *Rhizobium*-legume symbiosis (ABE et al. 1984, ANDREEVA 1984, DAZZO 1985, GATNER & GARDNER 1970). *Rhizobium* is an obligate aerobic

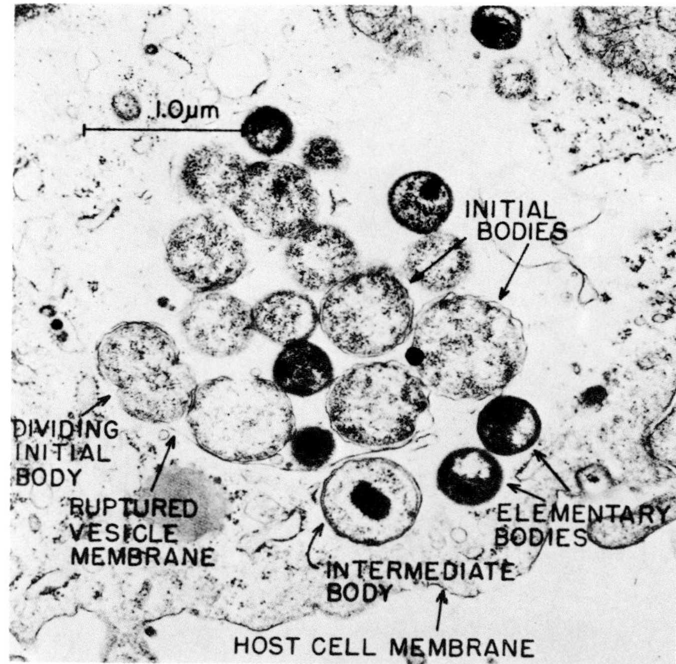

Fig. 232. Obligately intracellular bacteria; microcolony of *Chlamydia psittaci.* (From PAGE 1981, after CUTLIP)

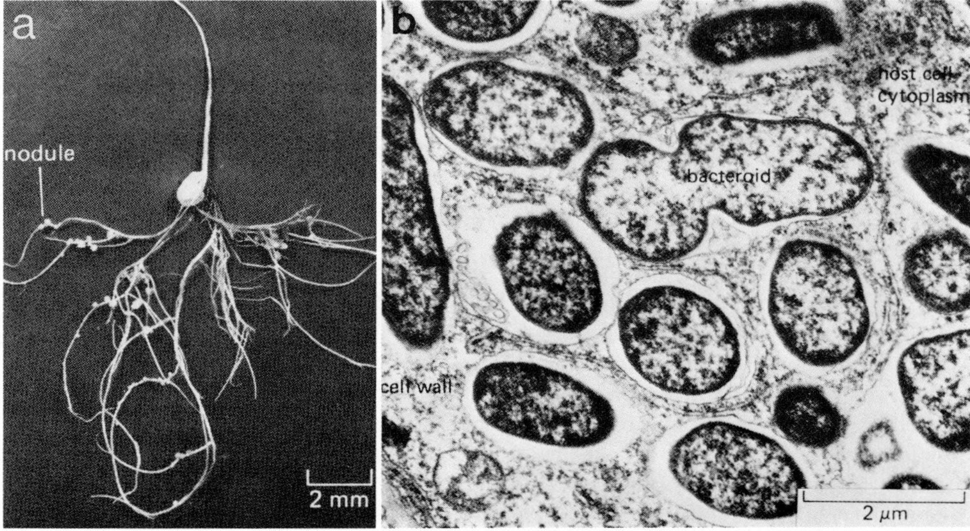

Fig. 233. a, a young pea plant in symbiotic association with the nitrogen-fixing bacterium *Rhizobium;* b, sectioned pea root nodule; the nitrogen-fixing *Rhizobium* bacteroids, surrounded by host cell-derived membranes, fill the host cell cytoplasm. (From ALBERTS et al. 1983; a, after JOHNSTON; b, after HUANG & MA)

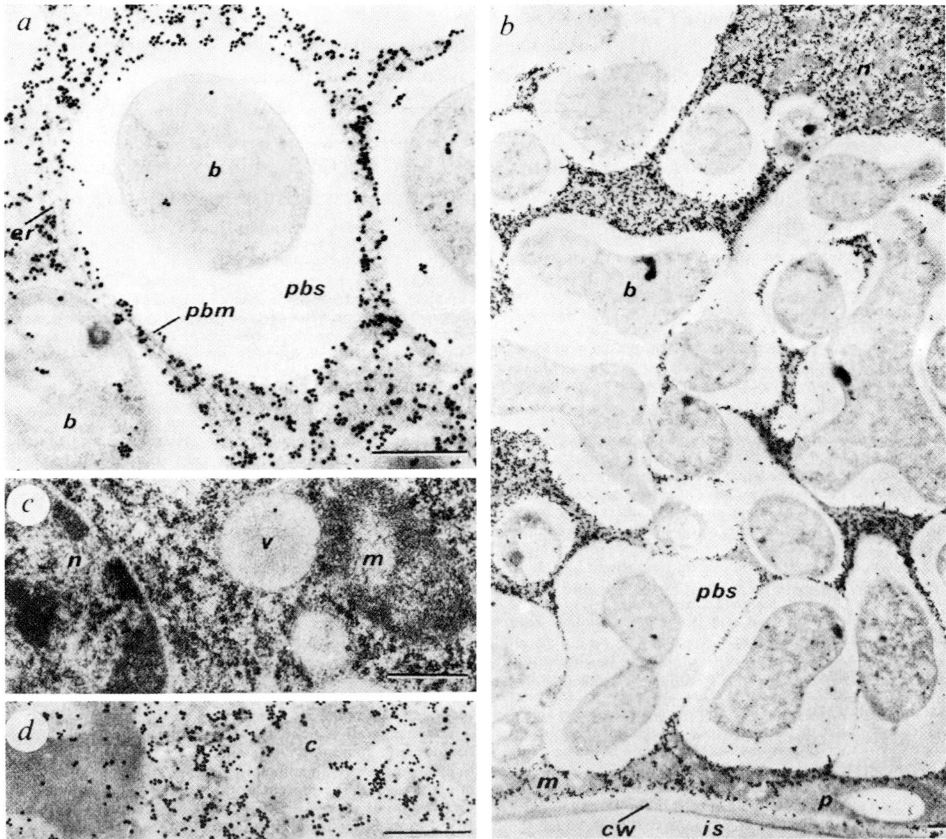

Fig. 234. Immuno-gold localization of leghaemoglobin in cytoplasm in nitrogen-fixing root nodules in pea. a, b, label in the cytoplasm; c, nucleus, vacuoles and mitochondria not labelled; d, chromatin not labelled. – b = bacteroid; c = chromatin; cw = cell wall; er = endoplasmic reticulum; is = intercellular space; m = mitochondria; n = nucleus; p = plastids; pbm = peribacteroid membrane; pbs = peribacteroid space; v = vacuole. Bars: 5 µm. (From ROBERTSON et al. 1984)

Gram-negative soil bacterium. It recognizes root hairs of legumes, infects them and induces the formation of root nodules. Usually a given strain of *Rhizobium* only infects a limited number of legume species (Fig. 233). The bacteria enter the root hairs and are usually found in structural differentiations called infection threads consisting of plant cell material. The threads branch and penetrate through cells within the cortex of the root. At the same time plant cell divisions in the inner root cortex occur. Finally, a root nodule is formed. Within the nodule the bacteria differentiate into bacteroids, and the bacterial genes involved in nitrogen fixation are expressed. A system involving leghaemoglobin (DAVENPORT 1960, ROBERTSON et al. 1984) is used by the partners for the regulation of the partial pressure of oxygen (Fig. 234). The plant takes care of the nutrient requirements of the symbionts. In turn, it receives fixed nitrogen from the bacteria. Nitrogen fixation is the enzymatic reduction of atmospheric dinitrogen to ammonia by the enzyme system nitrogenase. Recently, there has been increasing interest in the molecular genetics of *Rhizobium*

(Van den Bos et al. 1983, Pühler 1983, Simon et al. 1983; Werner et al. 1983; additional references can be found therein). From conjugation experiments it has been concluded that a megaplasmid harboured by *Rhizobium meliloti* plays an essential role in nodulation and symbiotic nitrogen fixation (summarized in Simon et al. 1983).

The formation of crown galls (or its stimulation; Silvere 1983) is the consequence of yet another type of interaction of a bacterium with a plant. Schell et al. (1983) (s. additional references therein) summarized the situation, described as "genetic colonization as an infectous mechanism", as follows: "Large plasmids in *Agrobacterium tumefaciens* (Ti) and *A. rhizogenes* (Ri) are known to endow these bacteria with the capacity to transfer a defined DNA fragment (T-DNA) into the plant cell nucleus and to covalently integrate this T-DNA segment in chromosomal DNA, thus creating a new locus at a number of possible sites. The mechanism of DNA transfer is still poorly understood. Genetic evidence indicates that a relatively large segment including the so-called *vir* region and the T-DNA region of Ti-plasmids are transferred. Under normal circumstances only the T-region is inserted in the chromosomal DNA and thus stably maintained. No functions located within the T-region are required for either transfer or integration. Large foreign DNA sequences of 50 kb or more, experimentally inserted within the T-region, are efficiently transferred and integrated in the plant chromosomal DNA. The plasmid-derived T-DNA was shown to consist of a number of well defined transcriptional units transcribed by the host polymerase II and coding for a number of different functions, e.g. enzymes involved in opine synthesis and functions involved in the inhibition of plant differentiation. Thus separate shoot suppressing and root suppressing functions have been identified. Removal of these tumor controlling genes does not affect DNA transfer or integration. Thus it was possible to design modified Ti-plasmids that can insert foreign genes in plant cells from which normal plants can be regenerated that express the foreign genes and transmit them sexually with normal mendelian segregation ratios. To be expressed in plants, the coding sequences of foreign genes have to be inserted behind plant promotor sequences. Several such constructed genes have been made and introduced in tabacco plants and their expression has been studied". These findings drastically illustrate the implications inherent in symbiosis between bacteria and higher organisms.

Intracellular bacteria in algae are known for a variety of algal genera belonging to the Chlorophyceae *(Volvox, Bryopsis, Penicillus)*, Raphidophyceae *(Gonyostomum)*, Tribophyceae *(Vaucheria)*, Chrysophyceae *(Chrysamoeba)*, and Dinophyceae *(Katodinium)* as summarized by Surek & Melkonian (1983; s. further references therein). According to ultrastructural investigations the bacteria are of the Gram-negative type. Usually the bacteria were not found to be enclosed within vacuolar host membranes. Two exceptions are the "intracellular" bacteria in *Penicillus capitatus* and in *Katodinium glandulum*. In some Euglenophyceae cytoplasmic as well as intranuclear bacteria have been found. Surek & Melkonian (1983) reported on the successful prolonged axenic culture of *Euglena mutabilis* with intracellular bacteria, and they provided further details on this presumptive symbiotic association. Further examples for interesting symbiotic associations in plants have been summarized by MacDonald (1983) and Nicolov et al. (1983).

Endocytobiosis of bacteria in ciliates has been investigated in a variety of associations, and the findings have been reviewed (s. Görtz 1983; further references therein). In most cases the bacteria were detected in the cytoplasm, with few exceptions where they were shown to be located in the nucleus (Görtz 1980, Görtz & Dieckmann 1980) of the host cell. In the cytoplasm they are often found in perisymbiotic vacuoles. While some bacteria obviously harm their host, others benefit them or may even be needed. The *omikron* particles of *Euplotes* are an example for an essential symbiont and have even been called "organelles" in order to emphasize their essentiallity. However, one should keep in mind the difference between cell organelle and an endocytobiont (NEUMÜLLER & SCHENK 1983): cell organelles are in metabolic, and additionally in an extensive or complete genetical dependency to the nucleocytoplasm. Contrary to it, and in spite of a more or less strong metabolic dependency the genetic autonomy of an organelle-like structure obviously refers to its endocytobiotic nature. One example for diverging views in this respect is the endocyanelle *Cyanocyta korschikoffiana* found in *Cyanophora paradoxa* (BOHNERT et al. 1983, NEUMÜLLER & SCHENK 1983, additional references therein), a colorless protozoon. The cyanelles function as photosynthetic entities which resemble free-living unicellular cyanobacteria (HERDMAN & STANIER 1977, KIES 1979, MARTENS et al. 1982). Like the latter the cyanelles possess unstacked thylakoids, phycobilisomes and – though rudimentary – a lysozyme-sensitive cell wall. The structural organization as well as size of their genome, however, suggest some relationship to chloroplasts (HEINHORST & SHIVELY 1983; additional references therein). Thus, HEINHORST & SHIVELY (1983) have addressed themselves the question of the cyanelles' evolutionary position by locating the genes for the large and the small subunit of RubisCO in the *Cyanophora paradoxa* system. The genes for both types of enzyme subunits were localized on the cyanelle genome, a finding supporting the view that the cyanelles are in fact endocytobionts, and not chloroplasts, which had found a balanced relation with their eukaryotic host. However, BOHNERT et al. (1983) proposed that cyanelles are a, probably ancient, form of plastids. They derived this view from studies on the size, segmental structure, and gene organization of the cyanelle DNA. Differences as compared to typical chloroplasts have been found supporting the impression gained from physiological and biochemical studies that cyanelles may be a "missing link" between cyanobacteria and plastids (GRAY & DOOLITTLE 1982). In this respect, the reader is referred to a discussion of principal questions related to the endosymbiotic origin for chloroplasts and mitochondria (FOX et al. 1980, SCHWEMMLER 1983), and the origin of the eukaryotic cell itself (CAVALIER-SMITH 1983; additional references therein), a topic which is outside the scope of this book.

Examples for symbiotic associations between bacteria and animals of various taxonomic positions have been described and are summarized, to some extent, by CAVANAUGH (1983), NARDON & WICKER (1983), GUPTA & PANT (1983) and GASSNER et al. (1983). A more general view in this respect has been discussed by TAYLOR & HARRISON (1983); these authors consider the host cell as a habitat or minute ecosystem, the "cytocosm".

References

ABAD-ZAPATERO, C., FOX, J. L. & HACKERT, M. L. (1977): The quaternary structure of a unique phycobiliprotein: B-phycoerythrin from *Porphyridium cruentum*. – Biochem. Biophys. Res. Commun. **78:** 266–272.

ABE, M., SHERWOOD, J. E., HOLLINGSWORTH, R. I. & DAZZO, F. B. (1984): Stimulation of clover root hair infection by lectin-binding Oligosaccharides from the capsular and extracellular Polysaccharides of *Rhizobium trifolii*. – J. Bacteriol. **160:** 517–520.

ABRAM, D. & KOFFLER, H. (1964): In vitro formation of flagella-like filaments and other structures from flagellin. – J. Molec. Biol. **9:** 168–185.

ABRAM, D., KOFFLER, H. & VATTER, A. E. (1965): Basal structure and attachment of flagella in cells of *Proteus vulgaris*. – J. Bacteriol. **90:** 1337–1354.

ABRAM, D., MITCHEN, J. R., KOFFLER, H. & VATTER, A. E. (1970): Differentiation within the bacterial flagellum and isolation of the proximal hook. – J. Bacteriol. **101:** 250–261.

ABRAM, D., VATTER, A. E. & KOFFLER, H. (1966): Attachment and structural features of flagella of certain bacilli. – J. Bacteriol. **91:** 2045–2068.

ACHTERRATH, M. & LICKFELD, K. G. (1972): Number, size and location of *Staphylococcus* mesosomes – a function of fixation parameters. – Proc. Fifth Eur. Congr. Electron Microscopy, Manchester. pp. 220–221. – Inst. of Physics, London.

ACHTMAN, M. (1975): Mating aggregates in *Escherichia coli* conjugation. – J. Bacteriol. **123:** 505–515.

ACHTMAN, M., KENNEDY, N. & SKURRAY, R. (1977): Cell-cell interactions in conjugating *Escherichia coli:* role of *traT* protein in surface exclusion. – Proc. Acad. Sci. USA **74:** 5104–5108.

ACHTMAN, M. & SKURRAY, R. (1977): A redefinition of the mating phenomenon in bacteria. – In: J. L. REISSIG (Ed.): Receptors and Recognition, Ser. B. vol. III, pp. 233–279. – Chapman & Hall, London.

ACKER, G. (1976): Intracelluläre Organisation phagenschwanzähnlicher Bacteriocine der Gruppe A in *Serratia marcenses*. – Arch. Microbiol. **111:** 175–183.

– (1977): The arrangement of lipopolysaccharides on the outer membrane of *Yersinia enterocolitica:* an electron microscopy study. – Zentralbl. Bakteriol. Hyg., 1. Abt. Orig. A **237:** 504–522.

ALAM, M. & OESTERHELT, D. (1984): Morphology, function and isolation of halobacterial flagella. – J. Molec. Biol. **176:** 459–475.

ALBERTS, B., BRAY, D., LEWIS, J., RAFF, M., ROBERTS, K. & WATSON, J. D. (1983): Molecular Biology of the Cell. – Garland, New York, London.

ALLEN, M. M. (1968): Ultrastructure of the cell wall and cell division of unicellular blue-green Algae. – J. Bacteriol. **96:** 842–852.

AMAHO, K. & UMEDA, A. (1977): Bacterial surfaces as revealed by the high resolution scanning electron microscope. – J. Gen. Microbiol. **98:** 297–299.

AMBROSE, E. J. (1961): The movements of fibrocytes. – Exp. Cell Res. Suppl. **8:** 54–73.

ANDERSON, T. F. (1949): On the mechanism of adsorption of bacteriophages to host cells. – In: A. A. MILES & N. W. PIRIE (Eds.): The Nature of the Bacterial Surface. pp. 76–95. – Blackwell Sci. Publ. Oxford.

ANDERSON, T. F., PREER, J. R., PREER, L. B. & BRAY, M. (1964): Studies on killing particles from *Paramecium:* The structure of refractile bodies from *Kappa* particles. – J. Microsc. **3:** 395–402.

ANDREESEN, M. & SCHLEGEL, H. G. (1974): A new coryneform hydrogen bacterium: *Corynebacterium autotrophicum* strain *7C II*. Isolation of a slime-free mutant. – Arch. Microbiol. **100:** 351–361.

ANDREEVA, I. N. (1984): Ultrastructure of symbiotic systems in legume root nodules. – Abstr. 8th Eur. Congr. Electron Microscopy. vol. III, pp. 2133–2134. Budapest.

ANDREOLI, A., SARANTO, J., BAECKER, P. A., SUEHIRO, S., ESCAMILLA, E. & STEINER, A. (1975): Biochemical properties of forespores isolated from *Bacillus cereus*. – In: P. GERHARDT, R. N. COSTILOW & H. L. SADOFF (Eds.) : Spores VI. pp. 418–424. – Amer. Soc. Microbiol., Washington D. C.

ARAGNO, M. & SCHLEGEL, H. G. (1977): *Alcaligenes ruhlandii* (PACKER & VISHNIAC) comb. nov., a peritrichous hydrogen bacterium previously assigned to *Pseudomonas*. – Internatl. J. System. Bacteriol. **27:** 279–281.

– (1981): The hydrogen-oxidizing bacteria. – In: M. P. STARR, H. STOLP, H. G. TRUEPER, A. BALOWS & H. G. SCHLEGEL (Eds.): The Prokaryotes. vol. I, pp. 865–893. – Springer, Berlin, Heidelberg, New York.

ARAGNO, M., WALTHER-MAURUSCHAT, A., MAYER, F. & SCHLEGEL, H. G. (1977): Micromorphology of Gram-negative hydrogen bacteria. I. Cell morphology and flagellation. – Arch. Microbiol. **114:** 93–100.

ARCHIBALD, A. R. (1980): Phage receptors in Gram-positive bacteria. – In: L. L. RANDALL & L. PHILLIPSON (Eds.): Virus Receptors. Part 1. Bacterial Viruses (receptors and recognition, Ser. B, vol. VII) pp. 7–26. – Chapman & Hall, London.

ARISTOVSKAYA, T. V. & HIRSCH, P. (1974): Genus *Pedomicrobium* Aristovsdaya 1961. – In: R. E. BUCHANAN & N. E. GIBBONS (Eds.): Bergey's Manual of Determinative Bacteriology, 8th ed., pp. 151–153. – Williams & Wilkins Co., Baltimore.

ARMBRUSTER, B. L., CARLEMALM, E., CHIOVETTI, R., GARVAVITO, M., HOBOT, J. A., KELLENBERGER, E. & VILLIGER, W. (1982): Specimen preparation for electron microscopy using low temperature embedding resins. – J. Microsc. **126:** 77–85.

ARONSON, A. I. & FITZ-JAMES, P. C. (1968): Biosynthesis of bacterial spore coats. – J. Molec. Biol. **33:** 199–212.

ASAKURA, S. (1970): Polymerization of flagellin and polymorphism of flagella. – Adv. Biophys. (Japan) **1:** 99–155.

ASTBURY, W. T., BEIGHTON, E. & WEIBULL, C. (1955): The structure of bacterial flagella. – Symp. Soc. Exp. Biol. **9:** 282–305.

AULING, G., MAYER, F. & SCHLEGEL, H. G. (1977): Isolation and partial characterization of normal and defective bacteriophages of Gram-negative hydrogen bacteria. – Arch. Microbiol. **155:** 237–249.

BAILLIE, E. & MURRELL, W. G. (1974): Some hydrodynamic properties of spore peptidoglycan. – Biochim. Biophys. Acta **37:** 23–31.

BALCH, W. E., FOX, G. E., MAGRUM, L. J., WOESE, C. R. & WOLFE, R. S. (1979): Methanogens: Reevaluation of a unique biological group. – Microbiol. Rev. **43:** 260–296.

BALKWILL, D. L., MARATEA, D. & BLAKEMORE, R. P. (1980): Ultrastructure of a magnetotactic spirillum. – J. Bacteriol. **141:** 1399–1408.

BARKER, H. A. (1940): Studies upon the methane fermentation. IV. The isolation and culture of *Methanobacterium omelianskii*. – Antonie van Leeuwenhoek J. Microbiol. Serol. **6:** 201–220.

BARNARD, S. D. & PARENTI, F. (1983): Ultrastructural morphology of the genus *Kineosporia*. – Current Microbiol. **8:** 173–176.

BARTHOLEMEW, J. W. & MITTWER, T. (1952): The Gram stain. – Bacteriol. Rev. **16:** 1–29.

BAULD, J. & TYLER, P. A. (1971): Taxonomic implications of reproductive mechanisms of *Hyphomicrobium*-facies and *Pedomicrobium*-facies of a pleomorphic budding bacterium. – Antonie van Leeuwenhoek J. Microbiol. Serol. **37:** 417–424.

BAULD, J., TYLER, P. A. & MARSHALL, K. C. (1971): Pleomorphy of a budding bacterium on various carbon sources. – Antonie van Leeuwenhoek J. Microbiol. Serol. **37:** 409–416.

BAUMEISTER, W. & KÜBLER, O. (1978): Topography of the cell surface of *Micrococcus radiodurans*. – Proc. Natl. Acad. Sci. USA **75:** 5525–5528.

BAYER, M. E. (1968a): Areas of adhesion between wall and membrane of *Escherichia coli*. – J. Gen. Microbiol. **53:** 395–404.

– (1968b): Adsorption of bacteriophages to adhesions between wall and membrane of *Escherichia coli*. – J. Virol. **2:** 346.

– (1975): Role of adhesion zones in bacterial cell surface function and biogenesis. – In: A. TZAGAGALOFF (Ed.): Membrane Biogenesis. pp. 393–427. – Plenum Publ. Corp., New York.

– (1979): The fusion sites between outer membrane and cytoplasmic membrane of bacteria: their role in membrane assembly and virus infection. – In: M. INOUYE (Ed.): Bacterial outer Membranes. Biogenesis and Functions. pp. 167–202. – John Wiley & Sons, Inc., New York.

BAYER, M. E. & THUROW, H. (1977): Polysaccharide capsule of *Escherichia coli*. Microscope study of its size, structure and sites of synthesis. – J. Bacteriol. **130:** 911–936.

BAYER, M. E. & REMSEN, C. C. (1970): Structure of *Escherichia coli* after freeze-etching. – J. Bacteriol. **101:** 304–313.

BAYER, M. H., COSTELLO, G. P. & BAYER, M. E. (1982): Isolation and partial characterization of membrane vesicles carrying markers of the membrane adhesion sites. – J. Bacteriol. **149:** 758–768.

BEAMAN, T. C., GREENAMYRE, J. T., CORNER, T. R., PANKRATZ, H. S. & GERHARDT, P. (1982): Bacterial spore heat resistance correlated with water content, wet density, and protoplast/sporoplast volume ratio. – J. Bacteriol. **150:** 870–877.

BEARD, J. P. & CONNOLLY, J. C. (1975): Detection of a protein, similar to the sex pilus subunit, in the outer membrane of *Escherichia coli* cells carrying a derepressed F-like R-factor. – J. Bacteriol. **122:** 59–65.

BEGG, K. J. & DONACHIE, W. D. (1977): Growth of the *Escherichia coli* cell surface. – J. Bacteriol. **129:** 1524–1536.

BERG, H. C. (1974): Dynamic properties of bacterial flagellar motors. – Nature **249:** 77–79.

– (1975): How bacteria swim. – Sci. Amer. **233:** 36–44.

BERG, H. C., MANSON, M. D. & CONLEY, M. P. (1982): Dynamics and energetics of flagellar rotation in bacteria. – In: W. B. AMOS & J. G. DUCKETT (Eds.): Prokaryotic and Eukaryotic Flagella. pp. 1–31. – Soc. Exp. Biol. Symp. No. XXXV. – University Press, Cambridge.

BERNS, D. S. & EDWARDS, M. R. (1965): Electron micrographic investigation of C-phycocyanin. – Arch. Biochem. Biophys. **110:** 511–616.

BERTHOLD, V. & GEIDER, D. (1976): Interaction of DNA with DNA-binding proteins; The characterization of protein HD from *Escherichia coli* and its nucleic acid complexes. – Eur. J. Biochem. **71:** 443–449.

BEUDEKER, R. F., CANNON, G. C., KUENEN, J. G. & SHIVELY, J. M. (1980): Relation between D-ribulose-1,5-bisphosphate carboxylase and CO_2 fixing capacity in the obligate chemolithotroph *Thiobacillus neapolitanus* grown under different limitations in the chemostat. – Arch. Microbiol. **124:** 185–189.

BEUDEKER, R. F., CODD, G. A. & KUENEN, J. G. (1981): Quantification and intracellular distribution of ribulose-1,5-bisphosphate carboxylase in *Thiobacillus neapolitanus,* as related to possible functions of carboxysomes. – Arch. Microbiol. **129:** 361–367.

BEUDEKER, R. F. & KUENEN, J. G. (1981): Carboxysomes: "Calvinosomes"? – FEBS Lett. **131:** 274–296.

BEVERIDGE, T. J. (1981): Ultrastructure, chemistry and function of the bacterial wall. – Internatl. Rev. Cytol. **72:** 229–317.

BEVERIDGE, T. J. & DAVIES, J. A. (1983): Cellular responses of *Bacillus subtilis* and *Escherichia coli* to the Gram stain. – J. Bacteriol. **156:** 846–858.

BEWLEY, J. D. (1979): Physiological aspects of desiccation tolerance. – Annual Rev. Plant Physiol. **30:** 195–238.

BIBERFELD, G. & BIBERFELD, P. (1970): Ultrastructural features of *Mycoplasma pneumoniae.* – J. Bacteriol. **102:** 855–861.

BIEBRICHER, C. K. & DÜKER, E.-M. (1984a): Light-microscopic visualization of F- and type l-pili. – J. Gen. Microbiol. **130:** 941–949.

– (1984b): F- and type l-piliation of *Escherichia coli.* – J. Gen. Microbiol. **130:** 951–957.

BIRDSELL, D. C. & COTA-ROBLES, E. H. (1967): Production and ultrastructure of lysozyme and ethylenediaminetetraacetatelysozyme spheroplasts of *Escherichia coli.* – J. Bacteriol. **93:** 427–437.

BIRDSELL, D. C., DOYLE, R. J. & MORGENSTERN, M. (1975): Organization of teichoic acid in the cell wall of *Bacillus subtilis.* – J. Bacteriol. **121:** 726–734.

BITTON, G. & MARSHALL, K. C. (Eds.) (1980): Adsorption of Microorganisms to Surfaces. – John Wiley & Sons, New York, Chichester, Brisbane, Toronto.

BLAKEMORE, R. P. & FRANKEL, R. B. (1981): Magnetic navigation in bacteria. – Sci. Amer. **245:** 42–49.

BLAND, C. E. & COUCH, J. N. (1981): The family Actinoplanaceae. – In: M. P. STARR, H. STOLP, H. G. TRUEPER, A. BALOWS & H. G. SCHLEGEL (Eds.): The Prokaryotes. vol. II, pp. 2004–2010. – Springer, Berlin, Heidelberg, New York.

BLAUROCK, A. E. (1975): Bacteriorhodopsin: A trans-membrane pump containing α-helix. – J. Molec. Biol. **93:** 139–158.

BLAUROCK, A. E. & KING, G. I. (1977): Asymmetric structure of the purple membrane. – Science **196:** 1101–1104.

BLAUROCK, A. E. & STOECKENIUS, W. (1971): Structure of the purple membrane. – Nature New Biol. **233:** 152–155.

BLAUROCK, A. E., STOECKENIUS, W., OESTERHELT, D. & SCHERPHOF, G. L. (1976): Structure of the cell envelope of *Halobacterium halobium.* – J. Cell Biol. **71:** 1–22.

BLOCK, S. M. & BERG, H. C. (1984): Successive incorporation of force-generating units in the bacterial rotary motor. – Nature **309:** 470–472.

BLOCK, S. M., SEGALL, J. E. & BERG, H. C. (1982): Impulse responses in bacterial chemotaxis. – Cell **31:** 215–226.

BOATMAN, E. S. & KENNY, G. E. (1971): Morphology and ultrastructure of *Mycoplasma pneumoniae* spherules. – J. Bacteriol. **106:** 1005–1015.

BOCHEM, H. P., SCHOBERTH, S. M., SPREY, B. & WENGLER, P. (1982): Thermophilic biomethanation of acetic acid: morphology and ultrastructure of a granular consortium. – Canad. J. Microbiol. **28:** 500–510.

BOCK, E. & HEINRICH, G. (1971): Struktur- und Funktionsänderungen in reaktivierenden Zellen von *Nitrobacter winogradskyi* Buch. – Arch. Mikrobiol. **77:** 349–65.

BODE, W. (1973): Die Bakteriengeißel (Flagella) und das Geißelprotein Flagellin. – Angew. Chemie **85:** 731–741.

BODE, W., ENGEL, J. & WINKELMAIR, D. (1972): A model of bacterial flagella based on small-angle X-ray scattering and hydrodynamic data which indicate an elongated shape of the flagellin protomer. – Eur. J. Biochem. **26:** 313–327.

BOGOMOLNI, R. A., BAKER, R. A., LOZIER, R. H. & STOECKENIUS, W. (1976): Light-driven proton translocations in *Halobacterium halobium.* – Biochim. Biophys. Acta **440:** 68–88.

– – – – (1980): Action spectrum and quantum efficiency for proton pumping in *Halobacterium halobium.* – Biochemistry **19:** 2152–2159.

BOGORAD, L. (1975): Phycobiliproteins and complementary adaptation. – Annual Rev. Plant Physiol. **26:** 369–401.

BOHNERT, H. J., MICHALOWSKI, C., KOLLER, B., DELIUS, H., MUCKE, H. & LÖFFELHARDT, W. (1983): The cyanelle genome from *Cyanophora paradoxa.* – In: H. E. A. SCHENK & W. SCHWEMMLER (Eds.): Endocytobiology. vol. II, pp. 433–448. – de Gruyter, Berlin, New York.

VAN DEN BOS, R. C., SCHETGENS, T. M. P., HONTELEZ, J. G. J., BAKKEREN, G., VAN DUN, C., BISSELING, T. & VAN KAMMEN, A. (1983): Expression of nodule-specific genes in both partners of the *Rhizobium* legume symbiosis. – In: H. E. A. SCHENK & W. SCHWEMMLER (Eds.): Endocytobiology. vol. II, pp. 573–581. – de Gruyter, Berlin, New York.

BOUBLIK, M. (1985): Ribosome structure. – In: NANNINGA (Ed.): Molecular Cytology of *Escherichia coli.* pp. 229–258. – Academic Press Inc., New York.

BOWIEN, B. & MAYER, F. (1978): Further studies on the quaternary structure of D-ribulose-1,5-bisphosphate carboxylase from *Alcaligenes eutrophus.* – Eur. J. Biochem. **88:** 97–107.

BOWIEN, B., MAYER, F., CODD, G. A. & SCHLEGEL, H. G. (1976): Purification, some properties and quaternary structure of the D-ribulose 1,5-diphosphate carboxylase of *Alcaligenes eutrophus.* – Arch. Microbiol. **110:** 157–166.

BOWMAN, C. M., SIDIKARO, J. & NOMURA, M. (1973): Mode of action of colicin *E3.* – In: L. P. HAGER (Ed.): Chemistry and Functions of Colicins. p. 87. – Academic Press Inc., New York, London.

BOYD, R. F. (1984): General Microbiology. – Times Mirror/Mosby, St. Louis/USA.

BRADLEY, D. E. (1967): Ultrastructure of bacteriophages and bacteriocins. – Bacteriol. Rev. **31:** 230.

BRADLEY, D. E. & DEWAR, C. A. (1967): Intracellular changes in cells of *Escherichia coli* infected with a filamentous bacteriophage. – J. Gen. Virol. **1:** 179.

BRANTON, D. (1966): Fracture faces of frozen membranes. – Proc. Natl. Acad. Sci. (Wash.) **55:** 1048–1056.

BRAUN, M., MAYER, F. & GOTTSCHALK, G. (1981): *Clostridium aceticum* (Wieringa), a microorganism producing acetic acid from molecular hydrogen and carbon dioxide. – Arch. Microbiol. **128:** 288–293.

BRAUN, V. (1975): Covalent lipoprotein from the outer membrane of *Escherichia coli.* – Biochim. Biophys. Acta **415:** 335–377.

BRAUN, V., SCHALLER, K. & WOLFF, H. (1973): A common receptor protein for phage *T5* and Colicin *M* in the outer membrane of *Escherichia coli B.* – Biochim. Biophys. Acta **323:** 87.

BRAUN, V. & WOLFF, H. (1973): Characterization of the receptor protein for phage *T5* and Colicin *M* in the outer membrane of *E. coli B.* – FEBS Lett. **34:** 77.

BREDT, W. (1973): Motility of mycoplasmas. – Ann. N. Y. Acad. Sci. **225:** 246–250.

BREITENREUTER, G., LOTTI, M., STÖFFLER-MEILICKE, M. & STÖFFLER, G. (1984): The use of monoclonal antibodies for the localization of proteins *S 3* and *S 7* on the ribosomal surface of *Escherichia coli.* – Abstr. 8th Eur. Congr. Electron Microscopy. vol. II, pp. 1553–1554. Budapest.

BRINTON, C. C. (1959): Non-flagellar appendages of bacteria. – Nature (Lond.) **183:** 782–786.

– (1965): The structure, function, synthesis and genetic control of bacterial pili and a molecular model for DNA and RNA transport in Gram-negative bacteria. – Trans. N. Y. Acad. Sci. 27: 1003–1054.

– (1971): The properties of sex pili, the viral nature of "conjugal" genetic transfer systems, and some possible approaches to the control of bacterial drug resistance. – Crit. Rev. Microbiol. **1:** 105–160.

BRINTON, C. C. & BEER, H. (1967): The interaction of male-specific bacteriophages with F-pili. – In: J. S. COLTER & W. PARANCHYCH (Eds.): The Molecular Biology of Viruses. pp. 251–289. – Academic Press, New York.

BRINTON, C. C., GEMSKI, P. & CARNAHAN, J. (1964): A new type of bacterial pilus genetically controlled by the fertility factor of *E. coli K-12* and its role in chromosome transfer. – Proc. Natl. Acad. Sci. USA **52:** 776–783.

BROCK, T. (1967): Life at high temperatures. – Science **158:** 1012–1019.

BROCK, T. (1981): Extreme termophiles of the genera *Thermus* and *Sulfolobus*. – In: M. P. STARR, H. STOLP, H. G. TRUEPER, A. BALOWS & H. G. SCHLEGEL (Eds.): The Prokaryotes. vol. I, pp. 978–984. – Springer, Berlin, Heidelberg, New York.

BROCKMAN, E. R. & TODD, R. L. (1974): Fruiting myxobacters as viewed with a scanning electron microscope. – Internatl. J. Syst. Bacteriol. **24:** 118–124.

BROWN, G. W. JR., PATE, J. L. & SUGIYAMA, H. (1971): Fine structure of protoplasts and L-forms of *Clostridium botulinum* types *A* and *E*. – J. Bacteriol. **105:** 1207–1210.

BRUNNER, H., KRAUSS, H., SCHAAR, H. & SCHIEFER, H. G. (1979): Electron microscopic studies on the attachment of *Mycoplasma pneumoniae* to Guinea pig erythrocytes. – Infect. Immun. **24:** 906–911.

BRYANT, D. A., COHEN-BAZIRE, G. & GLAZER, A. N. (1981): Characterization of the biliproteins of *Gloeobacter violaceus;* chromophore content of a cyanobacterial phycoerythrin carrying phycourobilin chromophore. – Arch. Microbiol. **129:** 190–198.

BRYANT, R. D., COSTERTON, J. W. & LAISHLEY, E. J. (1984): The role of *Thiobacillus albertis* glycocalyx in the adhesion of cells to elemental sulfur. – Canad. J. Microbiol. **30:** 81–90.

BRYANT, D. A., GLAZER, A. N. & EISERLING, F. A. (1976): Characterization and structural properties of the major biliproteins of *Anabaena* sp. – Arch. Microbiol. **110:** 61–75.

BRYANT, D. A., GUGLIELMI, G., TANDEAU DE MARSAC, N., CASTETS, A. & COHEN-BAZIRE, G. (1977): The structure of cyanobacterial phycobilisomes: a model. – Arch. Microbiol. **123:** 113–127.

BRYANT, M. P., WOLIN, E. A., WOLIN, M. J. & WOLFE, R. S. (1967): *Methanobacillus omelianskii,* a symbiotic association of two species of bacteria. – Arch. Microbiol. **59:** 20–31.

BURCHARD, R. P. (1981): Gliding motility of prokaryotes: Ultrastructure, physiology, and genetics. – Annual Rev. Microbiol. **35:** 497–529.

BURDETT, I. D. J. & MURRAY, R. G. L. (1974a): Septum formation in *Escherichia coli:* characterization of septal structure and the effects of antibiotics on cell division. – J. Bacteriol. **119:** 303–324.

– – (1974b): Electron microscope study of septum formation in *Escherichia coli* strains *B* and *B/r* during synchronous growth. – J. Bacteriol. **119:** 1039–1056.

BURDETT, I. D. J. & ROGERS, H. J. (1970): Modification of the appearance of mesosomes in sections of *Bacillus licheniformis* according to the fixation procedures. – J. Ultrastruct. Res. **30:** 354–367.

BURLEY, S. K. & MURRAY, R. G. E. (1983): Structure of the regular surface layer of *Bacillus polymyxa*. – Canad. J. Microbiol. **29:** 775–780.

BURMAN, L. G., RAICHLER, J. & PARK, J. T. (1983): Evidence for diffuse growth of the cylindrical portion of the *Escherichia coli* murein sacculus. – J. Bacteriol. **155:** 983–988.

BUTLER, R. D. & ALLSOPP, A. (1972): Ultrastructural investigations in the Stigonemataceae (Cyanophyta). – Arch. Mikrobiol. **82:** 283–299.

CAGLE, G. D. (1975): Fine structure and distribution of extracellular polymer surrounding selected aerobic bacteria. – Canad. J. Microbiol. **21:** 395–408.

CACLE, G. D., PFISTER, R. M. & VELA, R. G. (1972): Improved staining of extracellular polymer for electron microscopy: examination of *Azotobacter, Zoogloea, Leuconostoc,* and *Bacillus*. – Appl. Microbiol. **24:** 477–487.

CAGLE, G. D. & VELA, G. R. (1972): Cysts of *Azotobacter vinelandii* with double coats. – J. Bacteriol. **112:** 615–617.

CAIRNS, J. (1963a): The bacterial chromosome and its manner of replication as seen by autoradiography. – J. Molec. Biol. **6:** 208.

– (1963b): The chromosome of *Escherichia coli*. – Cold Spring Harbor Symp. Quant. Biol. **28:** 43–46.

CALENDAR, R. (1970): The regulation of phage development. – Annual Rev. Microbiol. **24:** 241.

CALLADINE, C. R. (1975): Construction of bacterial flagella. – Nature **255:** 121–124.

– (1982): Construction of bacterial flagellar filaments, and aspects of their conversion to different helical forms. – In: W. B. AMOS & J. G. DUCKETT (Eds.): Prokaryotic and Enkaryotic Flagella. pp. 33–51. Soc. Biol. Symp. No. XXXV. – Cambridge Univ. Press, Cambridge.

CAMPBELL, A. M. (1968): Techniques for studying defective bacteriophages. – In: K. MARAMOROSCH & H. KOPROWSKI (Eds.): Methods in Virology. vol. IV, p. 279. – Academic Press Inc., New York.

CANALE-PAROLA, E. (1978): Motility and chemotaxis of Spirochetes. – Annual Rev. Microbiol. **32:** 69–99.

– (1981): Free-living anaerobic and facultatively anaerobic spirochetes: the genus *Spirochaeta.* – In: M. P. STARR, H. STOLP, H. G. TRUEPER, A. BALOWS & H. G. SCHLEGEL (Eds.): The Prokaryotes. vol. I, pp. 538–547. – Springer, Berlin, Heidelberg, New York.

CANCEDDA, R. & SCHLESINGER, M. J. (1974): Localization of polyribosomes containing alkaline phosphatase nascent polypeptides on membranes of *Escherichia coli.* – J. Bacteriol. **117:** 290–301.

CANNO, G. C. & SHIVELY, J. M. (1983): Characterization of a homogeneous preparation of carboxysomes from *Thiobacillus neapolitanus.* – Arch. Microbiol. **134:** 52–59.

CARLEMALM, E. & KELLENBERGER, E. (1982): The reproducible observation of unstained embedded cellular material in thin sections: visualisation of an integral membrane protein by a new mode of imaging for STEM. – EMBO J. **1:** 63–67.

CARO, L. G. & SCHNÖS, M. (1966): The attachment of the male-specific bacteriophage *fl* to sensitive strains of *Escherichia coli.*– Proc. Natl. Acad. Sci. USA **56:** 126–132.

CARR, N. G. & SANDHU, G. R. (1966): Endogenous metabolism of polyphosphates in two photosynthetic micro-organisms. – Biochem. J. **99:** 29–30.

CASADIO, R. & STOECKENIUS, W. (1980): Effect of protein-protein interaction on light adaptation of bacteriorhodopsin. – Biochemistry **19:** 3374–3381.

CASSONE, A. & GARACI, E. (1977): The capsular network of *Klebsiella pneumoniae.* – Canad. J. Microbiol. **23:** 684–689.

CAVALIER-SMITH, T. (1983): Genetic symbionts and the origin of split genes and linear chromosomes. – In: H. E. A. SCHENK & W. SCHWEMMLER (Eds.): Endocytobiology. vol. II, pp. 29–45. – de Gruyter, Berlin, New York.

CAVANAUGH, C. M. (1983): Chemoautotrophic bacteria in marine invertebrates from sulfide-rich habitats: a new symbiosis. – In: H. E. A. SCHENK & W. SCHWEMMLER (Eds.): Endocytobiology, vol. II, pp. 699–708. – de Gruyter, Berlin, New York.

CHAI, N. C. & LARK, K. G. (1967): Segregation of deoxyribonucleic acid in bacteria: association of the segregating unit with the cell envelope. – J. Bacteriol. **94:** 415–421.

CHAMPNESS, J. N. (1971): X-ray and optical diffraction studies of bacterial flagella. – J. Molec. Biol. **56:** 295–310.

CHAO, L. & BOWEN, C. C. (1971): Purification and properties of glycogen isolated from a blue-green alga, *Nostoc muscorum.* – J. Bacteriol. **105:** 331–338.

DE CHASTELLIER, C., HELLIO, R. & RYTER, A. (1975 a): Study of cell wall growth of *Bacillus megaterium* by high resolution autoradiography. – J. Bacteriol. **123:** 1184–1196.

DE CHASTELLIER, C., FREHEL, C. & RYTER, A. (1975 b): Cell wall growth of *Bacillus megaterium:* cytoplasmic radioactivity after pulse-labeling with tritiated diaminopimelic acid. – J. Bacteriol. **123:** 1197–1207.

CHEESEMAN, P., TOMS-WOOD, A. & WOLFE, R. S. (1972): Isolation and properties of a fluorescent compound, $factor_{420}$, from *Methanobacterium* strain *M. o. H.* – J. Bacteriol. **112:** 527–531.

CHENG, K.-J., FAY, J. P., COLEMAN, R. N., MILLIGAN, L. P. & COSTERTON, J. W. (1981 a): Formation of bacterial microcolonies on feed particles in the rumen. – Appl. Environ. Microbiol. **41:** 298–305.

CHENG, K.-J., INGRAM, J. M. & COSTERTON, J. W. (1970): Alkaline phosphatase localization and spheroplast formation of *Pseudomonas aeruginosa.* – Canad. J. Microbiol. **16:** 1319–1324.

– – – (1971): Interactions of alkaline phosphatase and the cell wall of *Pseudomonas aeruginosa.* – J. Bacteriol. **107:** 325–336.

CHENG, K.-J., IRVIN, R. T. & COSTERTON, J. W. (1981 b): Autochthonous and pathogenic colonization of animal tissues by bacteria. – Canad. J. Microbiol. **27:** 461–490.

CHO, K. Y., DOY, C. H. & MERCER, E. H. (1967): Ultrastructure of the obligate halophilic bacterium *Halobacterium halobium.* – J. Bacteriol. **94:** 196–201.

CHOY, Y.-M., FEHMEL. F., FRANK, N. & STIRM, S. (1975): *Escherichia coli* capsule bacteriophages. VI. Primary structure of the bacteriophage 29 receptors, the *E. coli* serotype 29 capsular polysaccharide. – J. Virol. **16:** 581–590.

CHRYSOGELOS, S. & GRIFFITH, J. (1982): *Escherichia coli* single-strand binding protein organizes single-stranded DNA in nucleosome-like units. – Proc. Natl. Acad. Sci. USA **79:** 5803–5807.

CLARK-WALKER, G. D. (1969): Association of microcyst formation in *Spirillum itersonii* with the spontaneous induction of a defective bacteriophage. – J. Bacteriol. **97:** 885.

CLAUS, D., FAHMY, F., ROLF, H. J. & TOSUNOGLU, N. (1983): *Sporosarcina halophila* sp. nov., an obligate, slightly halophilic bacterium from salt marsh soils. – System. Appl. Microbiol. **4:** 496–506.

CLAYTON, R. K. & SISTROM, W. R. (Eds.) (1978): The Photosynthetic Bacteria. – Plenum Press, New York, London.

COETZEE, H. J., DE KLERK, H. C. & COETZEE, J. N. (1968): Bacteriophage-tail-like particles associated with intraspecies killing of *Proteus vulgaris*. – J. Gen. Virol. **2:** 29.

COHEN-BAZIRE, G., BEGUIN, S., RIMON, S., GLAZER, A. N. & BROWN, D. M. (1977): Physicochemical and immunological properties of allophycocyanins. – Arch. Microbiol. **111:** 225–238.

COHEN-BAZIRE, G. & LONDON, J. (1967): Base organelles of bacterial flagella. – J. Bacteriol. **94:** 458–465.

COHEN-BAZIRE, G. & SISTROM, W. R. (1966): The procaryotic photosynthetic apparatus. – In: L. P. VERNON & G. R. SEELY (Eds.): The Chlorophylls. pp. 313–341. – Academic Press Inc., New York.

COLLINS, J. F. & RICHMOND, M. H. (1962): Rate of growth of *Bacillus cereus* between divisions. – J. Gen. Microbiol. **28:** 15–33.

CONTI, S. F. & HIRSCH, P. (1965): Biology of budding bacteria. III. Fine structure of *Rhodomicrobium* and *Hyphomicrobium* spp. – J. Bacteriol. **89:** 503–512.

COOMBS, R. W., WILLISON, H. M. & EASTERBROOK, K. B. (1975): The morphological subunit of bacterial spinae. – Proc. Microsc. Soc. Canada **2:** 84–85.

COSNER, J. C. (1978): Phycobilisomes in spheroplasts of *Anacystis nidulans*. – J. Bacteriol. **135:** 1137–1140.

COSTERTON, J. W. (1984): Preservation of the dimensions and spatial arrangements of extracellular hydrated polysaccharide structures in scanning and transmission electron microscopy. – Abstr. 8th Eur. Congr. Electron Microscopy, vol. III, pp. 2375–2376. Budapest.

COSTERTON, J. W., GEESEY, G. G. & CHENG, K. J. (1978): How bacteria stick. – Sci. Amer. **238:** 86–95.

COSTERTON, J. W., INGRAM, J. M. & CHENG, K.-J. (1974): Structure and function of the cell envelope of Gram-negative bacteria. – Bacteriol. Rev. **38:** 87–110.

COSTERTON, J. W., IRVIN, R. T. & CHENG, K.-J. (1981): The bacterial glycocalix in nature and disease. – Annual. Rev. Microbiol. **35:** 299–324.

CRAWFORD, E. & GESTELAND, R. (1964): The adsorption of bacteriophage *R-17*. –Virology **22:** 165–167.

CRUDEN, D. L. & STANIER, R. Y. (1970): The characterization of *Chlorobium* vesicles and membranes isolated from green bacteria. – Arch. Mikrobiol. **72:** 115–134.

CULLIMORE, D. R., MCCANN, A. E. (1977): The identification, cultivation and control of iron bacteria in ground water. – In: F. A. SKINNER & J. M. SHEWAN (Eds.): Aquatic Microbiology. Soc. Appl. Bacteriol. Symp. No. 6, pp. 219–261. – Academic Press Inc., London.

CUMMINGS, D. J. (1965): Macromolecular synthesis during synchronous growth of *Escherichia coli B/r*. – Biochim. Biophys. Acta **85:** 341–349.

CURTISS, R. (1969): Bacterial conjugation. – Annual Rev. Microbiol. **23:** 69–136.

CURTISS, R., CARO, L. G., ALLISON, D. P. & STALLIONS, D. R. (1969): Early stages of conjugation in *Escherichia coli*. – J. Bacteriol. **100:** 1091–1104.

DAHLBACK, B., HERMANSSON, M., KJELLEBERG, S. & NORKRANS, B. (1981): The hydrophobicity of bacteria is an important factor in their initial adhesion at the air-water interface. – Arch. Microbiol. **128:** 267–270.

DAKER, W. D. & STACEY, M. (1938): CCLI. Investigation of a polysaccharide produced from sucrose by *Betabacterium vermiforme* (Ward-Meyer). – Biochem. J. **52:** 1946–1948.

DANEO-MOORE, L., DICKER, D. & HIGGINS, M. L. (1980): Structure of the nucleoid in cells of *Streptococcus faecalis*. – J. Bacteriol. **141:** 928–937.

DANIELS, M. J. & MEDDINS, B. M. (1974): Culture of *Spiroplasma* causing *Citrus* little leaf. – John Innes Inst. 64th Annual Rep. 1973, pp. 117–119. Norwich, U. K.

DANON, A. & STOECKENIUS, W. (1974): Photophosphorylation in *Halobacterium halobium*. – Proc. Natl. Acad. Sci. USA **71:** 1234–1238.

DAVENPORT, H. E. (1960): Haemoglobin in the root nodules of *Casuarina cunninghamiana*. – Nature **6:** 653–654.

DAVIES, J. A., ANDERSON, G. K., BEVERIDGE, T. J. & CLARK, H. C. (1983): Chemical mechanism of the Gram stain and synthesis of a new electron-opaque marker for electron microscopy which replaces the iodine mordant of the stain. – J. Bacteriol. **156:** 837–845.

DAVIS, M. J., WHITCOMB, R. F. & GILLASPIE JR., A. G. (1981): Fastidious bacteria of plant vascular tissue and invertebrates (including so-called *Rickettsia*-like bacteria). – In: M. P. STARR, H. STOLP, H. G. TRUEPER, A. BALOWS & H. G. SCHLEGEL (Eds.): The Prokaryotes. vol. II, pp. 2172–2188. – Springer, Berlin, Heidelberg, New York.

DAVIS, S. & WHITTENBURRY, R. (1970): Fine structure of methane and other hydrocarbon utilizing bacteria. – J. Gen. Microbiol. **61:** 227–232.

DAZZO, F. B. (1984): Bacterial adhesion to plant root surfaces. – In: K. C. MARSHALL (Ed.): Microbial Adhesion and Aggregation. pp. 85–93. – Springer, Berlin, Heidelberg, New York, Tokyo.

DIPPELL, R. V. (1958): The fine structure of *Kappa* in killer stock *51* of *Paramecium aurelia.* Preliminary observations. – J. Biophys. Biochem. Cytol. **4:** 125–128.

DOBLER, M., DOVERS, S. C., LAVES, E., BINDER, A. & ZUBER, H. (1972): Crystallization and preliminary crystal data of C-phycoyanin. – J. Molec. Biol. **71:** 785–787.

DOBSON, W. J. & MCCURDY, H. D. (1979): The function of fimbriae in *Myxococcus xanthus.* I. Purification and properties of *M. xanthus* fimbriae. – Canad. J. Microbiol. **25:** 1152–1160.

DOBSON, W. J., MCCURDY, H. D., MACRAE, T. H. (1979): The function of fimbriae in *Myxcoccus xanthus.* II. The role of fimbriae in cell-cell interactions. – Canad. J. Microbiol. **25:** 1359–1372.

DODDEMA, H. J. & VOGELS, G. D. (1978): Improved identification of methanogenic bacteria by fluorescence microscopy. – Appl. Environ. Microbiol. **36:** 752–754.

DODDEMA, H. J., CLAESEN, C. A., KELL, D. B., VAN DER DRIFT, C. & VOGELS. G. D. (1980): An adenine nucleotide translocase in the prokaryote *Methanobacterium thermoautotrophicum.* – Biochem. Biophys. Res. Commun. **95:** 1288–1293.

DODDEMA, H. J., HUTTEN, T. J., VAN DER DRIFT, C. & VOGELS, G. D. (1978): ATP hydrolysis and synthesis by the membrane-bound ATP synthetase complex of *Methanobacterium thermoautotrophicum.* – J. Bacteriol. **136:** 19–23.

DODDEMA, H. J., VAN DER DRIFT, C., VOGELS, G. D. & VEENHUIS, M. (1979): Chemiosmotic coupling in *Methanobacterium thermoautotrophicum:* hydrogen-dependent adenosine 5′-triphosphate synthesis by subcellular particles. – J. Bacteriol. **140:** 1081–1089.

DOMINGUE, G. J. (1983): Cell wall-deficient bacteria: Basic principles and clinical significance. – TIBS **8:** 338.

DOW, C. S., WESTMACOTT, D. & WHITTENBURRY, R. (1976): Ultrastructure of budding and prosthecate bacteria. – In: R. FULLER & D. W. LOVELOCK (Eds.): Microbial Ultrastructure. pp. 187–221. – Academic Press, London, New York, San Francisco.

DOYLE, R. J., KOCH, A. L. & CARSTENS, P. H. B. (1983): Cell wall-DNA association in *Bacillus subtilis.* – J. Bacteriol. **153:** 1521–1527.

DREWS, G. (1978): Structure and development of the membrane system of photosynthetic bacteria. – Curr. Topics Bioenerget. **8:** 161–207.

DREWS, G., WECKESSER, J. & MAYER, H. (1978): Cell envelopes. – In: R. K. CLAYTON & W. R. SISTROM (Eds.): The Photosynthetic Bacteria. pp. 61–77. – Plenum Press, New York, London.

VAN DRIEL, D., WICKEN, A. J., DICKSON, M. R. & KNOX, K. W. (1973): Cellular location of the lipoteichoic acids of *Lactobacillus fermenti* NCTC *6991* and *Lactobacillus casei* NCTC *6375*.– J. Ultrastruct. Res. **43:** 483–497.

DUBINIA, G. A. (1979): Mechanism of the oxidation of divalent iron and manganese by iron bacteria growing at neutral pH of the medium. – Microbiology **47:** 471–478.

DUBOCHET, J., MCDOWALL, A. W., MENGE, B., SCHMID, E. N. & LICKFELD, K. G. (1983): Electron microscopy of frozen-hydrated bacteria. – J. Bacteriol. **155:** 381–390.

DUCKWORTH, D. H. (1970): Biological activity of bacteriophage ghosts and "take-over" of host functions by bacteriophage. – Bacteriol. Rev. **34:** 344.

DUGUID, J. P., ANDERSON, E. S. & CAMPBELL, I. (1966): Fimbriae and adhesive properties in *Salmonella.* – J. Pathol. Bacteriol. **92:** 107–138.

DUGUID, J. P. & GILLIES, R. (1957): Fimbriae and adhesive properties in dysentery bacilli. – J. Pathol. Bacteriol. **74:** 397–411.

DUGUID, J. P. & OLD, D. C. (1980): Adhesive properties of Enterobacteriaceae. – In: E. H. BEACHEY (Ed.): Bacterial Adherence, Receptors and Recognition, ser. B. vol. VI, pp. 185–217. – Chapman & Hall, London.

DUGUID, J. P., SMITH, I. W., DEMPSTER, G. & EDMUNDS, P. N. (1955): Non-flagellar filamentous appendages ("fimbriae") and haemagglutinating activity in *Bacterium coli.* – J. Pathol. Bacteriol. **70:** 335–348.

DUNCAN, C. L., KING, G. J. & FRIEBEN, W. R. (1973): A paracrystalline inclusion formed during sporulation of enterotoxin-producing strains of *Clostridium perfringens* Type *A.* – J. Bacteriol. **144:** 845–859.

DUNLOP, W. F. & ROBARDS, A. W. (1973): Ultrastructural study of poly-β-hydroxybutyrate granules from *Bacillus cereus.* – J. Bacteriol. **114:** 1271–1280.

DWORKIN, M., KELLER, K. H. & WEISBERG, D. (1983): Experimental observations consistent with a surface tension model of gliding motility of *Myxococcus xanthus.* – J. Bacteriol. **155:** 1367–1371.

Easterbrook, K. B. (1974): The configuration of bacterial spines. – Proc. Microsc. Soc. Canada **1.**

Easterbrook, K. B. & Alexander, S. A. (1983): The initiation and growth of bacterial spinae. – Canad. J. Microbiol. **29:** 476–487.

Easterbrook, K. B. & Coombs, R. W. (1976): Spinin: the subunit protein of bacterial spinae. – Canad. J. Microbiol. **22:** 438–440.

Easterbrook, K. B., McGregor-Shaw, J. B. & McBride, R. P. (1973): Ultrastructure of bacterial spines. – Canad. J. Microbiol. **19:** 995–997.

Easterbrook, K. B., Willison, J. H. M. & Coombs, R. W. (1976): Arrangement of morphological subunits in bacterial spinae. – Canad. J. Microbiol. **22:** 619–629.

Eaton, M. W. & Ellar, D. J. (1974): Protein synthesis and break-down in the mother cell and forespore compartments during spore morphogenesis in *Bacillus megaterium.* – Biochem. J. **144:** 327–337.

Eberle, H. & Lark, K. G. (1966): Chromosome segregation in *B. subtilis.* – J. Molec. Biol. **22:** 183–186.

Ebersold, H. R., Cordier, J.-L. & Lüthy, P. (1981): Bacterial mesosomes: method-dependent artifacts. –Arch. Microbiol. **130:** 19–22.

Ebrahim-Nesbat, F. & Heitefuss, R. (1985): Rickettsien-ähnliche Bakterien (RLO) in Feinwurzeln erkrankter Fichten unterschiedlichen Alters. – Eur. J. Forest Pathol. *15:* 182–187.

Edelstein, E., Parks, L., Tsien, H.-C., Daneo-Moore, L. & Higgins, M. L. (1981): Nucleoid structure in freeze fractures of *Streptococcus faecalis:* effects of filtration and chilling. – J.Bacteriol. **146:** 798–803.

Edelstein, E. M., Rosenzweig, M. S., Daneo-Moore, L. & Higgins, M. L. (1980): Unit cell hypothesis for *Streptococcus faecalis.* – J. Bacteriol. **143:** 499–505.

Edwards, T. & McBride, B. C. (1975): New method for the isolation and identification of methanogenic bacteria. – Appl. Microbiol. **29:** 540–545.

Eickbush, T. H. & Moudrianakis, E. N. (1978): The compaction of DNA helices into either continuous supercoils or folded fiber-rods and toroids. – Cell **13:** 295–306.

Eigner, U. & Bock, E. (1971): Auf- und Abbau der Polyphosphatfraktion in Zellen von *Nitrobacter winogradskyi* Buch. – Arch. Mikrobiol. **81:** 367–378.

Eisenstein, B. I. (1981): Phase variation of type 1 fimbriae in *Escherichia coli* is under transcriptional control. – Science **214:** 337–339.

Eiserling, F. A. & Glazer, A. N. (1974): Blue-green algal proteins: Assembly forms of C-phycocyanin from *Synechococcus* sp. – J. Ultrastruct. Res. **47:** 16–25.

Elbein, A. D. & Mitchell, M. (1973): Levels of glycogen and trehalose in *Mycobacterium smegmatis* and the purification and properties of the glycogen synthetase. – J. Bacteriol. **113:** 863–873.

Ellar, D. J. & Eaton, M. W., Hogarth, C., Wilkinson, B. J., Deans, J. & La Nauze, J. (1975): Comparative biochemistry and function of forespore and mother-cell compartments during sporulation of *Bacillus megaterium* cells. – In: P. Gebhardt, R. N. Costilow & H. L. Sadoff (Eds.): Spores VI. pp. 425–433. – Amer. Soc. Microbiol., Washington D. C.

Ellar, D. & Lundgren, D. G. (1968): Morphology of poly-β-hydroxybutyrate granules. – J. Molec. Biol. **35:** 489–502.

Ellinger, A., Dworsky, P. & Weishäupl, V. (1982): Isolation of membrane-associated folded chromosomes from *Anacystis nidulans.* – Z. Allg. Mikrobiol. **22:** 17–27.

Emerson, S. U., Tokuyasu, K. & Simon, M. J. (1970): Bacterial flagella: Polarity of elongation. – Science **169:** 190–192.

Emeruwa, A. C. & Hawirko, R. Z. (1973): Poly-β-hydroxybutyrate metabolism during growth and sporulation of *Clostridium botulinum.* – J. Bacteriol. **116:** 989–993.

Engelhardt, H., Baumeister, W. & Saxton, W. O. (1983): Electron microscopy of photosynthetic membranes containing bacteriochlorophyll b. – Arch. Microbiol. **135:** 169–175.

Engelhardt, H., Engel, A. & Baumeister, W. (1984 a): STEM mass determination of the photosynthetic unit in native membranes of the bacterium *Ectothiorhodospirum halochloris.* – Abstr. 8th Eur. Congr. Electron Microscopy. vol. II, pp. 1503–1504. Budapest.

Engelhardt, H., Guckenberger, R. & Baumeister, W. (1984 b): Metal shadowing and decoration as tools for investigating the structure of bacterial photosynthetic membranes. – Abstr. 8th Eur. Congr. Electron Microscopy. vol. II, pp. 1279–1280. Budapest.

Van Ert, M. & Staley, J. T. (1971): Gas-vacuolated strains of *Microcyclus aquaticus.* – J. Bacteriol. **108:** 236–240.

ESHDAT, Y., OFEK, J., YASHOUVGAN, Y., SHARON, N. & MIRELMAN, D. (1978): Isolation of a mannose-specific lectin from *Escherichia coli* and its role in the adherence of the bacteria to epithelial cells. – Biochem. Biophys. Res. Commun. **85:** 1551–1559.

FATTOM, A. & SHILO, M. (1984): Hydrophobicity as an adhesion mechanism of benthic cyanobacteria. – Appl. Environ. Microbiol. **47:** 135–143.

FAURÉ-FREMIET, E. & ROUILLER, C. (1958): Etude au microscope électronique d'une bactérie sulfureuse, *Thiovulum majus* Hinze. – Exp. Cell Res. **14:** 29–46.

FEHMEL, F., FEIGE, U., NIEMANN, H. & STIRM, S. (1975): *Escherichia coli* capsule bacteriophages. VII. Bacteriophage *29* – host capsular polysaccharide interactions. – J. Virol. **16:** 591–601.

FELSENFELD, G. (1978): Chromatin. – Nature (Lond.) **271:** 115–122.

FERGUSON, T. J. & MAH, R. A. (1983): Isolation and characterization of an H_2-oxidizing thermophilic methanogen. – Appl. Environ. Microbiol. **45:** 265–274.

FIRSHEIN, W. (1972): The DNA membrane fraction of *Pneumococcus* contains a DNA replication complex. – J. Molec. Biol. **70:** 383–397.

FIRSOW, N. N. & DREWS, G. (1977): Differentiation of the intracytoplasmic membranes of *Rhodopseudomonas palustris* induced by variations of oxygen partial pressure or light intensity. – Arch. Microbiol. **115:** 299–306.

FITZ-JAMES, P. C. & YOUNG, E. (1969): Morphology of sporulation. – In: G. W. Gould & A. Hurst (Eds.): The Bacterial Spore. pp. 39–72. – Academic Press Inc., New York.

FIVES-TAYLOR, P. (1978): F-pili. – In: D. E. BRADLEY, E. RAIZEN, P. FIVES-TAYLOR & J. OU (Eds.): Pili. pp. 145–159. Internatl. Conf. on Pili, Washington D. C.

FLETCHER, M. & FLOODGATE, G. D. (1976): The adhesion of bacteria to solid surfaces. – In: R. FULLER & D. W. LOVELOCK (Eds.): Microbial Ultrastructure. pp. 101–107. – Academic Press, London.

FOLKHARD, W., LEONHARD, K. R., MALSEY, S., MARVIN, D. A., DUBOCHET, J., ENGEL, A., ACHTMANN, M. & HELMUTH, R. (1979): X-ray diffraction and electron microscope studies on the structure of bacterial F-pili. – J. Molec. Biol. **130:** 145–160.

FORMANEK, H., FORMANEK, S. & WAWRA, H. (1974): A three-dimensional atomic model of the murein layer of bacteria. – Eur. J. Biochem. **46:** 279–294.

FORSGREN, A. & SJÖQUIST, J. (1966): "Protein A" from *Staphylococcus aureus*. I. Pseudoimmune reaction with human γ-globulin. – J. Immunol. **97:** 822–827.

FOX, G. E., STACKENBRANDT, E., HESPELL, R. B., GIBSON, J., MANILOFF, J., DYER, T. A., WOLFE, R. S., BALCH, W. E., TANNER, R. S., MAGRUM, L. J., ZABLEN, L. B., BLAKEMORE, R., GUPTA, R., BONEN, L., LEWIS, B. J., STAHL, D. A., LUEHRSEN, K. R., CHEN, K. N. & WOESE, C. R. (1980): The phylogeny of prokaryotes. – Science **209:** 457–472.

FREDERICQ, P. (1963): On the nature of colicinogenic factors: a review. – J. Theoret. Biol. **4:** 159.

FREHEL, C., ROBBE, P., TINELLI, R. & RYTER, A. (1982): Relationship between biochemical and cytochemical results obtained on *Bacillus megaterium* and *Bacillus subtilis* cell-wall polysaccharides. – J. Ultrastruct. Res. **81:** 78–87.

FREHEL, C. & RYTER, A. (1969): Réversibilité de la sporulation chez *Bacillus subtilis*. – Ann. Inst. Pasteur Paris **117:** 297–311.

– – (1979): Peptidoglycan turnover during growth of a *Bacillus megaterium* Dap Lys mutant. – J. Bacteriol. **137:** 947–955.

– – (1980): *Bacillus megaterium* sporal peptidoglycan synthesis studied by high-resolution autoradiography. – J. Bacteriol. **144:** 789–799.

– – (1982): Electron microscope cytochemical study of cell-wall polysaccharides of *Bacillus subtilis* and two strains of *Bacillus megaterium*. – J. Ultrastruct. Res. **81:** 66–77.

FRIEDBERG, I. & AVIGAD, G. (1968): Structures containing polyphosphate in *Micrococcus lysodeikticus*. – J. Bacteriol. **96:** 544–553.

FRIEDRICH, B., HOGREFE, C. & SCHLEGEL, H. G. (1981): Naturally occurring genetic transfer of hydrogen-oxidizing ability between strains of *Alcaligenes eutrophus*. – J. Bacteriol. **147:** 198–205.

FROST, L. S., ARMSTRONG, G. D., FINLAY, B. B., EDWARDS, B. F. P. & PARANCHYCH, W. (1983): N-terminal amino acid sequencing of *EDP 208* conjugative pili. – J. Bacteriol. **153:** 950–954.

FUJITA, Y. & SHIMURA, S. (1974): Phycoerythrin of the marine blue-green alga *Trichodesmium thiebautii*. – Plant Cell Physiol. **15:** 939–942.

FULLER, R. & LOVELOCK, D. W. (Eds.) (1976): Microbial Ultrastructure. The Use of the Electron Microscope. – Academic Press, London, New York, San Francisco.

GANTT, E. (1969): Properties and ultrastructure of phycoerythrin from *Porphyridium cruentum.* – Plant Physiol. **44:** 1629–1638.
– (1977): Recent contributions in phycobiliproteins and phycobilisomes. – Photochem. Photobiol. **26:** 685–689.
– (1980): Structure and function of phycobilisomes: light harvesting pigment complexes in red and blue-green algae. – Internatl. Rev. Cytol. **66:** 45–80.
GANTT, E. & CONTI, S. F. (1969): Ultrastructure of blue-green algae. – J. Bacteriol. **97:** 1486–1493.
GANTT, E. & LIPSCHULTZ, C. A. (1977): Probing phycobilisome structure by immuno-electron microscopy. – J. Phycol. **13:** 185–192.
GANTT, E., LIPSCHULTZ, C. A., GRABOWSKI, J. & ZIMMERMANN, B. K. (1979): Phycobilisomes from blue-green and red algae: Isolation criteria and "dissociation" characteristics. – Plant Physiol. **63:** 615–620.
GARACI, E. & CASSONE, A. (1974): Capsular structures in *Klebsiella pneumoniae.* – Boll. Soc. Ital. Biol. Sperim. **50:** 1491–1495.
GARRO, A. J. & MARMUR, J. (1970): Defective bacteriophages. – J. Cell Physiol. **76:** 253.
GARTEN, W., HINDENNACH, I. & HENNING, U. (1975): The major proteins of the *Escherichia coli* outer cell envelope membrane. Characterization of proteins II* and III, comparison of all proteins. – Eur. J. Biochem. **59:** 215–221.
GASSNER, G., DUH, F.-M. & BROMEL, M. (1983): Chitinolytic activity: a prelude to a symbiotic relationship between bacteria and the screwworm fly? – In: H. E. A. SCHENK & W. SCHWEMMLER (Eds.): Endocytobiology. vol. II, pp. 801–811. – de Gruyter, Berlin, New York.
GATNER, E. M. S. & GARDNER, I. C. (1970): Observations on the fine structure of the root nodule endophyte of *Hippopaë rhamnoides* L. – Arch. Mikrobiol. **70:** 183–196.
GEBERS, R. & HIRSCH, P. (1978): Isolation and investigation of *Pedomicrobium* spp., heavy metal-depositing bacteria from soil habitats. – In: W. KRUMBEIN (Ed.): Environmental Biogeochemistry and Geomicrobiology. vol. III, pp. 911–922. – Ann Arbor Sci. Publ., Ann Arbor, Mich.
GEIDER, K. & HOFFMANN-BERLING, H. (1981): Proteins controlling the helical structure of DNA. – Annual Rev. Biochem. **50:** 233–260.
GEST, H. & FAVINGER, J. L. (1983): *Heliobacterium chlorum,* an anoxygenic brownish-green photosynthetic bacterium containing a "new" form of bacteriochlorophyll. – Arch. Microbiol. **136:** 11–16.
GHIORSE, W. C. (1980): Electron microscopic analysis of metal-depositing microorganisms in surface layers of Baltic Sea ferromanganese concretions. – In: P. A. TRUDINGER, M. R. WALTER & B. J. RALPH (Eds.): Biogeochemistry of Ancient and Modern Environments. pp. 345–354. – Austral. Acad. Sci., Canberra, and Springer-Verlag, New York.
GHIORSE, W. C. & HIRSCH P. (1978): Iron and manganese deposition by budding bacteria. – In: W. E. KRUMBEIN (Ed.): Environmental Biogeochemistry and Geomicrobiology. vol. III, pp. 897–909. – Ann Arbor Sci. Publ., Ann Arbor, Mich.
– – (1979): An ultrastructural study of iron and manganese deposition associated with extracellular polymers of *Pedomicrobium*-like budding bacteria. – Arch. Microbiol. **123:** 213–226.
– – (1982): Isolation and properties of ferromanganese-depositing budding bacteria from Baltic ferromanganese concretions. – Appl. Environ. Microbiol. **43:** 1464–1472.
GHOSH, B. K. (1974): The mesosome – a clue to the evolution of the plasma membrane. – Sub.-Cell. Biochem. **3:** 311–367.
GHOSH, B. K. & NANNINGA, N. (1976): Polymorphism of the mesosome in *Bacillus licheniformis* (749/C and 749). – J. Ultrastruct. Res. **56:** 107–120.
GHUYSEN, J. M. & SHOCKMAN, G. D. (1973): Biosynthesis of peptidoglycan. – In: L. LEIVE (Ed.): Bacterial Membranes and Walls. pp. 37–130. – Marcel Dekker, New York.
GIBBONS, N. E. (1974): Family V. Halobacteriaceae fam. nov. – In: R. E. BUCHANAN & N. E. GIBBONS (Eds.): Bergey's Manual of Determinative Bacteriology. 8th ed., pp. 269–273. – Williams & Wilkins Co., Baltimore, Md.
GIBSON, C. W., DANEO-MOORE, L. & HIGGINS, M. L. (1983): Cell wall assembly during inhibition of DNA synthesis in *Streptococcus faecium.* – J. Bacteriol. **155:** 351–356.
GIBSON JR., T. H. & STAEHELIN, L. A. (1978): Plasma membrane architecture of *Anabaena cylindrica:* occurrence of microplasmodesmata and changes associated with heterocyst development and the cell cycle. – Cytobiology **16:** 235–249.
– – (1979): Changes in thylakoid structure associated with the differentiation of heterocysts in the cyanobacterium *Anabaena cylindrica.* – Biochim. Biophys. Acta **546:** 373–382.

GIESBRECHT, P. (1972): Zur Morphogenese der Zellwand von Staphylokokken. – Mikroskopie **28:** 323–342.

GIESBRECHT, P. & DREWS, G. (1962): Elektronenmikroskopische Untersuchungen über die Entwicklung der "Chromatophoren" von *Rhodospirillum molischianum* Giesberger. – Arch. Mikrobiol. **43:** 152.

GIESBRECHT, P., LABISCHINSKI, H. & WECKE, J. (1985): A special morphogenetic wall defect and the subsequent activity of "murosomes" as the very reason for penicillin-induced bacteriolysis in staphylococci. – Arch. Microbiol. **141:** 315–324.

GILLERT, K.-E. & INST. WISS. FILM (1982): Aufbau und Verhalten beweglicher Kolonien von *Bacillus circulans.* – Film C 838 des IWF, Göttingen 1961. Publication: D. NIES & H. G. SCHLEGEL. Publ. Wiss. Film.: Sek. Med., Ser. 5, Nr. 23/ C 838.

GIVAN, A. L., GLASSEN, K., GREEN, R. S., LANG, W. K., ANDERSON, A. J. & ARCHIBALD, A. R. (1982): Relation between wall teichoic acid content of *Bacillus subtilis* and efficiency of adsorption of bacteriophages *SP 50* and *Φ 25*. – Arch. Microbiol. **133:** 318–322.

GLAGOLEVA, A. N. & SKULACHEV, V. P. (1978): The proton pump is a molecular engine of motile bacteria. – Nature **272:** 280–282.

GLAGOLEVA, T. N., GLAGOLEVA, A. N., GUSEV, M. V. & NIKITINA, K. A. (1980): Protonmotive force supports gliding in cyanobacteria. – FEBS Lett. **117:** 49–53.

GLATRON, M. F., LECADET, M. M. & DEDONDER, R. (1972): Structure of the parasporal inclusion of *Bacillus thuringiensis* Berliner: Characterization of a repetitive subunit. – Eur. J. Biochem. **30:** 330–338.

GLAUERT, A. M., KERRIDGE, D. & HORNE, R. W. (1963): The fine structure and mode of attachment of the sheated flagellum of *Vibrio metchnikkovii.* – J. Cell Biol. **18:** 327–336.

GLAUERT, A. M. & THORNLEY, M. J. (1969): The topography of the bacterial cell wall. – Annual Rev. Microbiol. **23:** 159–198.

GLAZER, A. N. (1976): Phycocyanins: Structure and function. – Photochem. Photobiol. Rev. **1:** 71–115.

– (1977): Structure and molecular organization of the photosynthetic accessory pigments of cyanobacteria and red algae. – Molec. Cell Biochem. **18:** 125–140.

GLAZER, A. N. & BRYANT, D. A. (1975): Allophycocyanin B (λ_{max} 671, 618 nm): a new cyanobacterial phycobiliprotein. – Arch. Microbiol. **104:** 15–22.

GLAZER, A. N. & FANG, S. (1973): Chromophore content of blue-green phycobiliproteins. – J. Biol. Chem. **248:** 659–622.

GLAZER, A. N. & HIXSON, C. S. (1975): Characterization of R-phycocyanin. Chromophore content of R-phycocyanin and C-phycoerythrin. – J. Biol. Chem. **250:** 5487–5495.

– – (1977): Subunit structure and chromophore composition of rhodophytan phycoerythrins. *Porphyridium cruentum* B-phycoerythrin and b-phycoerythrin. – J. Biol. Chem. **252:** 32–42.

GLAZER, A. N., WILLIAMS, R. C., YAMANAKA, G. & SCHACHMANN, H. K. (1979): Characteriziation of cyanobacterial phycobilisomes in zwitterionic detergents. – Proc. Natl. Acad. Sci. (Wash.) **76:** 6162–6166.

GLOSSMANN, H. & NEVILLE JR., D. D. (1971): Glycoproteins of cell surfaces. – J. Biol. Chem. **246:** 6339–6346.

GODING, J. W. (1978): Use of staphylococcal protein A as an immunological reagent. – J. Immunol. Meth. **20:** 241–253.

GÖRTZ, H.-D. (1980): Nucleus-specific symbionts in *Paramecium caudatum.* – In: W. SCHWEMMLER & H. E. A. SCHENK (Eds.): Endocytobiology, Endosymbiosis and Cell Biology. pp. 381–392. – de Gruyter, Berlin.

– (1983): Types of endonuclear symbiosis in ciliates. – In: H. E. A. SCHENK & W. SCHWEMMLER (Eds.): Endocytobiology. vol. II, pp. 517–522. – de Gruyter, Berlin, New York.

GÖRTZ, H.-D. & DIECKMANN, J. (1980): Life cycle and infectivity of *Holospora elagans* Haffkine, a micronucleus-specific symbiont of *Paramecium caudatum* (Ehrenberg). – Protistologica **16:** 591–603.

GOLECKI, J. R. (1977): Studies on ultrastructure and composition of cell walls of the cyanobacterium *Anacystis nidulans.* – Arch. Microbiol. **114:** 35–41.

– (1979): Ultrastructure of cell wall and thylakoid membranes of the thermophilic cyanobacterium *Synechococcus lividus* under the influence of temperature shifts. – Arch. Microbiol. **120:** 125–133.

GOLECKI, J. R. & DREWS, G. (1974): Zur Struktur der Blaualgen-Zellwand. Gefrierätzungsuntersuchungen an normalen und extrahierten Zellwänden von *Anabaena variabilis.* – Cytobiologie **8:** 213–227.

GOLECKI, J. R., DREWS, G. & BUHLER, R. (1979): The size and number of intramembrane particles in cells of the photosynthetic bacterium *Rhodopseudomonas capsulata* studied by freeze-fracture electron microscopy. – Cytobiologie **18:** 381–389.

GOLECKI, J. R. & OELZE, J. (1980): Differences in the architecture of cytoplasmic and intracytoplasmic membranes of the chemotrophically and phototrophically grown species of the Rhodospirallaceae. – J. Bacteriol. **144:** 781–788.

GOLECKI, J. R., SCHUMACHER, A. & DREWS, G. (1980): The differentiation of the photosynthetic apparatus and the intracytoplasmic membrane in cells of *Rhodopseudomonas capsulata* upon variation of light intensity. – Eur. J. Cell Biol. **23:** 1–5.

VAN GOOL, A. P., LAMBERT, R. & LAUDELOUT, H. (1969): The fine structure of frozen etched *Nitrobacter* cells. – Arch. Mikrobiol. **69:** 281–293.

GOTTSCHALK, G., ANDREESEN, J. R. & HIPPE, H. (1981): The genus *Clostridium* (nonmedical aspects). – In: M. P. STARR, H. STOLP, H. G. TRUEPER, A. BALOWS & H. G. SCHLEGEL (Eds.): The Prokaryotes. vol. II, pp. 1767–1803. – Springer, Berlin, Heidelberg, New York.

GOULBOURNE JR., E. A. & GREENBERG, E. P. (1981): Chemotaxis of *Spirochaeta aurantia:* Involvement of membrane potential in chemosensory signal transduction. – J. Bacteriol. **148:** 837–844.

GOULD, G. W. & HURST, A. (Eds.) (1969): The Bacterial Spore. – Academic Press, New York.

GOVAN, J. R. W. (1974): Studies on the pyocins of *Pseudomonas aeroginosa:* morphology and mode of action of contractile pyocins. – J. Gen. Microbiol. **80:** 1.

GRABOWSKY, J. & GANTT, E. (1978): Photophysical properties of phycobiliproteins from phycobilisomes: Fluorescence lifetimes, quantum yields, and polarization spectra. – Photochem. Photobiol. **39–45.**

GRÄF, W. (1965): Bewegungsorganellen bei Myxobakterien. – Arch. Hyg. **149:** 518.

GRAY, B. H. & GANTT, E. (1975): Spectral properties of phycobiliproteins from the blue-green alga *Nostoc* sp. – Photochem. Photobiol. **21:** 121–128.

GRAY, M. W. & DOOLITTLE, W. F. (1982): Has the endosymbiont hypothesis been proven? – Microbiol. Rev. **46:** 1–42.

GREENBERG, E. P. & CANALE-PAROLA, E. (1977): Motility of flagellated bacteria in viscous environments. – J. Bacteriol. **132:** 356–358.

GRIEBEL, R. J. & MERRICK, J. M. (1971): Metabolism of poly-β-hydroxybutyrate: Effect of mild alkaline extraction on native poly-β-hydroxybutyrate granules. – J. Bacteriol. **108:** 782–789.

GRIEBEL, R. J., SMITH, Z. & MERRICK, J. M. (1968): Metabolism of poly-β-hydroxybutyrate. Purification, composition, and properties of native poly-β-hydroxybutyrate granules from *Bacillus megaterium.* – Biochemistry **7:** 3676–3681.

GRIEF, C., HILLS, G. J. & SHAW, P. J. (1984): 3-dimensional structure of cell wall components of methanogenic bacteria. – Abstr. 8th Eur. Congr. Electron Microscopy. vol. II, pp. 1511–1512. Budapest.

GRIFFITH, J. D. (1976): Visualization of prokaryotic DNA in a regularly condensed chromatin-like fiber. – Proc. Natl. Acad. Sci. USA **73:** 563–567.

GROSS, H., MUELLER, T., WILDHABER, I. & WINKLER, H. P. (1984): Recent progress in high resolution shadowing. – Abstr. 8th Eur. Congr. Electron Microscopy. vol. II, pp. 1293–1294. Budapest.

GUGLIELMI, G., COHEN-BAZIRE, G. & BRYANT, D. A. (1981): The structure of *Gloebacter violaceus* and its phycobilisomes. – Arch. Microbiol. **129:** 181–189.

GUINAND, M., VACHERON, M. J., MICHEL, G. & TIPPER, D. J. (1979): Location of peptidoglycan lytic enzymes in *Bacillus sphaericus.* – J. Bacteriol. **138:** 126–132.

GUNSALUS, R. P. & WOLFE, R. S. (1978): Chromophoric factors F_{342} and F_{430} of *Methanobacterium thermoautotrophicum.* – FEMS Microbiol. Lett. **3:** 191–193.

GUPTA, M. & PANT, N. C. (1983): Symbiotes of *Dacus cucurbitae* and their in vitro physiology. - In: H. E. A. SCHENK & W. SCHWEMMLER (Eds.): Endocytobiology. vol. II, pp. 739–749. – de Gruyter, Berlin, New York.

GYSI, J. & ZUBER, H. (1976): Allophycocyanin I – a second cyanobacterial allophycocyanin? Isolation, characterization, and comparison with allophycocyanin II from the same alga. – FEBS Lett. **68:** 49–54.

HAGEAGE JR., G. J., EANES, E. D. & GHERNA, R. L. (1973): X-ray diffraction studies of the sulfur globules accumulated by *Chromatium* species. – J. Bacteriol. **101:** 464–469.

HAMILTON, R. C., BOVER, F. G. & MASON, T. J. (1975): An association between fimbriae and and pores in the wall of *Fusiformis nodosus.* – J. Gen. Microbiol. **91:** 421–424.

HAMILTON, L. D. & GETTNER, M. E. (1958): Fine structure of *Kappa* in *Paramecium aurelia.* – Biophys. Biochem. Cytol. **4:** 122–123.

HAMKALO, B. A. & MILLER JR., O. L. (1973): Electronmicroscopy of genetic activity. – Annual Rev. Biochem. **42:** 379–396.

HANERT, H. H. (1981): The genus *Siderocapsa* (and other iron- or manganese-oxidizing eubacteria). – In: M. P. STARR, H. STOLP, H. G. TRUEPER, A. BALOWS & H. G. SCHLEGEL (Eds.): The Prokaryotes. vol. I, pp. 1049–1060. – Springer, Berlin, Heidelberg, New York.

HANSON, R. S., PETERSON, J. A. & YOUSTEN, A. Ă. (1970): Unique biochemical events in bacterial sporulation. – Annual Rev. Microbiol. **24:** 53–90.

HARIHARAN, I. K., CZOLIJ, R. & WAKE, R. G. (1982): Conformation and segregation of nucleoids accompanying cell length extension after completion of a single round of DNA replication in germinated and outgrowing *Bacillus subtilis* spores. – J. Bacteriol. **150:** 861–869.

HAROLD, F. M. (1963): Inorganic polyphosphate of high molecular weight from *Aerobacter aerogenes.* – J. Bacteriol. **86:** 885–887.

– (1966): Inorganic polyphosphates in biology: Structure, metabolism, and function. – Bacteriol. Rev. **30:** 772–794.

HASE, T. (1983): Assembly of *Rickettsia tsutsugamuski* progeny in irradiated L cells. – J. Bacteriol. **154:** 976–979.

VAN HEEL, M. (1984): Multivariate statistical analysis of images of biological macromolecules: Powerful new methods. – Abstr. 8th Eur. Congr. Electron Microscopy. vol. II, pp. 1317–1325. Budapest.

HEGERL, R. , HOPPE, W., KNAUER, V. & TYPKE, D. (1984): Some aspects of the 3 D reconstruction of individual objects. – Abstr. 8th Eur. Congr. Electron Microscopy. vol. II, pp. 1363–1373. Budapest.

HEIDELBERGER, M., KENDALL, F. E. & SCHERP, H. W. (1936): The specific polysaccharides of types *I, II* and *III Pneumococcus.* – J. Exp. Med. **64:** 577–584.

HEIDRICH, H. G. & OLSEN, W. L. (1975): Deoxyribonucleic acid-envelope complexes from *Escherichia coli.* – J. Cell Biol. **67:** 444–460.

HEINHORST, S. & SHIVELY, J. M. (1983): The genes for the two subunits of ribulose-bisphosphate carboxylase in *Cyanophora paradoxa.* – In: H. E. A. SCHENK & W. SCHWEMMLER (Eds.): Endocytobiology. vol. II, pp. 449–450. – de Gruyter, Berlin, New York.

HELMSTETTER, C. E. & COOPER, S. (1968): DNA synthesis during the division cycle of rapidly growing *Escherichia coli B/r.* – J. Molec. Biol. **31:** 507–518.

HENDERSON, R., JUBB, J. S. & WHYTOCK, S. (1978): Specific labelling of the protein and lipid on the extracellular surface of purple membrane, – J. Molec. Biol. **123:** 259–274.

HENDERSON, R. & UNWIN, P. N. T. (1975): Three-dimensional model of purple membrane obtained by electron microscopy. – Nature **257:** 28–32.

HENNING, U. & SCHWARZ, U. (1973): Determination of cell shape. – In: L. LEIVE (Ed.): Bacterial Membranes and Walls. pp. 413–438. – Dekker, New York.

HEPPEL, L. A. (1969): The effect of osmotic shock on release of bacterial proteins and on active transport. – J. Gen. Physiol. **54:** 95–109.

– (1971): The concept of periplasmic enzymes. – In: L. I. ROTHFIELD (Ed.): Structure and Function of Biological Membranes. pp. 223–247. – Academic Press Inc., New York.

HEPTINSTALL, S., ARCHIBALD, A. R. & BADDILEY, J. (1970): Teichoic acids and membrane function in bacteria. – Nature **225:** 519–521.

HERBERT, B. N., GOULD, H. J. & CHAIN, E. B. (1971): Crystal protein of *Bacillus thuringiensis* var. *tolworthi.* – Eur. J. Biochem. **24:** 366–375.

HERDMAN, M. & STANIER, R. Y. (1977): The cyanelle: chloroplast or endosymbiotic prokaryote? – FEMS Microbiol. Lett. **1:** 7–12

HEUMANN, W. (1968): Conjugation in star-forming *Rhizobium lupini.* – Molec. Gen. Genet. **102:** 132–144.

HEUMANN, W. & MARX, R. (1964): Feinstruktur und Funktion der Fimbrien bei dem sternbildenden Bacterium *Pseudomonas echinoides.* – Arch. Mikrobiol. **47:** 325–337.

HIGGINS, M. L. & DANEO-MOORE, L. (1974): Factors influencing the frequency of mesosomes observed in fixed and unfixed cells of *Streptococcus faecalis.* – J. Cell Biol. **61:** 288–300.

HIGGINS, M. L., TSIEN, H. C. & DANEO-MOORE, L. (1976): Organization of mesosomes in fixed and unfixed cells. – J. Bacteriol. **127:** 1519–1523.

HIGHTON, P. J. (1976): The analysis of mesosomes in *Bacillus* species by sectioning and negative staining. – In: R. FULLER & D. W. LOVELOCK (Eds.): Microbial Ultrastructure. pp. 161–173. – Academic Press, London, New York, San Francisco.

HILL, S., DROZD, J. W. & POSTGATE, J. R. (1972): Environmental effects on the growth of nitrogen-fixing bacteria. – J. Appl. Chem. Biotechnol. **22:** 541–558.

HINDENNACH, I. & HENNING, U. (1975): The major proteins of the *Escherichia coli* outer cell envelope membrane. Preparative isolation of all major membrane proteins. – Eur. J. Biochem. **59:** 207–213.

HIRSCH, P. (1968): Biology of budding bacteria. IV. Epicellular deposition of iron by aquatic budding bacteria. – Arch. Mikrobiol. **60:** 201–216.

– (1974): Budding bacteria. – Annual Rev. Microbiol. **28:** 391–444.

HIRSCH, P. & CONTI, S. F. (1964): Biology of budding bacteria. I. Enrichment, isolation and morphology of *Hyphomicrobium* spp. – Arch. Mikrobiol. **48:** 339–357.

HIRSCH, P. & RHEINHEIMER, G. (1968): Biology of budding bacteria. V. Budding bacteria in aquatic habitats: occurrence, enrichment and isolation. – Arch. Mikrobiol. **62:** 289–306.

HITCHINS, V. M. & SADOFF, H. L. (1970): Morphogenesis of cysts in *Azotobacter vinelandii.* – J. Bacteriol. **104:** 492–498.

– – (1973): Sequential metabolic events during encystment of *Azotobacter vinelandii.* – J. Bacteriol. **113:** 1273–1279.

HOARE, D. S., INGRAM, L. O., THURSTON, E. L. & WALKUP, R. (1971): Dark heterotrophic growth of an endophytic blue-green alga. – Arch. Mikrobiol. **78:** 310–321.

HOBOT, J. A., CARLEMALM, E., VILLIGER, W. & KELLENBERGER, E. (1984): Periplasmic gel: new concept resulting from the reinvestigation of bacterial cell envelope ultrastructure by new methods. – J. Bacteriol. **160:** 143–152.

HOCHKEPPEL, H.-K., GORDON, J. & BRACK, C. (1977): RNA of the small subunit of the *Escherichia coli* ribosome with additional protein-binding sites. – FEBS Lett. **77:** 277–280.

HÖLTJE, J. V. & SCHWARZ, U. (1985): Structure and biosynthesis of the murein layer. – In: N. NANNINGA (Ed.): Molecular Cytology of *Escherichia coli.* pp. 77–119. – Academic Press Inc., New York.

HOENINGER, J. F. M., VAN ITERSON, W. & VAN ZANTEN, E. N. (1966): Basal bodies of bacterial flagella in *Proteus mirabilis.* II. Electron microscopy of negatively stained material. – J. Cell Biol. **18:** 603–618.

HOGREFE, C. & FRIEDRICH, B. (1984): Isolation and characterization of megaplasmid DNA from lithoautotrophic bacteria. – Plasmid **12:** 161–169.

HOLT, S. C. & BEVERIDGE, T. J. (1982): Electron microscopy: its development and application to microbiology. – Canad. J. Microbiol. **28:** 1–53.

HOLT, S. C., GAUTHIER, J. J. & TIPPER, D. J. (1975): Ultrastructural studies of sporulation in *Bacillus sphaericus.* – J. Bacteriol. **122:** 1322–1338.

HOLT, S. C. & LEADBETTER, E. R. (1969): Comparative ultrastructure of selected aerobic spore-forming bacteria: a freeze-etching study. – Bacteriol. Rev. **33:** 346–378.

HOMMA, J. Y., GOTO, S. & SHIONOYA, H. (1967): Relationship between pyocine and temperate phage of *Pseudomonas aeruginosa.* II. Isolation of pyocines from strain *P 1-III* and their characteristics. – Japan. J. Exp. Med. **37:** 373.

HOMMA, J. Y. & SHIONOYA, H. (1967): Relationship between pyocine and temperate phage of *Pseudomonas aeruginosa.* III. Serological relationship between pyocines and temperate phages. – Japan. J. Exp. Med. **37:** 395.

HORISBERGER, M. (1969): Structure of the dextran of the "Tibi"-grains. – Carbohyd. Res. **10:** 379–385.

HOROWITZ, S., DOYLE, R. J., YOUNG, F. E. & STREIPS, U. N. (1979): Selective association of the chromosome with membrane in a stable *L*-form of *Bacillus subtilis.* – J. Bacteriol. **138:** 915–922.

HOUWINK, A. C. & VAN ITERSON, W. (1950): Electron microscopical observations on bacterial cytology. II. A study of flagellation. – Biochim. Biophys. Acta **5:** 10–44.

HOVIND-HOUGEN, K. (1976): Determination by means of electron microscopy of morphological criteria of value for classification of some spirochetes, in particular treponemes. – Acta Pathol. Microbiol. Scand. Ser. B, Suppl. **255:** 1–41.

HOYLE, B. D. & BEVERIDGE, T. J. (1984): Metal binding by the peptidoglycan sacculus of *Escherichia coli K-12.* – Canad. J. Microbiol. **30:** 204–211.

HUIS IN 'T VELD, J. H. J. & LINSSEN, W. H. (1973): The localization of streptococcal group and type antigens: an electron microscopic study using ferritin-labelled antisera. – J. Gen. Microbiol. **74:** 315–324.

HUMPHREY, B. A., DICKSON, M. R. & MARSHALL, K. C. (1979): Physicochemical and in situ observations on the adhesion of gliding bacteria to surfaces. – Arch. Microbiol. **120:** 231–238.

HUSER, B. A., WUHRMANN, K. & ZEHNDER, A. J. B. (1982): *Methanothrix soehngenii* gen. nov. sp. nov., a new acetotrophic non-hydrogen-oxidizing methane bacterium. – Arch. Microbiol. **132:** 1–9.

HYAMS, J. S. & BORING, G. G. (1975): Flagellar coordination in *Chlamydomonas reinhardtii:* Isolation and reactivation of the flagellar apparatus. – Science **189:** 891–893.

IINO, T. (1969a): Polarity of flagellar growth in *Salmonella.* – J. Gen. Microbiol. **56:** 227–239.

– (1969b): Genetics and chemistry of bacterial flagella. – Bacteriol. Rev. **33:** 454–475.

– (1977): Genetics of structure and function of bacterial flagella. – Annual Rev. Genet. **11:** 161–182.

IKEDA, K. & EGAMI, F. (1969): Receptor substance for pyocin *R I*. Partial purification and chemical properties. – J. Biochem. (Tokyo) **65:** 603.

– – (1973): Lipopolysaccharide of *Pseudomonas aeruginosa* with special reference to pyocin *R* receptor activity. – J. Gen. Appl. Microbiol. **19:** 115.

IKEDA, T., KAMIYA, R. & YAMAGUCHI, S. (1983): Excretion of flagellin by a short-flagella mutant of *Salmonella typhimurium*. – J. Bacteriol. **153:** 506–510.

IMAE, Y., STROMINGER, M. B. & STROMINGER, J. L. (1976): Electron microscope studies of conditional spore cortexless mutants of *Bacillus sphaericus*. – J. Bacteriol. **127:** 1568–1570.

INGRAHAM, J. L., MAALOE, O. & NEIDHARDT, F. C. (1983): Growth of the Bacterial Cell. – Sinauer Assoc. Inc., Sunderland/USA.

INOUYE, H. & BECKWITH, J. (1977): Synthesis and processing of an *Escherichia coli* alkaline phosphatase precursor in vitro. – Proc. Natl. Acad. Sci. USA **74:** 1440–1444.

ISHIHARA, A., SEGALL, J. E., BLOCK, S. M. & BERG, H. C. (1983): Coordination of flagella on filamentous cells of *Escherichia coli* – J. Bacteriol. **155:** 228–237.

ISMAIL, G. & SOM, F. M. (1982): Haemagglutination reaction and epithelial cell adherence activity of *Serratia marcescens*. – J. Gen. Appl. Microbiol. **28:** 161–168.

VAN ITERSON, W., HOENIGER, J. F. M. & VAN ZANTEN, E. N. (1966): Basal bodies of bacterial flagella in *Proteus mirabilis*. I. Electron microscopy of sectioned material. – J. Cell Biol. **31:** 585–601.

ITO, K., SATO, T. & YURA, T. (1977): Synthesis and assembly of the membrane protein in *E. coli*. – Cell **11:** 551–559.

ITO, S. & KAGEYAMA, M. (1970): Relationship between pyocine and a bacteriophage in *Pseudomonas aeruginosa*. – J. Gen. Appl. Microbiol. **16:** 231.

IVANOVICS, G. (1962): Bacteriocins and bacteriocin-like substances. – Bacteriol. Rev. **26:** 108.

JANNASCH, H. W. & WIRSEN, C. O. (1981): Morphological survey of microbial mats near deep-sea thermal vents. – Appl. Environ. Microbiol. **41:** 528–538.

JARCHAU, T. (1985): Untersuchungen zum Vorkommen und zur Häufigkeit von Proteinvorstufen bei der Bildung von 1- und F-Pili in *Escherichia coli*. – Doctoral thesis, Göttingen.

JAROSCH, R. (1962): Gliding. – In: R. A. LEWIN (Ed.): Physiology and Biochemistry of Algae. pp. 573–581. – Academic Press, New York.

JAVOR, B., REQUADT, C. & STOECKENIUS, W. (1982): Box-shaped halophilic bacteria. – J. Bacteriol. **151:** 1532–1542.

JAYAWARDENE, A. & FARKAS-HIMSLEY, H. (1970): Mode of action of vibriocin. – J. Bacteriol. **102:** 382.

JENSEN, T. E. (1968): Electron microscopy of polyphosphate bodies in a blue-green alga, *Nostoc pruniforme*. – Arch. Mikrobiol. **62:**–152.

– (1970): Comparative ultrastructure of polyphosphate bodies in several blue-green algae. – J. Cell Biol. **47:** 241 A-242 A.

JENSEN, T. E. & SICKO, L. M. (1971): Fine structure of poly-β-hydroxybutyric acid granules in a blue-green alga, *Chlorogloea fritschii*. – J. Bacteriol. **106:** 683–686.

– – (1973): The fine structure of *Chlorogloea fritschii* cultured in sodium acetate enriched medium. – Cytologia **38:** 381–391.

JOHANNSSEN, W., SCHUETTE, H., MAYER, H. & MAYER, F. (1984): Structural analysis of EcoRI-DNA complexes as revealed by electron microscopy. – Arch. Microbiol. **140:** 265–270.

JOHNSON, H. H. & STOKES, J. C. (1966): Manganese oxidation by *Sphaerotilus discophorus*. – J. Bacteriol. **91:** 1543–1547.

JONES, G. W. (1977): The attachment of bacteria to the surface of animal cells. – In: J. L. REISSIG (Ed.): Microbial Interactions. Receptors and Recognition Ser. B. vol. III, pp. 139–176. – Chapman & Hall, London.

JONES, G. W. & ISAACSON, R. E. (1983): Proteinaceous bacterial adhesions and their receptors. – Crit. Rev. Microbiol. **10:** 229–260.

JONES, W. J., LEIGH, J. A., MAYER, F., WOESE, C. R. & WOLFE, R. S. (1983): *Methanococcus jannaschii* sp. nov., an extremely thermophilic methanogen from a submarine hydrothermal vent. – Arch. Microbiol. **136:** 254–261.

KAGAWA, H., MORISHITA, H. & ENOMOTO, M. (1981): Reconstruction in vitro of flagellar filaments onto hook structures attached to bacterial cells. – J. Molec. Biol. **153:** 465–470.

KAGAWA, H., NISHIYAMA, T. & YAMAGUCHI, S. (1983): Motility development of *Salmonella typhimurium* cells with fla *V* mutations after addition of exogenous flagellin. – J. Bacteriol. **155:** 435–437.

KAGEYAMA, M. (1970 a): Genetic mapping of a bacteriocinogenic factor in *Pseudomonas aeruginosa*. I. Mapping of pyocin *R2* factor by conjugation, – J. Gen. Appl. Microbiol. **16:** 523.
– (1970 b): Genetic mapping of a bacteriocinogenic factor in *Pseudomonas aeroginosa*. II. Mapping of pyocin *R 2* factor by transduction with phage *F 116*. – J. Gen. Appl. Microbiol. **16:** 531.
KAGEYAMA, M., IKEDA, K. & EGAMI, F. (1964): Studies of a pyocin. III. Biological properties of the pyocin. – J. Biochem. (Tokyo) **55:** 59.
KAHANE, I., BANAI, M., RAZIN, S. & FELDNER, J. (1982): Attachment of mycoplasmas to host cell membranes. – Rev. Infect. Dis. **4** (Suppl.): 185–192.
KAISER, G. E. & DOETSCH, R. N. (1976): The effects of alkalinity and hypertonicity on the morphology and motility of *Leptospira interrogans (biflexa)* strain *B 16*. – J. Gen. Microbiol. **96:** 25–33.
KAMIYA, R. & ASAKURA, S. (1976): Flagellar transformations at alkaline pH. – J. Molec. Biol. **108:** 513–518.
KANDLER, O. (1979): Zellwandstrukturen bei Methan-Bakterien. – Naturwissensch. **66:** 95–105.
– (1981): Archaebakterien und Phylogenie der Organismen. – Naturwissensch. **68:** 183–192.
KAO, O. H. W., EDWARDS, M. R. & BERNS, D. S. (1975): Physical-chemical properties of C-phycocyanin isolated from an acido-thermophilic eukaryote, *Cyanidium caldarium*. – Biochem. J. **147:** 63–70.
KAPLAN, S. (1978): Control and kinetics of photosynthetic membrane development. – In: R. CLAYTON & W. SISTROM (Eds.): The Photosynthetic Bacteria. pp. 809–839. – Plenum Publ. Corp., New York.
– (1981): Development of the membranes of photosynthetic bacteria. – Photochem. Photobiol. **34:** 769–774.
KAPLAN, S. & ARNTZEN, C. J. (1982): Photosynthetic membrane structure and function. – In: S. GOVINJEE (Ed.): Photosynthesis: Comparative Aspects of Bacteria and Green Plants. pp. 65–153. – Academic Press Inc., New York.
KAUFMANN, F. (1944): Zur Serologie der *Coli*-Gruppe. – Acta Pathol. Microbiol. Scand. **21:** 20–45.
KAUFMANN, N., REIDL, H.-H., GOLECKI, J. R., GARCIA, A. F. & DREWS, G. (1982): Differentiation of the membrane system in cells of *Rhodopseudomonas capsulata* after transition from chemotrophic to phototrophic growth conditions. – Arch. Microbiol. **131:** 313–322.
KAWATA, T., MASUDA, K., YOSHINO, K. & FUJIMOTO, M. (1974): Regular array in the cell wall of *Lactobacillus fermenti* as revealed by freeze-etching and negative staining. – Japan. J. Microbiol. **18:** 469–476.
KAZIRO, Y. & TANAKA, M. (1965 a): Studies on the mode of action of pyocin. I. Inhibition of macromolecular synthesis in sensitive cells. – J. Biochem. (Tokyo) **57:** 689.
– – (1965 b): Studies on the mode of action of pyocin. II. Inactivation of ribosomes. – J. Biochem. (Tokyo) **58:** 357.
KAZIRO, Y., TANAKA, M. & SHIMAZONO, N. (1964): Mode of action of pyocin: Inactivation of ribosomes in supporting poly U-directed incorporation of phenylalanine. – Biochem. Biophys. Res. Commun. **17:** 624.
KEDDIE, R. M. & JONES, D. (1981): Saprophytic, aerobic coryneform bacteria. – In: M. P. STARR, H. STOLP, H. G. TRUEPER, A. BALOWS & H. G. SCHLEGEL (Eds.): The Prokaryotes. vol. II, pp. 1838–1878. Springer, Berlin, Heidelberg, New York.
KELL, D. B., DODDEMA, H. J., MORRIS, J. G. & VOGELS, G. D. (1981): Energy coupling in methanogens. – In: H. DALTON (Ed.): Microbial, Growth on C_1 Compounds. Proc. Internatl. Symp. 1981. pp. 159–170. – Heyden & Sons Ltd., London.
KELLENBERGER, E. & RYTER, A. (1958): Cell wall and cytoplasmic membrane of *Escherichia coli*. – J. Biophys. Biochem. Cytol. **4:** 323–326.
KELLENBERGER, E., RYTER, A. & SÉCHAUD, J. (1958): Electron-microscopic study of DNA containing plasms. II. Vegetative and mature phage DNA as compared with normal bacterial nucleoids in different physiological states. – J. Biophys. Biochem. Cytol. **4:** 671–678.
KELLER, K. H., GRADY, M. & DWORKIN, M. (1983): Surface tension gradients: Feasible model for gliding motility of *Myxococcus xanthus*. – J. Bacteriol. **155:** 1358–1366.
KERRIDGE, D., HORNE, R. W. & GLAUERT, A. M. (1962): Structural components of flagella from *Salmonella typhimurium*. – J. Molec. Biol. **4:** 227–238.
KESSEL, M. (1978): A unique crystalline wall layer in the cyanobacterium *Microcystis marginata*. – J. Ultrastruct. Res. **62:** 203–212.
KESSEL, M. & COHEN, Y. (1982): Ultrastructure of a square bacterium from a brine pool in Southern Sinai. – J. Bacteriol. **150:** 851–860.
KESSEL, M., MACCOLL, R., BERNS, D. S. & EDWARDS, M. R. (1973): Electron microscope and pysical chemical characterization of C-phycocyanin from fresh extracts of two blue-green algae. – Canad. J. Microbiol. **19:** 831–836.

KHAK-MUN, T. (1968): The biological nature of iron-mangenese crusts of soil-forming rocks in Sakhalin mountain soils. – Mikrobiologiya **37:** 749–753. (In Russian).

KIES, L. (1979): Zur schematischen Einordnung von *Cyanophora paradoxa, Gloeochaete wittrockiana* und *Glaucocystis nostichinearium.* – Ber. Deutsch. Bot. Ges. **92:** 445–454.

KIM, K. S. & BARKSDALE, L. (1969): Crystalline inclusions of bacterium *22 M.* – J. Bacteriol. **98:** 1390–1394.

KLUG, A. (1967): The design of self-assembling systems of equal units. – Symp. Internatl. Soc. Cell Biol. **6:** 1–18.

KOBAYSHI, T., RINKER, J. N. & KOFFLER, H. (1959): Purification and chemical properties of flagellin. – Arch. Biochem. Biophys. **84:** 342–362.

KOBAYASHI, Y., SIEGELMAN, H. W. & HIRS, C. H. W. (1972): C-phycocyanin from *Phormidium luridium.* Isolation of subunits. – Arch. Biochem. Biophys. **152:** 187–198.

KOCH, A. L. (1984): How bacteria get their shapes: The surface stress theory. – Comments Molec. Cell. Biophys. **2:** 179–196.

KOCH, A. L., HIGGINS, M. L. & DOYLE, R. J. (1981): Surface tension-like forces determine bacterial shapes: *Streptococcus faecium.* – J. Gen. Microbiol. **123:** 151–161.

KODAKA, H., ARMFIELD, A. Y., LOMBARD, G. L. & DOWELL, V. R. (1982): Practical procedure for demonstrating bacterial flagella. – J. Clin. Microbiol. **16:** 948–952.

KÖNIG, H. & STETTER, O. (1982: Isolation and characterization of *Methanolobus tindarius* sp. nov., a coccoid methanogen growing only on methanol and methylamines. – Zentralbl. Bakteriol. Hyg. 1. Abt. Orig. Reihe C **3:** 478–490.

KOFFLER, H. & KOBAYASHI, T. (1975): Purification of flagella and flagellin with ammonium sulfate. – Arch. Biochem. Biophys. **67:** 246–248.

KOHRING, G. W., MAYER, F. & MAYER, H. (1985): Immunoelectron microscopic localization of the restriction endonuclease *EcoRI* in *Escherichia coli BS 5.* – Eur. J. Cell Biol. **37:** 1–6.

KOLLER, K. P., WEHRMEYER, W. & MÖRSCHEL, E. (1978): Biliprotein assembly in the disc-shaped phycobilisomes of *Rhodella violacea.* – Eur. J. Biochem. **91:** 57–63.

KOLLER, K. P., WEHRMEYER, W. & SCHNEIDER, H. (1977): Isolation and characterization of disc-shaped phycobilisomes from the red alga *Rhodella violacea.* – Arch. Microbiol. **112:** 61–67.

KOPPES, L. J. H., MEIJER, M., OONK, H. B., DE JONG, M. A. & NANNINGA, N. (1980): Correlation between size and age at different events in the cell division cycle of *Escherichia coli.* – J. Bacteriol. **143:** 1241–1252.

KOPPES, L. J. H. & NANNINGA, N. (1980): Positive correlation between size at initiation of chromosome replication in *Escherichia coli* and size at initiation of cell constriction. – J. Bacteriol. **143:**89–99.

KOPPES, L. J. J., OVERBEEKE, N. & NANNINGA, N. (1978 a): DNA replication pattern and cell wall growth in *Escherichia coli PAT 84.* – J. Bacteriol. **133:** 1053–1061.

KOPPES, L. J. H., WOLDRINGH, C. L. & NANNINGA, N. (1978 b): Size variations and correlation of different cell cycle events in slow-growing *Escherichia coli.* – J. Bacteriol. **134:** 423–433.

KORHONEN, T. K., TARKKA, E., RANTA, H. & HAAHTELA, K. (1983): Type 3 fimbriae of *Klebsiella* sp.: Molecular characterization and role in bacterial adhesion to plant roots. – J. Bacteriol. **155:** 860–865.

KOSHLAND JR., D. E. (1977): A response regulator model in a simple sensory system. – Science **196:** 1055–1063.

– (1979): A model regulatory system: Bacterial chemotaxis. – Physiol. Rev. **59:** 811–862.

KROL, P. M. & FARKAS-HIMSLEY, H. (1972): Mode of action of vibriocin: effects on membrane permeability and transport. – Microbios **6:** 199.

KÜHLWEIN, H. (1969): Some aspects of morphogenesis and fruiting body formation in myxobacteria. – J. Appl. Bacteriol. **32:** 19–21.

KUHN, D. A. (1981): The genera *Simonsiella* and *Alysiella.* – In: M. P. STARR, H. STOLP, H. G. TRUEPER, A. BALOWS & H. G. SCHLEGEL (Eds.): The Prokaryotes. vol. I, pp. 390–399. – Springer, Berlin, Heidelberg, New York.

KUHN, H.-M., MEIER, U. & MAYER, H. (1984): *ECA,* das gemeinsame Antigen der Enterobacteriaceae – Stiefkind der Mikrobiologie. – Forum Mikrobiol. **7:** 274–285.

KUNKEL, D. D. (1982): Thylakoid centers: Structures associated with the cyanobacterial photosynthetic membrane system. – Arch. Mictrobiol. **133:** 97–99.

KUPFER, D. & ZUSMAN, D. R. (1984): Changes in cell-suface hydrophobicity of *Myxococcus xanthus* are correlated with sporulation-related events in the developmental program. – J. Bacteriol. **159:** 776–779.

KUSHNAREV, V. M. & SMIRNOVA, T. S. (1966): Electron microscopy of alkaline phosphatase of *Escherichia coli.* – Canad. J. Microbiol. **12:** 605–608.

KUTZNER, H. J. (1981): The family Streptomycetaceae. – In: M. P. STARR, H. STOLP, H. G. TRUEPER, A. BALOWS & H. G. SCHLEGEL (Eds.): The Prokaryotes. vol. II, pp. 2028–2090. – Springer, Berlin, Heidelberg, New York.

KYLE, D. J., STAEHELIN, L. A. & ARNTZEN, C. J. (1983): Lateral mobility of the light-harvesting complex in chloroplast membranes controls excitation energy distribution in higher plants. – Arch. Biochem. Biophys. **222:** 527–541.

LAISHLEY, E. J., MACALISTER, T. J., CLEMENTS, I. & YOUNG, C. (1973): Isolation and morphology of native intracellular polyglucose granules from *Clostridium pasteurianum.* – Canad. J. Microbiol. **19:** 991–994.

LALUCAT, J., ALVAREZ, A., PARES, R. & SCHLEGEL, H. G. (1980): R-bodies in *Pseudomonas.* – In: W. SCHWEMMLER & H. E. A. SCHENK (Eds.): Endocytobiology, Endosymbiosis and Cell Biology. pp. 409–416. – Walter de Gruyter, Berlin.

LALUCAT, J. & MAYER, F. (1978): "Spiral-bodies" – intracytoplasmic membraneous structures in a hydrogen-oxidizing bacterium. – Z. Allg. Mikrobiol. **18:** 422–427.

LALUCAT, J., MEYER, O., MAYER, F., PARES, R. & SCHLEGEL, H. G. (1979): R-bodies in newly isolated free-living hydrogen-oxidizing bacteria. – Arch. Microbiol. **121:** 9–15.

LALUCAT, J., PARES, R. & SCHLEGEL, H. G. (1982): *Pseudomonas taeniospiralis* sp. nov., an R-body-containing hydrogen bacterium. – Internatl. J. Syst. Bacteriol. **32:** 332–338.

LAMED, R., SETTER, E. & BAYER, E. A. (1983): Characterization of a cellulose-binding, cellulase-containing complex in *Clostridium thermocellum.* – J. Bacteriol. **156:** 828–836.

LAMPEN, J. O. (1974): Movement of extracellular enzymes across membranes. – In: M. A. SLEIGH & D. H. JENNINGS (Eds.): Symp. Soc. Exp. Biol. vol. XXVIII, pp. 351–374. – Cambridge Univ. Press, England.

LANG, N. J. (1968): The fine structure of blue-green algae. – Annual Rev. Microbiol. **22:** 15–46.

LANG, N. J., SIMON, R. D. & WOLK, C. P. (1972): Correspondence of cyanophycin granules with structured granules in *Anabaena cylindrica.* – Arch. Mikrobiol. **83:** 313–320.

LANGENBERG, K. F., BRYANT, M. P. & WOLFE, R. S. (1968): Hydrogen-oxidizing methane bacteria. II. Electron microscopy. – J. Bacteriol. **95:** 1124–1129.

LANYI, J. K. (1978): Light energy conversion in *Halobacterium halobium.* – Microbiol. Rev. **42:** 682–706.

LARSEN, H. (1962): Halophilism. – In: I. C. GUNSALUS & R. Y. STANIER (Eds.): Bacteria: A treatise on structure and function. vol. IV, pp. 297–342. – Academic Press Inc., New York.

– (1967): Biochemical aspects of extreme halophilism. – Adv. Microbiol. Physiol. **1:** 97–132.

LARSEN, S. H., READER, R. W., KANT, E. N., TSO, W.-W. & ADLER, J. (1974): Change in direction of flagellar rotation is the basic of the chemotactic response in *Escherichia coli.* – Nature **249:** 75–77.

LAUTERBORN, R. (1915): Die sapropelische Lebewelt (Ein Beitrag zur Biologie des Faulschlamms natürlicher Gewässer). – Verh. Naturhist.-Med. Vereins **13:** 395–480.

LECADET, M. M. & DEDONDER, R. (1971): Biogenesis of the crystalline inclusion of *Bacillus thuringiensis* during sporulation. – Eur. J. Biochem. **23:** 282–294.

LEDERBERG, J. (1956): Conjugal pairing in *Escherichia coli.* – J. Bacteriol. **71:** 497–498.

LEDERBERG, J., CAVALLI, L. L. & LEDERBERG, E. M. (1952): Sex compatibility in *Escherichia coli.* – Genetics **37:** 720–730.

LEE, D. R., SCHNAITTMAN, C. A. & PUGSLEY, A. P. (1979): Chemical heterogeneity of major outer membrane pore proteins of *Escherichia coli.* – J. Bacteriol. **138:** 861–870.

LEFORT-TRAN, M., COHEN-BAZIRE, G. & POUPHILE, M. (1973): Les membranes photosynthétiques des algues à biliprotéines observées après cryodécapage. – J. Ultrastruct. Res. **44:** 199–209.

LEIFSON, E. (1960): Atlas of Bacterial Flagellation. – Academic Press Inc., New York, London.

– (1964): *Hyphomicrobium neptunium* sp. n. – Antonie van Leeuwenhoek J. Microbiol. Serol. **30:** 249–256.

LEMASSON, C., TANDEAU DE MARSAC, N. & COHEN-BAZIRE, G. (1973): Role of allophycocyanin as a light-harvesting pigment in cyanobacteria. – Proc. Natl. Acad. Sci. (Wash.) **70:** 3130–3133.

LENNETTE, E. H., SPAULDING, E. H. & TRUANT, J. P. (Eds.) (1974): Manual of Clinical Microbiology. 2nd ed. – Amer. Soc. Microbiol., Washington D. C.

LEWIN, R. A. (1981): The Prochlorophytes. – In: M. P. STARR, H. STOLP, H. G. TRUEPER, A. BALOWS & H. G. SCHLEGEL (Eds.): The Prokaryotes. vol. I, pp. 257–266. – Springer, Berlin, Heidelberg, New York.

LEY, A. C., BUTLER, W. L., BRYANT, D. A. & GLAZER, A. N. (1977): Isolation and function of allophycocyanin B of *Porphyridium cruentum.* – Plant Physiol. **59:** 974–980.

LICHTLÉ, C. & THOMAS J. C. (1976): Etude ultrastructurale des thylacoïdes des algues à phycobiliprotéines, comparison des résultats obtenus par fixation classique et cryodécapage. – Phycologia **15:** 393–404.

LICKFELD, K. G. & ACHTERRATH, M. (1972): Polymorphismus des *Staphylococcus aureus*-Mesosoms. – Cytobiologie, Z. Exp. Zellforsch. **6:** 74–85.
LIN, L. P., PANKRATZ, S. & SADOFF, H. L. (1978): Ultrastructural and physiological changes occurring upon germination and outgrowth of *Azotobacter vinelandii* cysts. – J. Bacteriol. **135:** 641–646.
LIN, L. P. & SADOFF, H. L. (1968): Encystment and polymer production by *Azotobacter vinelandii* in the presence of β-hydroxybutyrate. – J. Bacteriol. **95:** 2336–2343.
– – (1969): Chemical composition of *Azotobacter vinelandii* cysts. – J. Bacteriol. **100:** 480–486.
LINDLEY, E. V. & MACDONALD, R. E. (1979): A second mechanism for sodium extrusion in *Halobacterium halobium:* a lightdriven sodium pump. – Biochem. Biophys. Res. Commun. **88:** 491–499.
LINDSAY, S. S., WHEELER, B., SANDERSON, K. E. & COSTERTON, J. W. (1973): The release of alkaline phosphatase and lipopolysaccharide during the growth of rough and smooth strains of *Salmonella typhimurium.* – Canad. J. Microbiol. **19:** 335–343.
LINNETT, P. E. & TIPPER, D. J. (1976): Transcriptional control of peptidoglycan precursor synthesis during sporulation in *Bacillus sphaericus.* – J. Bacteriol. **125:** 565–574.
LOPERFIDO, B. & SADOFF, H. L. (1973): Germination of *Azotobacter vinelandii* cysts: Sequence of macromolecular synthesis and nitrogen fixation. – J. Bacteriol. **113:** 841–846.
LOTZ, W. (1976): Defective bacteriophages: The page tail-like particles. – In: F. E. HAHN (Ed.): Progress in Molecular and Subcellular Biology. vol. IV, pp. 53–102. – Springer, Berlin, Heidelberg, New York.
LOWY, J. (1965): Structure of the proximal ends of bacterial flagella. – J. Molec. Biol. **14:** 297–299.
LOWY, J. & HANSON, J. (1965): Electron microscope studies of bacterial flagella. – J. Molec. Biol. **11:** 293–313.
LOWY, J. & SPENCER, M. (1968): Structure and function of bacterial flagella. – Symp. Soc. Exp. Biol. **22:** 215–236.
LOZIER, R. H. (1982): Rapid kinetic optical absorption spectroscopy of bacteriorhodopsin photocycles. – Meth. Enzymol. **88:** 133–162.
LÜNSDORF, H., EHRIG, K., FRIEDL., P. & SCHAIRER, H. U. (1984): Use of monoclonal antibodies in immunoelectron microscopy for the determination of subunit stoichiometry in oligomeric enzymes. – J. Molec. Biol. **173:** 131–136.
LUFT, J. H. (1971): Ruthenium red and violet. I. Chemistry, purification, methods of use for electron microscopy and mechanisms of action. – Anat. Rec. **171:** 347–368.
LUGTENBERG, B. & VAN ALPHEN, L. (1983): Molecular architecture and functioning of the outer membrane of *Escherichia coli* and other Gram-negative bacteria. – Biochim. Biophys. Acta **737:** 51–115.
LURIA, S. E. (1964): On the mechanism of action of colicines. – Ann. Inst. Pasteur **107** (Suppl. **5**): 67.
LUTZ, L. (1899): Recherches biologiques sur la constitution du "Tibi". – Bull. Soc. Mycol. France **15:** 68–72.
MACALISTER, T. J., IRVIN, R.T. & COSTERTON, J. W. (1977a): Cell surface-localized alkaline phosphatase of *Escherichia coli* as visualized by reaction product deposition and ferritin-labelled antibodies. – J. Bacteriol. **130:** 318–328.
– – – (1977b): Immunocytological investigation of protein synthesis in *Escherichia coli.* – J. Bacteriol **130:** 329–338.
MACDONALD, R. E., GREENE, R. V., CLARK, R. D. & LINDLEY, E. V. (1979): Characterization of the lightdriven sodium pump of *Halobacterium halobium.* – J. Biol. Chem. **254:** 11831–11838.
MCDONALD, R. M. (1983): Bacterium-like organelles in V. A. mycorrhizal fungi. – In: H. E. A. SCHENK & W. SCHWEMMLER (Eds.): Endocytobiology. vol. II, pp. 595–603. – de Gruyter, Berlin, New York.
MACKEY, B. M. & MORRIS, J. G. (1971): Ultrastructural changes during sporulation of *Clostridium pasteurianum.* – J. Gen. Microbiol. **66:** 1–13.
MACKIE, E. B., BROWN, K. N., LAM, J. & COSTERTON, J. W. (1979): Morphological stabilization of capsules of group *B* streptococci, types *Ia, Ib, II,* and *III,* with specific antibody. – J. Bacteriol. **138:** 609–617.
MACNAB, R. M. (1983): Bacterial motility: Energization and switching of the flagellar motor. – In: H. SUND & C. VEEGER (Eds.): Mobility and Recognition in Cell Biology. pp. 499–516. – de Gruyter, Berlin, New York.
MACRAE, I. C. & EDWARDS, J. F. (1972): Adsorption of colloidal iron by bacteria. – Appl. Microbiol. **24:** 819–823.
MACRAE, T. H. & MCCURDY, H. D. (1975): Ultrastructural studies of *Chondromyces crocatus* vegetative cells. – Canad. J. Microbiol. **21:** 1815–1826.
MADOFF, S. (1981): The L-forms of bacteria. – In: M. P. STARR, H. STOLP, H. G. TRUEPER, A. BALOWS & H. G. SCHLEGEL (Eds.):The Prokaryotes. vol. II, pp. 2225–2237. – Springer, Berlin, Heidelberg, New York.

MAH, R. A. & SMITH, M. R. (1981): The methanogenic bacteria. – In: M. P. STARR, H. STOLP, H. G. TRUEPER, A. BALOWS & H. G. SCHLEGEL (Eds.): The Prokaryotes. vol. I, pp. 948–977. – Springer, Berlin, Heidelberg, New York.

MANILOFF, J. (1979): *Mycoplasma* gliding may be a biased brownian motion. – J. Theor. Biol. **81:** 617–620.

MANN, S., FRANKEL, R. B. & BLAKEMORE, R. P. (1984): Structure, morphology and crystal growth of bacterial magnetite. – Nature **130:** 405–407.

MANNING, P. A. & ACHTMANN, M. (1979): Cell-to-cell interactions in conjugating *Escherichia coli:* the involvement of the cell envelope. – In: M. INOUYE (Ed.): Bacterial Outer Membranes. Biogenesis and Functions. pp. 409–447. – Wiley, New York.

MANNING, P. A., BEUTIN, L. & ACHTMANN, M. (1980): Outer membrane of *Escherichia coli:* properties of the F sex factor *traT* protein which is involved in surface exclusion. – J. Bacteriol. **142:** 285–294.

MARAMOROSCH, K. (1963): Arthropod transmission of plant viruses. – Annual Rev. Entomol. **8:** 369–414.

– (Ed.) (1969): Viruses, Vectors, and Vegetation. – Wiley (Interscience), New York.

– (1979): How *Mycoplasmas* and *Rickettsias* induce plant disease. – In: J. HORSFALL & E. COWLING (Eds.): Plant Disease. vol. IV, pp. 203–217. – Academic Press, New York, San Francisco, London.

MARCUS, L. & KANESHIRO, T. (1972): Lipid composition of *Azotobacter vinelandii* in which the internal membrane network is induced or repressed. – Biochim. Biophys. Acta (Amsterdam) **288:** 296–303.

MARGULIS, L., TO, L. P. & CHASE, D. G. (1981): The genera *Pillotina, Hollandina,* and *Diplocalyx.* – In: M. P. STARR, H. STOLP, H.G. TRUEPER, A. BALOWS & H. G. SCHLEGEL (Eds.): The Prokaryotes. vol. I, pp. 548–554. – Springer, Berlin, Heidelberg, New York.

MARTENS, S., BRANDT, P. & WIESSNER, W. (1982): On the developmental dependence between *Cyanophora paradoxa* and *Cyanocyta korschikoffiana* in symbiosis. Host dependent development of endocyanelles. – Planta **155:** 190–192.

MARVIN, D. A. & HOHN, B. (1969): Filamentous bacterial viruses. – Bacteriol. Rev. **33:** 172–209.

MASUDA, K. & KAWATA, T. (1979): Ultrastructure and partial characterization of a regular array in the cell wall of *Lactobacillus brevis.* – Microbiol. Immunol. **23:** 941–953.

– – (1980): Reassembly of the regularly arranged subunits in the cell wall of *Lactobacillus brevis* and their reattachment to cell walls. – Microbiol. Immunol. **24:** 299–308.

– – (1981): Characterization of a regular array in the wall of *Lactobacillus buchneri* and its reattachment to the other wall components. – J. Gen. Microbiol. **124:** 81–90.

– – (1983): Distribution and chemical characterization of regular arrays in the cell walls of strains of the genus *Lactobacillus.* – FEMS Microbiol. Lett. **20:** 145–150.

MAUNSBACH, A. B. (1984): A survey of recent improvements of EM methods in cell biology. – Abstr. 8th Eur. Congr. Electron Microscopy. vol. III, pp. 2401–2408. Budapest.

MAYER, F. (1969): Die Fimbrien von *Rhizobium lupini 1/50, sta⁻.* – Arch. Mikrobiol. **68:** 179–186.

– (1971): Elektronenmikroskopische Untersuchung der Fimbrienkontraktion bei dem sternbildenden Bodenbakterium *Pseudomonas echinoides.* – Arch. Mikrobiol. **76:** 166–173.

– (1978): Vergleichende Feinstruktur-Untersuchungen an Bakterien mit den Methoden Negativ-Kontrastisierung, Ultradünnschnitt-Technik und Gefrierätzung. – Mikroskopie (Wien) **34:** 39–43.

MAYER, F., KALL, S. & SCHMITT, R. (1974): Untersuchung der morphologischen Grundlagen für einen möglichen Übertragungsweg der DNA bei der Konjugation sternbildender Bodenbakterien. – Z. Allg. Mikrobiol. **14:** 221–228.

MAYER, F., LURZ, R. & SCHOBERTH, S. (1977): Electron microscopic investigation of the hydrogen-oxidizing acetate-forming anaerobic bacterium *Acetobacterium woodii.* – Arch. Microbiol. **115:** 207–213.

MAYFIELD, C. I. & INNIS, W. E. (1977): A rapid, simple method for staining bacterial flagella. – Canad. J. Microbiol. **23:** 1311–1313.

MAYR-HARTING, A., HEDGES, A. J. & BERKELEY, R. C. W. (1972): Methods for studying bacteriocins. – In: J. R. NORRIS & D. W. RIBBONS (Eds.): Methods in Microbiology. vol. VIIA, pp. 315–422. – Academic Press, New York, London.

MCCOWAN, R. P., CHENG, K. J. & COSTERTON, J. W. (1980): Adherent bacterial populations on the bovine rumen wall: Distribution patterns of adherent bacteria. – Appl. Environ. Microbiol. **39:** 233–241.

MCCOY, R. E. (1981): Wall-free prokaryotes of plants and invertebrates. – In: M. P. STARR, H. STOLP, H. G. TRUEPER, A. BALOWS, & H. G. SCHLEGEL (Eds.): The Prokaryotes. vol. II, pp. 2238–2246. – Springer, Berlin, Heidelberg, New York.

MCGREGOR-SHAW, J. B., EASTERBROOK, K. B. & MCBRIDE, R. P. (1973): A bacterium with echinulate (nonprosthecate) appendages. – Internatl. J. Syst. Bacteriol. **23:** 267–270.

McMichael, J. C. & Ou, J. T. (1979): Structure of common pili from *Escherichia coli*. – J. Bacteriol. **138:** 969–975.

Mejer, M., De Jong, M. A., Woldringh, C. L. & Nanninga, N. (1976a): Factors affecting the release of folded chromosomes from *Escherichia coli*. – Eur. J. Biochem. **63:** 469–475.

– – – – (1976b): Significance of folded chromosomes released from amino-acid-starved *Escherichia coli* cells. – Eur. J. Biochem. **65:** 409–414.

Mendel, L., Timm, B. & Subramanian, A. R. (1978): Primary structures of two homologous ribosome-associated DNA-binding proteins of *Escherichia coli*. – FEBS Lett. **95:** 395–398.

Meng, K. E. & Pfister, R. M. (1980): Intracellular structures of *Mycoplasma pneumoniae* revealed after membrane removal. – J. Bacteriol. **144:** 390–399.

Mescher, M. F. & Strominger, J. L. (1976): Structural (shape maintaining) role of the cell surface glycoprotein of *Halobacterium salinarium*. – Proc. Natl. Acad. Sci. USA **73:** 2687–2691.

Mett, H., Kloetzlen, L. & Vosbeck, K. (1983): Properties of pili from *Escherichia coli SS142* that mediate mannose-resistant adhesion to mammalian cells. – J. Bacteriol. **153:** 1038–1044.

Miano, A., Losso, M. A., Gianfranceschi, G. L. & Gualerzi, C. O. (1982): Proteins from the prokaryotic nucleoid. I. Effect of NS 1 and NS 2 (HU) proteins on the thermal stability of DNA. – Biochem. Internatl. **5:** 415–422.

Miller, J., Plapp, R. & Kandler, O. (1966): The amino acid sequence of the serine containing murein of *Butybacterium Rettgeri*. – Biochim. Biophys. Res. Commun. **25:** 415–420.

Miller, J. R. & Kline, B. C. (1979): Biochemical characterization of nonintegrated plasmid-folded chromosome complexes: Sex factor F and the *Escherichia coli* nucleoid. – J. Bacteriol. **137:** 885–890.

Miller, K. R. (1979): Structure of a bacterial photosynthetic membrane. – Proc. Natl. Acad. Sci. (USA) **76:** 6415–6419.

– (1982): Three-dimensional structure of a photosynthetic membrane. – Nature (London) **300:** 53–55.

Miller, M. M. & Lang, N. J. (1968): The fine structure of akinete formation and germination in *Cylindrospermum*.– Arch. Mikrobiol. **60:** 303–313.

Miller, Jr., O. L., Hamkalo, B. A. & Thomas, Jr., C. A. (1970): Visualization of bacterial genes in action. – Science **169:** 392–395.

Minkley, E. G., Polen, S., Brinton, C. C. & Ippen-Ihler, K. (1976): Identification of the structural gene for F-pilin. – J. Molec. Biol. **108:** 111–121.

Miorner, H., Johansson, G. & Kronvall, G. (1983): Lipoteichoic acid is the major cell-wall component responsible for surface hydrophobicity of group *A* streptococci. – Infect. Immun. **39:** 336–343.

Mitchell, P. (1984): Bacterial flagellar motors and osmoelectric molecular rotation by an axially transmembrane well and turnstile mechanism. – FEBS Lett. **176:** 287–294.

Mitruka, B. M., Costilow, R. N., Black, S. H. & Pepper, R. E. (1967): Comparisons of cells, refractile bodies, and spores of *Bacillus popilliae*. – J. Bacteriol. **94:** 759–765.

Mitsui, Y., Dyer, F. P. & Langridge, R. (1973): X-ray diffraction studies of bacterial pili. – J. Molec. Biol. **79:** 57–64.

Mizushima, S. (1985): Structure and assembly of the outer membrane. – In: N. Nanninga (Ed.): Molecular Cytology of *Escherichia coli*. pp. 39–75. – Academic Press Inc., New York.

Möller, B., Ossmer, R., Howard, B. H., Gottschalk, G. & Hippe, H. (1984): *Sporomusa*, a new genus of Gram-negative anaerobic bacteria including *Sporomusa sphaeroides* spec. nov. and *Sporomusa ovata* spec. nov. – Arch. Microbiol. **139:** 388–396.

Mörschel, E., Koller, K. P., Wehrmeyer, W. & Schneider, H. (1977): Biliprotein assembly on the disc-shaped phycobilisomes of *Rhodella violacea*. I. Electron microscopy of phycobilisomes in situ and analysis of their architecture after isolation and negative staining. – Cytobiologie **16:** 118–129.

Moinas, M., Horisberger, M. & Bauer. H. (1980): The structural organization of the "Tibi" grain as revealed by light, scanning and transmission microscopy. – Arch. Microbiol. **128:** 157–161.

Moll, G. & Ahrens, R. (1970): Ein neuer Fimbrientyp. – Arch. Microbiol. **70:** 361–368.

Mollenhauer, H. H. & Hopkins, D. L. (1974): Ultrastructural study of Pierce's disease bacterium in grape xylem tissue. – J. Bacteriol. **119:** 612–618.

Moore, D., Sowa, B. A. & Ippen-Ihler, K. (1981): Location of an F-pilin pool in the inner membrane. – J. Bacteriol. **146:** 251–259.

Moore, R. L. (1981): The genera *Hyphomicrobium, Pedomicrobium,* and *Hyphomonas*. – In: M. P. Starr, H. Stolp, H. G. Trueper, A. Balows & H. G. Schlegel (Eds.): The Prokaryotes. vol. I, pp. 480–487. – Springer, Berlin, Heidelberg, New York.

MOORE, R. L. & HIRSCH, P. (1973a): Nuclear apparatus of *Hyphomicrobium.* – J. Bacteriol. **116:** 1447–1455.
– – (1973b): First generation synchrony of isolated *Hyphomicrobium* swarmer populations. – J. Bacteriol. **116:** 418–423.
MOOREHOUSE, R., WINTER, W. T., ARNOTT, S. & BAYER, M. E. (1977): Conformation and molecular organization in fibers of capsular polysaccharide from *Escherichia coli M 41* mutant. – J. Molec. Biol. **109:** 373–391.
MUCKLE, G., OTTO, J. & RÜDIGER, W. (1978): On the linkages between chromophore and protein in biliproteins. VII. Amino acid sequence in the chromophore regions of C-phycoerythrin from *Pseudanabaena W 1173* and *Phormidium persicinum.* – Hoppe-Seyler's Z. Physiol. Chem. **359:** 345–355.
MUCKLE, G. & RÜDIGER, W. (1977): Chromophore content of C-phycoerythrin from various cyanobacteria. – Z. Naturforsch. **32e:** 957–962.
MUDD, S., HEINMETS, F. & ANDERSON, T. F. (1943): The pneumococcal capsular swelling reaction, studied with the aid of the electron microscopy. – J. Exp. Med. **78:** 327–332.
MULDER, E. G. & DEINEMA, M. H. (1981): The sheathed bacteria. – In: M. P. STARR, H. STOLP, H. G. TRUEPER, A. BALOWS & H. G. SCHLEGEL (Eds.): The Prokaryotes. vol. I, pp. 425–440. – Springer, Berlin, Heidelberg, New York.
MULLAKHANBHAI, M. F. & LARSEN, H. (1975): *Halobacterium volcanii* spec. nov., a Dead Sea bacterium with a moderate salt requirement. – Arch. Microbiol. **104:** 207–214.
MUÑOZ, E. (1982): Polymorphism and conformational dynamics of F1-ATPase from bacterial membranes: A model for the regulation of these enzymes on the basis of molecular plasticity. – Biochim. Biophys. Acta **650:** 233–265.
MURPHY, J. R., GIRARD, A. E. & TILTON, R. C. (1974): Ultrastructure of a marine *Thiobacillus.* – J. Gen. Microbiol. **85:** 130–138.
MURRAY, R. G. E. & BIRCH-ANDERSEN, A. (1963): Specialized structure in the region of the flagella tuft in *Spirillum serpens.* – Canad. J. Microbiol. **9:** 393–401.
MURRAY, R. G. E. & DOUGLAS, H. C. (1950): The reproductive mechanism of *Rhodomicrobium vanielii* and the accompanying nuclear changes. – J. Bacteriol. **59:** 157–167.
MURRAY, R. G.E. & ELDER, E. H. (1949): The predominance of counterclockwise rotation during swarming of *Bacillus* species. – J. Bacteriol. **58:** 351–359.
MURRAY, R. G. E., STEED, P. & ELSON, H. E. (1965): The location of the mucopeptide in sections of the cell wall of *Escherichia coli* and other Gram-negative bacteria. – Canad. J. Microbiol. **11:** 547–560.
MURRELL, W. G., OHYE, D. F. & GORDON, R. A. (1969): Cytological and chemical structure of the spore. – In: L. L. CAMPBELL (Ed.): Spores IV. pp. 1–19. – Amer. Soc. Microbiol., Bethesda, Md.
MURRY, R. G. E. & WATSON, S. W. (1965): Structure of *Nitrosocystis oceanus* and *Nitrobacter.* – J. Bacteriol. **89:** 1594–1609.
VON MUTO, T. (1904): Ein eigentümlicher *Bacillus,* welcher sich schneckenartig bewegende Kolonien bildet (*B. helicoides*). – Centralbl. Bakteriol. **37**, Part I.: Originale: 325.
NANNINGA, N. (1968): Structural features of mesosomes (chondrioids) of *Bacillus subtilis* after freeze-etching. – J. Cell Biol. **39:** 251–263.
– (1970a): Mesosome induction by chemical fixation as verified by freeze-fracturing. – In: P. FAVARD (Ed.): Seventh Internatl. Congr. Electron Microsc., Grenoble. vol. III, pp. 349–350. – Soc. Fr. Microsc. Electron., Paris.
– (1970b): Ultrastructure of the cell envelope of *Escherichia coli B* after freeze-etching. – J. Bacteriol. **101:** 297–303.
– (1971): The mesosome of *Bacillus subtilis* as affected by chemical and physical fixation. – J. Cell. Biol. **48:** 219–224.
– (1973a): Freeze-fracturing of microorganisms: Physical and chemical fixation of *Bacillus subtilis.* – In: L. BENEDETTI & P. FAVARD (Eds.): Freeze-etching, Techniques and Applications, pp. 151–180. – Soc. Fr. Microsc. Electron., Paris.
– (1973b): Structural aspects of ribosomes. – Internatl. Rev. Cytol. **35:** 135–188.
– (1975): Mesosomal cytology. – In: T. HASEGAWA (Ed.): Proc. First Intersect. Congr. IAMS. vol. I, pp. 527–538. – Sci. Council of Japan.
NANNINGA, N., BRAKENHOFF, G. J., MEIJER, M. & WOLDRINGH, C. L. (1984): Bacterial anatomy in retrospect and prospect. – Antonie van Leeuwenhoek J. Microbiol. Serol. **50:** 433–460.

NANNINGA, N., GARRETT, R. A., STÖFFLER, G. & KLOTZ, G. (1972): Ribosomal proteins. XXXVIII. Electron microscopy of ribosomal protein S4-16S RNA complexes of *Escherichia coli.* – Molec. Gen. Genet. **119:** 175–184.

NANNINGA, N. & WOLDRINGH, C. L. (1981): The interpretation of chemically fixed and freeze-fractured bacterial nucleoplasm. – Acta Histochem. **23:** 39–53.

– – (1985): Cell growth, genome duplication and cell division. – In: N. NANNINGA (Ed.): Molecular Cytology of *Escherichia coli.* pp. 259–318. – Academic Press Inc., New York.

NANNINGA, N., WOLDRINGH, C. L. & KOPPES, L. J. H. (1982): Growth and division of *Escherichia coli.* – In: C. NICOLINI (Ed.): Cell Growth. pp. 225–270. – Plenum Publ. Corp., New York.

NARDON, P. & WICKER, C. (1983): Genetic control of symbiotes by the host in the insect *Sitophilus oryzae* L. (Coleoptera Curculionidae). – In: H. E. A. SCHENK & W. SCHWEMMLER (Eds.): Endocytobiology. vol. II, pp. 727–731. – de Gruyter, Berlin, New York.

NEIMARK, H. C. (1977): Extraction of an actin-like protein from the prokaryote *Mycoplasma pneumoniae.* – Proc. Natl. Acad. Sci. USA **74:** 4041–4045.

NEU, H. C. & HEPPEL, L. A. (1965): The release of enzymes from *Escherichia coli* by osmotic shock and during the formation of spheroplasts. – J. Biol. Chem. **240:** 3685–3692.

NEUGEBAUER, D. CH. & ZINGSHEIM, H. P. (1978): The two faces of the purple membrane; structural differences revealed by metal decoration. – J. Molec. Biol. **123:** 235–246.

NEU-MÜLLER, M. & SCHENK, H. E. A. (1983): On the biosynthesis of photosynthetic pigments in *Cyanophora paradoxa.* – In: H. E. A. SCHENK & W. SCHWEMMLER (Eds.): Endocytobiology. vol. II, pp. 451–463. – de Gruyter, Berlin, New York.

NEUSHUL, M. (1971): Uniformity of the thylakoid structure in a red, a brown, and two blue-green algae. – J. Ultrastruct. Res. **37:** 532–543.

NEWMAN, C. N. & KUBITSCHEK, H. E. (1978): Variation in periodic replication of the chromosome in *Escherichia coli B/rTT.* – J. Molec. Biol. **121:** 461–471.

NEWMAN, H. N. (1976): Dental plaque. – In: R. FULLER & D. W. LOVELOCK (Eds.): Microbial Ultrastructure. pp. 224–263. – Academic Press, London, New York, San Francisco.

NICHOLSON, D. E. & FOX, G. E. (1983): Molecular evidence for a close phylogenetic relationship among box-shaped halophilic bacteria, *Halobacterium vallismortis,* and *Halobacterium marismortui.* – Canad. J. Microbiol. **29:** 52–59.

NICOLOV, N. N., KOSTOVA, K. P. & VAKLINOVA, S. G. (1983): Relationship of symbiosis between unicellular green algae and nitrogen fixation bacteria. – In: H. E. A. SCHENK & W. SCHWEMMLER (Eds.): Endocytobiology. vol. II, pp. 623–630. – de Gruyter, Berlin, New York.

NICOLSON, G. L. & SCHMIDT, G. L. (1971): Structure of the *Chromatium* sulfur particle and its protein membrane. – J. Bacteriol. **105:** 1142–1148.

NIEDERMAN, R. A. & GIBSON, K. D. (1978): Isolation and physicochemical properties of membranes from purple photosynthetic bacteria. – In: R. CLAYTON & W. SISTROM (Eds.): The Photosynthetic Bacteria. pp. 78–118. – Plenum Publ. Corp., New York.

NIEDERMAN, R. A., HUNTER, C. N., INAMINE, G. S. & MALLON, D. E. (1981): Development of the bacterial photosynthetic apparatus. – In: G. AKOYUNOGLOU (Ed.): Photosynthesis. V. Chloroplast Development. pp. 663–674. – Balaban Internatl. Sci. Services, Philadelphia.

NIEMEYER, R. & RICHTER, G. (1969): Schnellmarkierte Polyphosphate und Metaphosphate bei der Blaualge *Anacystis nidulans.* – Arch. Mikrobiol. **69:** 54–59.

NIENHAUS, F. (1976): Rickettsien, Mykoplasmen und Spiroplasmen als Erreger von Pflanzenkrankeiten. – Ber. Deutsch. Bot. Ges. **89:** 531–545.

– (1979): Lärchen-Degeneration durch Rickettsienähnliche Bakterien. – Allgem. Forstz. **6:** 130–132.

NIENHKAUS, F., BRÜSSEL, H. & SCHINZER, U. (1976): Soil-borne transmission of *Rickettsia*-like organisms found in stunted and wiches' broom-diseased larch trees *(Larix decidua).* – Z. Pflanzenkrankh. Pflanzenschutz **83:** 309–316.

NIENHAUS, F. & SIKORA, R. A. (1979): Mycoplasmas, spiroplasmas, and *Rickettsia*-like organisms as plant pathogens. – Annual Rev. Phytopathol. **17:** 37–58.

NIERZWICKI, S. A., MARATEA, D., BALKWILL, D. L. HARDIE, L. P., MEHTA, V. B. & STEVENS JR., S. E. (1982): Ultrastructure of the cyanobacterium *Mastigocladus laminosus.* – Arch. Microbiol. **133:** 11–19.

NISHIYAMA, Y., KIMURA, S. & EDA, T. (1984): Electron microscopic studies on reproductive process in bacterial L-forms. – Abstr. 8th Eur. Congr. Electron Microscopy. vol. III, pp. 2377–2378. Budapest.

Nokhal, T.-H. & Mayer, F. (1979): Structural analysis of four strains of *Paracoccus denitrificans.* – Antonie van Leeuwenhoek J. Microbiol. Serol. **45:** 185–197.

Nomura, M. (1963): Mode of action of colicines. – Cold Spring Harbor Symp. Quant. Biol. **28:** 315.

– (1967): Colicins and related bacteriocins. – Annual Rev. Microbiol. **21:** 257.

Norris, J. R. (1969): Spores. – In: L. L. Campbell (Ed.): Spores **4:** pp. 45–58. – Amer. Soc. Microbiol., Washington D. C.

Norris, J. R. & Proctor, H. M. (1969): Crystalline inclusions in *Bacillus thuringiensis.* – J. Bacteriol. **98:** 824–826.

Novick, N. J. & Tyler, M. E. (1985): Isolations and characterization of *Alteromonas luteoviolacea* strains with sheathed flagella. – Internatl. J. System. Bacteriol. **35:** 111–113.

Novotny, C. P. & Fives-Taylor, P. (1974): Retraction of F-pili. – J. Bacteriol. **117:** 1306–1311.

Oberti, J., Caravano, R. & Roux, J. (1982): Demonstration of an external layer in several species of the genus *Brucella.* – Canad. J. Microbiol. **28:** 1300–1303.

O'Brien, E. J. & Bennett, P. M. (1972): Structure of straight flagella from a mutant *Salmonella.* – J. Molec. Biol. **70:** 133–152.

O'Carra, P. & O'Heocha, C. (1976): Algal biliproteins and phycobilins. – In: T. W. Goodwin (Ed.): Chemistry and Biochemistry of Plant Pigments. 2nd ed., pp. 28–376. – Academic Press, New York.

Oesterhelt, D. & Stoeckenius, W. (1973): Functions of a new photoreceptor membrane. – Proc. Natl. Acad. Sci. USA **70:** 2853–2857.

– – (1974): Isolation of the cell membrane of *Halobacterium halobium* and its fractionation into red and purple membranes. – Meth. Enzymol. **31:** 667–678.

Ofek, I., Whitnack, E. & Beachey, E. M. (1983): Hydrophobic interaction of group *A* streptococci with hexadecane droplets. – J. Bacteriol. **154:** 139–145.

Olsen, W. L., Heidrich, H. G., Henning, K. & Hofschneider, P. H. (1974): Desoxyribonucleic acid-envelope complexes isolated from *Escherichia coli* by free-flow electrophoresis: Biochemical and electron microscope characterization. – J. Bacteriol. **118:** 646–653.

Oppenheim, J. & Marcus, L. (1970): Correlation of ultrastructure in *Azotobacter vinelandii* with nitrogen source for growth. – J. Bacteriol. **101:** 286–291.

Orpin, C. G. (1973): The intracellular polysaccharide of the rumen bacterium "Eadie's Oval". – Arch. Mikrobiol. **90:** 247–254.

Ou, J. T. & Anderson, T. F. (1970): Role of pili in bacterial conjugation. – J. Bacteriol. **102:** 648–654.

Ozeki, H. (1968): Methods for the study of colicine and colicinogeny. – In: K. Maramorosch & H. Koprowski (Eds.): Methods in Virology. vol. IV, p. 565. – Academic Press Inc., New York.

Pachas, N. & Currid, V. R. (1974): L-form induction, morphology, and development in two related strains of *Erysipelothrix rhusiopathiae.* — J. Bacteriol. **119:** 576–582.

Paci, M., Pon, C. L., Losso, M. A. & Gualerzi, C. (1984): Proteins from the prokaryotic nucleoid. High-resolution ^{1}H NMR spectroscopic study of *Escherichia coli* DNA-binding proteins NS 1 and NS 2. – Eur. J. Biochem. **138:** 193–200.

Page, L. A. (1981): Obligately intracellular bacteria: the genus *Chlamydia.* – In: M. P. Starr, H. Stolp, H. G. Trueper, A. Balows & H. G. Schlegel (Eds.): The Prokaryotes. vol. II, pp. 2210–2222. – Springer, Berlin, Heidelberg, New York.

De Phamphilis, M. L. & Adler, J. (1971 a): Purification of intact flagella from *Escherichia coli* and *Bacillus subtilis.* – J. Bacteriol. **105:** 376–383.

– – (1971 b): Fine structure and isolation of the hook-basal-body-complex of flagella from *Escherichia coli* and *Bacillus subtilis.* – J. Bacteriol. **105:** 384–395.

– – (1971 c): Attachment of flagellar basal-bodies to the cell envelope: Specific attachment to the outer, lipopolysaccharid membrane and the cytoplasmic membrane. – J. Bacteriol. **105:** 396–407.

Pangborn, J., Marr, A. G. & Robrish, S. A. (1962): Localization of respiratory enzymes in intracytoplasmic membranes of *Azotobacter agilis.* – J. Bacteriol. **84:** 669–678.

Pantskhava, E. S. (1977): Role of low molecular weight factors from *Methanobacillus kuzneceovii* in an activating effect of visible light on methane formation and reduction of pyridine nucleotides. – Biokhimiya **42:** 549–559.

Parameswaran, N. (1983): *Rickettsia*-like organisms in *Cryptococcus fagisuga* on beech. – Eur. J. Forest Pathol. **13:** 376–380.

Pardee, A. B., Wu, P. C. & Zusman, D. R. (1983): Bacterial division and the cell envelope. – In: L. Leive (Ed.): Bacterial Membranes and Walls. pp. 357–412. – Dekker, New York.

PARKER, N. D. & MUNN, C. B. (1974): Increased cell-surface hydrophobicity associated with possession of an additional surface protein by *Aeromonas salmonicida.* – FEMS Microbiol. Lett. **21:** 233–237.

PARKES, K. & WALSBY, A. E. (1981): Ultrastructure of a gas-vacuolate square bacterium. – J. Gen. Microbiol. **126:** 503–506.

PATE, J. L. & CHANG, L.-Y. E. (1979): Evidence that gliding motility in prokaryotic cells is driven by rotary assemblies in the cell envelopes. – Curr. Microbiol. **2:** 59–64.

PATE, J. L., SHAH, V. K. & BRILL, W. J. (1973): Internal membrane control in *Azotobacter vinelandii.* – J. Bacteriol. **114:** 1346–1350.

PATTERSON, M. J. & HAFEEZ, A. E. B. (1976): Group *B* streptococci in human disease. – Bacteriol. Rev. **40:** 774–792.

PAYNTER, M. J. B. & HUNGATE, R. E. (1968): Characterization of *Methanobacterium mobilis* sp. n., isolated from bovine rumen. – J. Bacteriol. **95:** 1943–1951.

PETERSON, J. E. (1969): The fruiting myxobacteria: Their properties, distribution and isolation. – J. Appl. Bacteriol. **32:** 5–12.

PETERSON, J. E., RODWELL, A. W. & RODWELL, E. S. (1973): Occurrence and ultrastructure of a variant *(rho)* form of *Mycoplasma.* – J. Bacteriol. **115:** 411–425.

DE PETRIS, S. (1976): Ultrastructure of the cell wall of *Escherichia coli* and chemical nature of its constituent layers. – J. Ultrastruct. Res. **19:** 45–83.

PETTIJOHN, D. E. & SINDEN, R. R. (1985): Structure of the isolated nucleoid. – In: N. NANNINGA (Ed.): Molecular Cytology of *Escherichia coli.* pp. 199–227. – Academic Press Inc., New York.

PETZOLD, H., MARWITH, R. & KUNZE, L. (1973): Elektronenmikroskopische Untersuchungen über intrazelluläre rickettsienähnliche Bakterien in triebsuchtkranken Äpfeln – Phytopathol. Z. **78:** 170–181.

PFENNIG, N. (1967): Photosynthetic bacteria. – Annual Rev. Microbiol. **21:** 285–324.

PFENNIG, N. & TRUEPER, H. G. (1974a): Genus *Rhodomicrobium* Duchow and Douglas 1949. – In: R. E. BUCHANAN & N. E. GIBBONS (Eds.): Bergey's Manual of Determinative Bacteriology. pp. 33–34. – Williams & Wilkins, Baltimore.

– – (1974b): The phototrophic bacteria. – In: R. E. BUCHANAN & N. E. GIBBONS (Eds.): Bergey's Manual of Determinative Bacteriology, 8th Ed., pp. 24–64. – Williams & Wilkins, Baltimore.

PFENNIG, N., WIDDEL, F. & TRUEPER, H. G. (1981): The dissimilatory sulfate-reducing bacteria. – In: M. P. STARR, H. STOLP, H. G. TRUEPER, A. BALOWS & H. G. SCHLEGEL (Eds.): The Prokaryotes. vol. I, pp. 926–940. – Springer, Berlin, Heidelberg, New York.

PIETSCHMANN, K. (1942): Über die Begeißelung der Bakterien. – Arch. Mikrobiol. **12:** 377–472.

PITEL, D. W. & GILVARG, C. (1970): Mucopeptide metabolism during growth and sporulation in *Bacillus megaterium.* – J. Biol. Chem. **245:** 6711–6717.

POPE, L. M., HOARE, D. S. & SMITH, A. J. (1969): Ultrastructure of *Nitrobacter agilis* grown under autotrophic and heterotrophic conditions. – J. Bacteriol. **97:** 936–939.

POPE, L. M., YOLTON, D. P. & RODE, L. J. (1968): Crystalline inclusions of *Clostridium cochlearium.* – J. Bacteriol. **96:** 1895–1868.

POST, E., GOLECKI, J. R. & OELZE, J. (1982): Morphological and ultrastructural variations in *Azotobacter vinelandii* growing in oxygen-controlled continuous culture. – Arch. Microbiol. **133:** 75–82.

POST, E., VAKALOPOULOU, E. & OELZE, J. (1983): On the relationship of intracytoplasmic to cytoplasmic membranes in nitrogen-fixing *Azotobacter vinelandii.* – Arch. Microbiol. **134:** 265–269.

POSTGATE, J. R. & KELLY, D. P. (Eds.) (1982): Sulphur Bacteria. – Roy. Soc., London.

PREER, J. R., PREER, L. B. & JURAND, A. (1974): *Kappa* and other endosymbionts in *Paramecium aurelia.* – Biol. Rev. **38:** 113–163.

PREER, J. R. & STARK, P. (1953): Cytological oberservations on the cytoplasmic factor *"Kappa"* in *Paramecium aurelia.* – Exp. Cell Res. **5:** 478–491.

PREER, L. B. (1981): Prokaryotic symbionts of *Paramecium.* – In: M. P. STARR, H. STOLP, H. G. TRUEPER, A. BALOWS & H. G. SCHLEGEL (Eds.): The Prokaryotes. Vol. II, pp. 2127–2130. – Springer, Berlin, Heidelberg, New York.

PRINGSHEIM, E. G. (1951): The Vitreoscillacaea: a family of colourless, gliding filamentous organisms. – J. Gen. Microbiol. *5:* 124–149.

PROCTOR, H. M., NORRIS, J. R. & RIBBONS, D. W. (1969): Fine structure of methane-utilizing bacteria. – J. Appl. Bacteriol. **32:** 118.

PROGULSKE, A. & HOLT, S. C. (1980): Transmission-scanning electron microscopic observations of selected *Eikenella corrodens* strains. – J. Bacteriol. **143:** 1003–1018.

PÜHLER, A. (Ed.) (1983): Molecular Genetics of the Bacteria/plant Interaction. – Springer, Berlin, Heidelberg, New York, Tokyo

QUACKENBUSH, R. L. & BURBACH, J. A. (1983): Cloning and expression of DNA sequences associated with the killer trait of *Paramecium tetraurelia* Stock 47. – Proc. Natl. Acad. Sci. USA **80**: 250–254.

RASCH, M., SAXTON, W. O. & BAUMEISTER, W. (1984): Structure of the regular surface layer of *Acetogenium kivui*. – Abstr. 8th Eur. Congr. Electron Microscopy. vol. II, pp. 1509–1510. Budapest.

RASKA, I., MAYER, F., EDELBLUTH, C. & SCHMITT, R. (1976): Structure of plain and complex flagellar hooks of *Pseudomonas rhodos*. – J. Bacteriol. **125**: 679–688.

RASTOGI, N., FREHEL, C. & DAVID, H. L. (1984): Evidence for taxonomic utility of periodic acid-thiocarbohydrazide-silver proteinate cytochemical staining for electron microscopy. – Internatl. J. System. Bacteriol. **34**: 293–299.

RASTOGI, N., FREHEL, C., RYTER, A., OHAYON, H., LESOURD, M. & DAVID, H. L. (1981): Multiple drug resistance of *Mycobacterium avium:* Is the wall architecture responsible for the exclusion of antimicrobial agents? – Antimicrobiol Agents Chemother. **20**: 666–677.

RAUVALA, H. (1983): Cell surface carbohydrates and cell adhesion. – Trends Biochem. Sci. **8**: 323–325.

RAZIN, S. (1973): Physiology of mycoplasmas. – Adv. Microbiol. Physiol. **10**: 1–80.

– (1975): The *Mycoplasma* membrane. – Progr. Surf. Membr. Sci. **9**: 257–312.

– (1978): The mycoplasmas. – Microbiol. Rev. **42**: 414–470.

REED, D. W. & RAVEED, D. (1972): Some properties of the ATPase from chromathophores of *Rhodopseudomonas sphaeroides* and its structural relationship to the bacteriochlorophyll proteins. – Biochim. Biophys. Acta **283**: 79–91.

REEVES, P. (1972): The Bacteriocins. (Molecular Biology, Biochemistry, and Biophysics. vol. XI) – Springer, Berlin, Heidelberg, New York.

REICHENBACH, H. (1965): Untersuchungen an *Archangium violaceum*. Ein Beitrag zur Kenntnis der Myxobakterien. – Arch. Mikrobiol. **52**: 376–403.

– (1974): Die Biologie der Myxobakterien. – Biol. unserer Zeit. **4**: 33–45.

REICHENBACH, H. & DWORKIN, M. (1981 a): Introduction to the gliding bacteria. – In: M. P. STARR, H. STOLP, H. G. TRUEPER, A. BALOWS & H. G. SCHLEGEL (Eds.): The Prokaryotes. vol. I, pp. 315–327. – Springer, Berlin, Heidelberg, New York.

– – (1981 b): The order Cytophagales (with addenda on the genera *Herpetosiphon, Saprospira*, and *Flexithrix*). – In: M. P. STARR, H. STOLP, H. G. TRUEPER, A. BALOWS & H. G. SCHLEGEL (Eds.): The Prokaryotes. vol. I, pp. 356–379. – Springer, Berlin, Heidelberg, New York.

REICHLE, R. E. & LEWIN, R. A. (1968): Purification and structure of rhapidosomes. – Canad. J. Microbiol. **14**: 211.

REMSEN, C. C. (1968): Fine structure of the mesosome and nucleoid in frozen-etched *Bacillus subtilis*. – Arch. Mikrobiol. **61**: 40–47.

– (1978): Comparative subcellular architecture of photosynthetic bacteria. – In: R. K. CLAYTON & W. R. SISTROM (Eds.): The Photosynthetic Bacteria. pp. 31–60. – Plenum Press, New York, London.

REMSEN, C. C. & LUNDGREN, D. G. (1966): Electron microscopy of the cell envelope of *Ferrobacillus ferrooxidans* prepared by freeze-etching and chemical fixation techniques. – J. Bacteriol. **92**: 1765–1771.

REMSEN, C. C., WATSON, S. W., WATERBURY, J. B. & TRUEPER, H. G. (1968): Fine structure of *Ectothiorhodospira mobilis* Pelsh. – J. Bacteriol. **95**: 2392.

DI RENZO, J. M., NAKAMURA, K. & INOUYE, M. (1978): The outer membrane proteins of Gram-negative bacteria: biosynthesis, assembly and functions. – Annual Rev. Biochem. **47**: 481–532.

RIPPKA, R., DERUELLES, J., WATERBURY, J. B., HERDMAN, M. & STANIER, R. Y. (1979): Generic assignments, strain histories and properties of pure cultures of cyanobacteria. – J. Gen. Microbiol. **111**: 1–61.

RIPPKA, R., WATERBURY, J. & COHEN-BAZIRE, G. (1974): A cyanobacterium which lacks thylakoids. – Arch. Microbiol. **100**: 419–436.

ROBERTSON, J. G., LYTTLETON, P., WILLIAMSON, K. I. & BATT, R. D. (1975): The effect of fixation procedures on the electron density of polysaccharide granules in *Nocardia corallina*. – J. Ultrastruct. Res. **50**: 321–332.

ROBERTSON, J. G., WELLS, B., BISSELING, T., FARNDEN, K. J. F. & JOHNSTON, A. W. B. (1984): Immuno-gold localization of leghaemoglobin in cytoplasm in nitrogen-fixing root nodules of pea. – Nature **311**: 254–256.

ROGERS, H. J. (1983): Bacterial Cell Structure. – Van Nostrand Reinhold (UK) Co. Ltd.

ROGERS, H. J., PERKINS, H. R. & WARD, J. B. (1980): Microbial Cell Walls and Membranes. – Chapman & Hall, London, New York.

ROHDE, M., MAYER, F. & MAYER, O. (1984): Immunocytochemical localization of carbon monoxide oxidase in *Pseudomonas carboxydovorans* – the enzyme is attached to the inner aspect of the cytoplasmic membrane. – J. Biol. Chem. **259:** 14788–14792.

ROMANENKO, V. M. (1984): Electron microscopic study of polyphosphate inclusions in cells of *Pseudomonas holci* and *Pseudomonas phaseolicola.* – Abstr. 8th Eur. Congr. Electron Microscopy. vol. III, pp. 2383–2384. Budapest.

ROMESSER, J. A., WOLFE, R. S., MAYER, F., SPIESS, E. & WALTHER-MAURUSCHAT, A. (1979): *Methanogenium*, a new genus of marine methanogenic bacteria, and characterization of *Methanogenium cariaci* sp. nov. and *Methanogenium marisnigri* sp. nov. – Arch. Microbiol. **121:** 147–153.

ROSENBERG, E. (Ed.) (1984): Myxobacteria, Development and Cell Interactions. – Springer, Berlin, Heidelberg, New York.

ROSENBERG, E., KAPLAN, N., PINES, O., ROSENBERG, M. & GUTNICK, D. (1983): Capsular polysaccharides interfere with adherence of *Acinetobacter calcoaceticus* to hydrocarbons. – FEMS Microbiol. Lett. **17:** 157–160.

ROSENBERG, M., GUTNICK, D. & ROSENBERG, E. (1980): Adherence of bacteria to hydrocarbons: a simple method for measuring cell-surface hydrophobicity. – FEMS Microbiol. Lett. **9:** 29–33.

ROSENBERG, M. & ROSENBERG, E. (1981): Role of adherence in growth of *Acinetobacter calcoaceticus RAG-1* on hexadecane. – J. Bacteriol. **148:** 51–57.

ROSENBERG, M., ROSENBERG, E., JUDES, H. & WEISS, E. (1983): Bacterial adherence to hydrocarbons and to surfaces in the oral cavity. – FEMS Microbiol. Lett. **20:** 1–5.

ROSENBUSCH, J. P. (1974): Characterization of the major envelope protein from *Escherichia coli* – J. Biol. Chem. **249:** 8019–8029.

ROSENBUSCH, J. P., STEVEN, A. C., ALKAN, M. & REGENASS, M. (1980): Matrix porin: a periodically arranged protein in the outer membrane of *Escherichia coli* – In: W. BAUMEISTER & W. VOGELL (Eds.): Electron Microscopy at Macromolecular Dimensions. pp. 1–10. – Springer, Berlin, Heidelberg, New York.

ROUVIÈRE-YANIV, J. & GROS, F. (1975): Characterization of a novel, low-molecular-weight DNA-binding protein from *Escherichia coli.* – Proc. Natl. Acad. Sci. USA **72:** 3428–3432.

ROWLES, C. R., PARTON, R. & JEYNES, M. H. (1976): Some aspects of the cell walls of *Vibrio* spp. – In: R. FULLER & D. W. LOVELOCK (Eds.): Microbial Ultrastructure. pp. 109–115. – Academic Press, London.

RUCINSKY, T. E. & COTA-ROBLES, E. H. (1973): The intracellular organization of bacteriophage tail-like particles in cells of *Chromobacterium violaceum* following mitomycin C treatment. – J. Ultrastruct. Res. **43:** 260.

RYTER, A. (1968): Association of the nucleus and the membrane of bacteria: A morphological study. – Bacteriol. Rev. **32:** 39–54.

RYTER, A. & CHANG, A. (1975): Localization of transcribing genes in the bacterial cell by means of high resolution autoradiography. – J. Molec. Biol. **98:** 797–810.

RYTER, A., HIROTA, Y. & SCHWARZ, U. (1973): Process of cellular division in *Escherichia coli.* Growth pattern of *E. coli* murein. – J. Molec. Biol. **78:** 185–195.

RYTER, A. & KELLENBERGER, E. (1958): Etude au microscope électronique des plasmas contenant de l'acide désoxyribonucléique. I. Le nucléide de bactéries en croissance active. – Z. Naturforsch. **13b:** 597–605.

RYTER, A., SHUMAN, H. & SCHWARTZ, M. (1975): Integration of the receptor for bacteriophage *lambda* in the outer membrane of *Escherichia coli:* coupling with cell division. – J. Bacteriol. **122:** 295–301.

SABET, S. F. & SCHNAITMAN, C. A. (1973): Chemistry of the colicin *E* receptor. – In: L. P. HAGER (Ed.): Chemistry and Functions of Colicins. p. 59. – Academic Press, New York, London.

SAGLIO, P., L'HOSPITAL, M., LAFLÈCHE, D., DUPONT, G., BOVÉ, J. M., TULLY, J. G. & FREUNDT, G. A. (1973): *Spiroplasma citri* gen. and sp. nov.: A *Mycoplasma*-like organism associated with "stubborn" disease of *Citrus.* – Internatl. J. System. Bacteriol. **23:** 191–204.

SALIT, I. E. & GOTSCHLICH, E. C. (1977): Haemagglutination by purified type 1 *Escherichia coli* pili. – J. Exp. Med. **146:** 1182–1194.

SALTON, M. R. J. (1963): The relationship between the nature of the cell wall and the Gram stain. – J. Gen. Microbiol. **30:** 223–235.

– (1964): The Bacterial Cell Wall. – Elsevier Publ. Co., Amsterdam.

SALTON, M. R. J. & OWEN, P. (1976): Bacterial membrane structure. – Annual Rev. Microbiol. **30:** 451–482.

SANTO, L. Y. & DOI, R. H. (1973): Crystal formation by a ribonucleic acid polymerase mutant of *Bacillus subtilis.* – J. Bacteriol. **116:** 479–482.

SARGENT, M. G. (1979): Surface extension and the cell cycle in prokaryotes. – Adv. Microbiol. **18:** 105–176.

SAUER, F. D., ERFLE, J. D. & MAHADEVAN, S. (1980): Methane production by the membranous fraction of *Methanobacterium thermoautotrophicum.* – Biochem. J. **190:** 177–182.

– – – (1981): Evidence for an internal electrochemical proton gradient in *Methanobacterium thermoautotrophicum.* – J. Biol. Chem. **256:** 9843–9848.

SAYLES JR., V. B., ARONSON, J. N. & ROSENTHAL, A. (1970): Small polypeptide components of the *Bacillus thuringiensis* parasporal crystalline inclusion. – Biochem. Biophys. Res. Commun. **41:** 1126–1133.

SCHELL, J., VAN MONTAGU, M., HERNALSTEENS, J. P., WILLMITZER, L., LEEMANS, J., JOOS, H., OTTEN, L., DE GREVE, H., HOLSTERS, M., ZAMBRYSKI, P., HERRERA, L. & DEPICKER, A. (1983): Crown gall: Genetic colonization as an infectious mechanism. – In: H. E. A. SCHENK & W. SCHWEMMLER (Eds.): Endocytobiology. vol. II, pp. 593–594. – de Gruyter, Berlin, New York.

SCHENK, H. E. A., SCHWEMMLER, W. (1983): Endocytobiology. vol. II. – de Gruyter, Berlin, New York.

SCHERRER, R. (1963): Cell structure and quantitative Gram stain of *Bacillus megaterium.* – J. Gen. Microbiol. **31:** 135–145.

SCHIEFER, H.-G., GERHARDT, U., BRUNNER, H. & KRÜPE, M. (1974): Studies with lectins on the surface carbohydrate structures of *Mycoplasma* membranes. – J. Bacteriol. **120:** 81–88.

SCHIEFER, H.-G., KRAUSS, H., BRUNNER, H. & GERHARDT, U. (1975): Ultrastructural visualization of anionic sites on *Mycoplasma* membranes by polycationic ferritin. – J. Bacteriol. **127:** 461–468.

SCHLAEPPI, J.-M. & KARAMATA, D. (1982): Cosegregation of cell wall and DNA in *Bacillus subtilis.* – J. Bacteriol. **152:** 1231–1240.

SCHLAEPPI, J.-M., POOLEY, H. M. & KARAMATA, D. (1982): Identification of cell wall subunits in *Bacillus subtilis* and analysis of their segregation during growth. – J. Bacteriol. **149:** 329–337.

SCHLEGEL, H. G. (1981): Allgemeine Mikrobiologie. – Thieme, Stuttgart, New York.

SCHLEGEL, H. G., GOTTSCHALK, G. & VON BARTHA, R. (1961): Formation and utilization of poly-β-hydroxybutyric acid by knallgas bacteria *(Hydrogenomonas).* – Nature (Lond.) **191:** 463

SCHLEIFER, K. H. & KANDLER, O. (1972): Peptiodoglycan types of bacterial cell walls and their taxonomic implications. – Bacteriol. Rev. **36:** 407–477.

SCHMIDT, G. (1972): Basalstrukturen von Lipopolysacchariden verschiedener Enterobacteriaceen. Genetische und serologische Untersuchungen an *R*-Mutanten. – Zentralbl. Bakteriol. Parasitenkd. Infektionskr. Hyg. 1. Abt.: Orig. Reihe A, **A220:** 472–476.

SCHMIDT, G., JANN, B. & JANN, K. (1970): Immunochemistry of *R* lipopolysaccharides of *Escherichia coli.* – Eur. J. Biochem. **16:** 382–392.

SCHMIDT, G. L., NICOLSON, G. L. & KAMEN, M. D. (1971): Composition of the sulfur particle of *Chromatium vinosum* strain *D.* – J. Bacteriol. **105:** 1137–1141.

SCHMIDT, J. M. (1981): The genera *Caulobacter* and *Asticcacaulis.* – In: M. P. STARR, H. STOLP, H. G. TRUEPER, A. BALOWS & H. G. SCHLEGEL (Eds.): The Prokaryotes. vol. I, pp. 466–476. – Springer, Berlin, Heidelberg, New York.

SCHMIDT, J. M. & STARR, M. P. (1981): The *Blastocaulis-Planctomyces* group of budding and appendaged bacteria. – In: M. P. STARR, H. STOLP, H. G. TRUEPER, A. BALOWS & H. G. SCHLEGEL (Eds.): The Prokaryotes. vol. I, pp. 496–504. – Springer, Berlin, Heidelberg, New York.

SCHMIDT, J. M. & SWAFFORD, J. R. (1981): The genus *Seliberia.* – In: M. P. STARR, H. STOLP, H. G. TRUEPER, A. BALOWS & H. G. SCHLEGEL (Eds.): The Prokaryotes. vol. I, pp. 516–519. – Springer, Berlin, Heidelberg, New York.

SCHMIDT, K. (1980): A comparative study on the composition of chlorosomes (*Chlorobium* vesicles) and cytoplasmic membranes from *Chloroflexus aurantiacus* strain *OK-70-fl* and *Chlorobium limicola* f. *thiosulfatophilum* strain *6230.* – Arch. Microbiol. **124:** 21–31.

SCHMITT, R. (1972): Bakteriengeißeln: Bewegliche Biopolymere. – Biol. unserer Zeit **2:** 83–91.

SCHMITT, R., RASKA, I. & MAYER, F. (1974): Plain and complex flagella of *Pseudomonas rhodos:* Analysis of fine structure and composition. – J. Bacteriol. **117:** 844–857.

SCHUMACHER, A. & DREWS, G. (1979): Effects of light intensity on membrane differentiation in *Rhodopseudomonas capsulata.* – Biochim. Biophys. Acta **547:** 417–428.

SCHWARZ, U., ASMUS, A. & FRANK, H. (1969): Autolytic enzymes and cell division of *Escherichia coli.* – J. Molec. Biol. **41:** 419–429.

SCHWARZ, U., RYTER, A., RAMBACH, A., HELLIO, R. & HIROTA, Y. (1975): Process of cellular division in *Escherichia coli:* Differentiation of growth zones in the sacculus. – J. Molec. Biol. **98:** 749–760.

SCHWARZMANN, S. & BORING III, J. R. (1971): Antiphagocytic effect of slime from a mucoid strain of *Pseudomonas aeruginosa.* – Infect. Immun. **3:** 762–767.

SCHWEISFURTH, R., ELEFTHERIADIS, D., GUNDLACH, H., JACOBS, M. & JUNG, W. (1978): Microbiology of the precipitation of manganese. – In: W. E. KRUMBEIN (Ed.): Environmental Biogeochemistry and Geomicrobiology. vol. III, pp. 923–928. – Ann Arbor Sci., Ann Arbor, Mich.

SCHWEMMLER, W. (1983): Endocytobiosis as an intracellular ecosystem. – In: H. E. A. SCHENK & W. SCHWEMMLER (Eds.): Endocytobiology. vol. II, pp. 363–411. – de Gruyter, Berlin, New York.

SCHWEMMLER, W. & SCHENK, H. E. A. (eds.) (1980): Endocytobiology. vol. I. – de Gruyter, Berlin, New York.

SCOTLAND, S. M., RICHMOND, J. E. & ROWE, B. (1983): Adhesion of enteropathogenic strains of *Escherichia coli (EPEC)* to *HEp-2* cells is not dependent on the presence of fimbriae. – FEMS Lett. **20:** 191–195.

SHINN, L. E. (1940): The motion of colonies of *Bacillus alvei* as shown by lapse-time cinematography. – J. Bacteriol. **39:** 22.

SHINOMIYA, T. (1972): Studies on biosynthesis and morphogenesis of R-type pyocins of *Pseudomonas aeruginosa.* II. Biosynthesis of antigenic proteins and their assembly into pyocin particles in mitomycin C-induced cells. – J. Biochem. (Tokyo) **72:** 39.

SHIOZAWA, J., CUENDET, P., ZUBER, H. & DREWS, G. (1981): The light-harvesting B800-650 complex of *Rhodopseudomonas capsulata:* Analysis of the native complex and its subunits. – In: G. AKOYUNOGLU (Ed.): Photosynthesis. III. Structure and Molecular Organization of the Photosynthetic Apparatus. pp. 427–434. – Balaban Internatl. Sci. Services, Philadelphia.

SHIOZAWA, J., WELTE, W., HODAPP, N. & DREWS, G. (1982): Studies on the size and composition of isolated light-harvesting B800-850 pigment-protein complex of *Rhodopseudomonas capsulata.* – Arch. Biochem. Biophys. **213:** 473–485.

SHIRAKIHARA, Y. & WAKABAYASHI, T. (1979): Three-dimensional image reconstruction of straight flagella from a mutant *Salmonella typhimurium.* – J. Molec. Biol. **131:** 485–507.

SHIVELY, J. M. (1974): Inclusion bodies of prokaryotes. – Annual Rev. Microbiol. **28:** 167–187.

SHIVELY, J. M., BALL, F., BROWN, D. H. & SAUNDERS, R. E. (1973a): Functional organelles in prokaryotes: Polyhedral inclusions (carboxysomes) of *Thiobacillus neapolitanus.* – Science **182:** 584–586.

SHIVELY, J. M., BALL, F. & KLINE, B.W. (1973b): Electron microscopy of the carboxysomes (polyhedral bodies) of *Thiobacillus neapolitanus.* – J. Bacteriol. **116:** 1405–1411.

SHIVELY, J. M., DECKER, G. L. & GREENAWALT, J. W. (1970): Comparative ultrastructure of the Thiobacilli. – J. Bacteriol. **101:** 618–627.

SHUKLA, J. S. & IYER, V. N. (1961): Aerobic spore-forming bacilli showing the phenomen of colony migration. – Arch. Mikrobiol. **39:** 298–312.

SHUR, B. D. & ROTH, S. (1975): Cell surface glycosyltransferases. – Biochim. Biophys. Acta **415:** 473–512.

SILVERE, A. P. (1983): Cytological interaction of *Agrobacterium tumefaciens* with plant cells on storage tissue discs. – In: H. E. A. SCHENK & W. SCHWEMMLER (Eds.): Endocytobiology. vol. II, pp. 587–592. – de Gruyter, Berlin, New York.

SILVERMAN, M. R. & SIMON, M. I. (1972): Flagellar assembly mutants in *Escherichia coli.* – J. Bacteriol. **112:** 986–993.

– – (1974): Flagellar rotation and the mechanism of bacterial motility. – Nature (Lond.) **249:** 73–74.

– – (1977): Bacterial flagella. – Annual Rev. Microbiol. **31:** 397–419.

SIMON, L. D. (1969): The infection of *Escherichia coli* by *T2* and *T4* bacteriophages as seen in the electron microscope. III. Membrane-associated intracellular bacteriophages. – Virology **38:** 285.

SIMON, L. D. & ANDERSON, T. F. (1967): The infection of *Escherichia coli* by *T2* and *T4* bacteriophages as seen in the electron microscope. I. Attachment and penetration. – Virology **32:** 279.

SIMON, M., SILVERMAN, M., MATSUMURA, P., RIDGWAY, H., KOMEDU, Y. & HILMEN, M. (1978): Structure and function of bacterial flagella. – In: R. Y. STAINIER, H. J. ROGERS & B. J. WAND (Eds.): Relation between Structure and Function in the Procaryotic Cell. pp. 271–284. – Symp. Soc. Gen. Microbiol. **28,** Univ.-Press, London, New York, Melbourne.

SIMON, R., MÜLLER, P., PRIEFER, U., WEBER, G. & PÜHLER, A. (1983): Genetics of the *Rhizobium meliloti/ Medicago sativa* symbiosis. – In: H. E. A. SCHENK & W. SCHWEMMLER (Eds.): Endocytobiology. vol. II, pp. 557–572. – de Gruyter, Berlin, New York.

SIMON, R. D. (1971): Cyanophycin granules from the blue-green alga *Anabaena cylindrica:* A reserve material consisting of copolymers of aspartic acid and arginine. – Proc. Natl. Acad. Sci. USA **68:** 265–367.

– (1973a): The effect of chloramphenicol on the production of cyanophycin granule polypeptide in the blue-green alga *Anabaena cylindrica.* – Arch. Mikrobiol. **92:** 115–122.

– (1973b): Measurement of the cyanophycin granule polypeptide contained in the blue-green alga *Anabaena cylindrica.* – J. Bacteriol. **114:** 1213–1216.

Singer, S. J. & Nicolson, G. L. (1972): The fluid mosaic model of the structure of cell membranes. – Science **175:** 720–731.

Skerman, V. B. D., Sly, L. J. & Williamson, M.-L. (1983): *Conglomeromonas largomobilis* gen. nov., sp. nov., a sodium-sensitive, mixed-flagellated organism from fresh waters. – Internatl. J. Syst. Bacteriol. **33:** 300–308.

Sleytr, U. B. (1978): Regular arrays of macromolecules on bacterial cell walls: Structure, chemistry, assembly, and function. – Internatl. Rev. Cytol. **53:** 1–64.

Sleytr, U. B. & Glauert, A. M. (1973): Evidence for an empty core in a bacterial flagellum. – Nature (Lond.) **241:** 542–543.

– – (1975): Analysis of regular arrays of subunits on bacterial surfaces: Evidence for a dynamic process of assembly. – J. Ultrastruct. Res. **50:** 103–116.

– – (1976): Ultrastructure of the cell walls of two closely related clostridia that possess different regular arrays of subunits. – J. Bacteriol. **126:** 869–882.

Sleytr, U. B. & Messner, P. (1983): Crystalline surface layers on bacteria. – Annual Rev. Microbiol. **37:** 311–339.

Sloof, P., Maagdelijn, A. & Boswinkel, E. (1983): Folding of prokaryotic DNA. Isolation and characterization of nucleoids from *Bacillus licheniformis.* – J. Molec. Biol. **163:** 277–297.

Smit, J. A., Hugo, N. & De Klerk, H. C. (1969): A receptor for a *Proteus vulgaris* bacteriocin. – J. Gen. Virol. **5:** 33.

Smit, J. & Nikaido, H. (1978): Outer membrane of Gram-negative bacteria XVIII. Electron microscopic studies on porin insertion sites and growth of cell surface of *Salmonella typhimurium.* – J. Bacteriol. **135:** 687–702.

Smith, H. (1977): Microbial surfaces in relation to pathogenicity. – Bacteriol. Rev. **41:** 475–500.

Smith. P. F. (1971): The Biology of Mycoplasmas. – Academic Press, New York, London.

Smith, P. H. & Hungate, R. E. (1958): Isolation and characterization of *Methanobacterium ruminantium* n. sp. – J. Bacteriol. **75:** 713–718.

Smith, R. W. & Koffler, H. (1971): Bacterial flagella. – Adv. Microbiol. Physiol. **6:** 219–339.

Smith, U. & Ribbons, D. (1970): Fine structure of *Methanomonas methanooxidans.* – Arch. Mikrobiol. **74:** 110–122.

Socolofsky, M. D. & Wyss, O. (1961): Cysts of *Azotobacter.* – J. Bacteriol. **81:** 946–954.

– – (1962): Resistance of the *Azotobacter* cyst. – J. Bacteriol. **84:** 119–124.

Sommerville, H. J. (1971): Formation of the parasporal inclusion of *Bacillus thuringiensis.* – Eur. J. Biochem. **18:** 226–237.

Somerville, H. J., Delafield, F. P. & Rittenberg, S. C. (1968): Biochemical homology between crystal and spore protein of *Bacillus thuringiensis.* – J. Bacteriol. **96:** 721–726.

Sowa, B. A., Moore, D. & Ippen-Ihler, K. (1983): Physiology of F-pilin synthesis and utilization. – J. Bacteriol. **153:** 962–968.

Spiess, E. (1979): Structure of the 50S ribosomal subunit from *Escherichia coli.* Investigation of the intact subunit and core particles by electron microscopy and analogue image processing. – Eur. J. Cell Biol. **19:** 120–130.

Spormann, A. M. & Wolfe, R. S. (1984): Chemotactic, magnetotactic and tactile behaviour in a magnetic spirillum. – FEMS Lett. **22:** 171–177.

Sprague, S. G., Staehelin, L. A., Di Bartolomeis, M. J. & Fuller, R. C. (1981a): Isolation and development of chlorosomes in the green bacterium *Chloroflexus aurantiacus.* – J. Bacteriol. **147:** 1021–1031.

Sprague, S. G., Staehelin, L. A. & Fuller, R. C. (1981b): Semiaerobic induction of bacteriochlorophyll synthesis in the green bacterium *Chloroflexus aurantiacus.* – J. Bacteriol. **147:** 1032–1039.

Springer, E. L. & Roth, I. L. (1973a): The ultrastructure of the capsules of *Diplococcus pneumoniae* and *Klebsiella pneumoniae* stained with red ruthenium. – J. Gen. Microbiol. **74:** 21–31.

– – (1973b): Ultrastructure of the capsule of *Klebsiella pneumoniae* and slime of *Enterobacter aerogenes* as revealed by freeze-etching. – Arch. Mikrobiol. **93:** 277–286.

SPROTT, G. D., SOWDEN, L. C., COLVIN, J. R. & JARRELL, K. F. (1984): Methanogenesis in the absence of intracytoplasmic membranes. – Canad. J. Microbiol. **30:** 594–604.

STAEHELIN, L. A. (1976): Reversible particle movements associated with unstacking and restacking of chloroplast membranes. – J. Cell Biol. **71:** 136–158.

– (1981): Freeze-fracture studies of green plant and *Prochloron* thylakoids. A status report. – In: G. AKOYUNOGLOU (Ed.): Photosynthesis. III. Structure and Molecular Organization of the Photosynthetic Apparatus. pp. 3–14. – Balaban Internatl. Sci. Services, Philadelphia.

STAEHELIN, L. A., ARMOND, P. A. & MILLER, K. R. (1977): Chloroplast membrane organization at the supramolecular level and its functional implications. – Brookhaven Symp. Biol. **28:** 278–315.

STAEHELIN, L. A., GIDDINGS, T. H., BADAMI, P. & KRZYMOWSKI, W. W. (1978a): A comparison of the supramolecular architecture of photosynthetic membranes of blue-green, red and green algae and of higher plants. – In: D. W. DEAMER (Ed.): Light Transducing Membranes, Structure, Function and Evolution. pp. 335–355. – Academic Press, New York.

STAEHELIN, L. A., GOLECKI, J. R. & DREWS, G. (1980): Supramolecular organization of chlorosomes (*Chlorobium* vesicles) and of their membrane attachment sites in *Chlorobium limicola*. – Biochim. Biophys. Acta **589:** 30–45.

STAEHELIN, L. A., GOLECKI, J. R., FULLER, R. C. & DREWS, G. (1978b): Visualization of the supramolecular architecture of chlorosomes (*Chlorobium*-type vesicles) in freeze-fractured cells of *Chloroflexus aurantiacus*. – Arch. Microbiol. **119:** 269–277.

STAHL, S. J., STEWART, K. R. & WILLIAMS, F. D. (1983): Extracellular slime associated with *Proteus mirabilis* during swarming. – J. Bacteriol. **154:** 930–937.

STALEY, J. T. & BAULD, J. (1981): The genus *Planctomyces*. – In: M. P. STARR, H. STOLP, H. G. TRUEPER, A. BALOWS & H. G. SCHLEGEL (Eds.): The Prokaryotes. vol. I, pp. 505–508. – Springer, Berlin, Heidelberg, New York.

STALLIONS, D. R. & CURTISS, R. (1972): Bacterial conjugation under anaerobic conditions. – J. Bacteriol. **111:** 294–295.

STANIER, R. Y., ADELBERG, E. A., INGRAHAM, J. C. & WHEELIS, M. L. (1979): Introduction to the Microbial World. – Prentice-Hall, Englewood Cliffs, New Jersey.

STANIER, R. Y. & COHEN-BAZIRE, G. (1977): Phototrophic prokaryotes: The cyanobacteria. – Annual Rev. Microbiol. **31:** 225–274.

STANIER, R. Y., KUNISAWA, R., MANDEL, M. & COHEN-BAZIRE, G. (1971): Purification and properties of unicellular blue-green algea (Order Chroococcales). – Bacteriol. Rev. **35:** 171–205.

STANIER, R. Y., PFENNIG, N. & TRUEPER, H. G. (1981): Introduction to the phototrophic bacteria. – In: M. P. STARR, H. STOLP, H. G. TRUEPER, A. BALOWS & H. G. SCHLEGEL (Eds.): The Prokaryotes. vol. I, pp. 197–211. – Springer, Berlin, Heidelberg, New York.

STARR, M. P. (1981): The genus *Lampropedia*. – In: M. P. STARR, H. STOLP, H. G. TRUEPER, A. BALOWS & H. G. SCHLEGEL (Eds.): The Prokaryotes. vol. II, pp. 1530–1536: – Springer, Berlin, Heidelberg, New York.

STARR, M. P., SAYRE, R. M. & SCHMIDT, J. M. (1983): Assignment of ATCC *27377* to *Planctomyces staleyi* sp. nov. and conservation of *Pasteuria ramosa* Metchnikoff 1888 on the basis of type descriptive material. – Internatl. J. System. Bacteriol. **33:** 666–671.

STEENSLAND, H. & LARSEN, H. (1969): A study of the cell envelope of the Halobacteria. – J. Gen. Microbiol. **55:** 325–336.

STETTER, K. O., THOMM, M., WINTER, J., WILDGRUBER, G., HUBER, H., ZILLIG, W., JANÉCOVIC, D., KÖNIG H., PALM, P. & WUNDERL, S. (1981): *Methanothermus fervidus,* sp. nov., a novel extremely thermophilic methanogen isolated from an Icelandic hot spring. – Zentralbl. Bakteriol. Hyg., 1. Abt. Orig. **C2:** 166–178.

STEWART, J. R. & BROWN, R. M. (1969): *Cytophaga* that kills or lyses algae. – Science **164:** 1523–1525.

STIRM, S. & FREUND-MÖLBERT, E. (1971): *Escherichia coli* capsule bacteriophages. II. Morphology. – J. Virol. **8:** 330–342.

STOCKDALE, H., RIBBONS, D. W. & DAWES, E. A. (1968): Occurrence of poly-β-hydroxybutyrate in the Azotobacteriaceae. – J. Bacteriol. **95:** 1798–1803.

STOECKENIUS, W. (1981): Walsby's square bacterium: fine structure of an orthogonal prokaryote. – J. Bacteriol. **148:** 352–360.

STOECKENIUS, W. & BOGOMOLNI, R. A. (1982): Bacteriorhodopsin and related pigments of halobacteria. – Annual Rev. Biochem. **52:** 587–615.

STOECKENIUS, W. & KUNAU, W. H. (1968): Further characterization of particulate fractions from lysed cell envelopes of *Halobacterium halobium* and isolation of gas vacuole membranes. – J. Cell Biol. **38:** 337–357.

STOECKENIUS, W., LOZIER, R. H. & BOGOMOLNI, R. A. (1979): Bacteriorhodopsin and the purple membrane of halobacteria. – Biochim. Biophys. Acta **505:** 215–278.

STOECKENIUS, W. & ROWEN, R. (1967): A morphological study of *Halobacterium halobium* and its lysis in a media of low salt concentration. – J. Cell Biol. **34:** 365–393.

STÖFFLER, G., LOTTI, M., NOAH, M. & STÖFFLER-MEILICKE, M. (1984): The localization of proteins L5 and L10 on the surface of 50 S subunits of *Escherichia coli.* – Abstr. 8th Eur. Congr. Electron Microscopy. vol. II, pp. 1551–1552. Budapest.

STÖFFLER-MEILICKE, M. & STÖFFLER, G. (1984): The distribution of risobosomal proteins on the surface of the 30 S subunit of *Escherichia coli* – Abstr. 8th Eur. Congr. Electron Microscopy. vol. II, pp. 1525–1536. Budapest.

STOLP, H. (1968): *Bdellovibrio bacteriovorus* – ein räuberischer Bakterienparasit. – Naturwissensch. **55:** 57–63.

STONINGTON, O. G. & PETTIJOHN, D. E. (1971): The folded genome of *Escherichia coli* isolated in a protein-DNA-RNA complex. – Proc. Natl. Acad. Sci. USA **68:** 6–9.

STRYER, L. (1981): Biochemistry. – Freeman, San Francisco.

STUTZER, A. & HARTLEB, R. (1898): Untersuchungen über die bei der Bildung von Salpeter beobachteten Mikroorganismen. – Mitt. Landwirtsch. Inst. Univ. Breslau **1:** 75–100.

SUREK, B. & MELKONIAN, M. (1983): Intracellular bacteria in the Euglenophyceae: Prolonged axenic culture of an algal-bacterial system. – In: H. E. A. SCHENK & W. SCHWEMMLER (Eds.): Endocytobiology. vol. II, pp. 475–486. – de Gruyter, Berlin, New York.

SURYANARYANA, T. & SUBRAMANIAN, A. R. (1978): Specific associaton of two homologous DNA-binding proteins to the native 30 S ribosomal subunits of *Escherichia coli.* – Biochim. Biophys. Acta **520:** 342–357.

SUTHERLAND, I. W. (1972): Bacterial exopolysaccharides. – In: A. H. ROSE & D. W. TEMPEST (Eds.): Advances in Microbial Physiology. vol. VIII, pp. 143–213. – Academic Press, London, New York.

SUZUKI, T. & IINO, T. (1977): Appearance of straight flagellar filaments in the presence of p-fluorophenyl-alanine in *Pseudomonas aeruginosa.* – J. Bacteriol. **129:** 527–529.

SWANEY, L. M., LUI, Y.-P., TO, C.-M., TO, C.-C., IPPE-IHLER, K. & BRINTON, C. C. (1977): Isolation and characterization of *Escherichia coli* phase variants and mutants deficient in type 1 pilus production. – J. Bacteriol. **130:** 495–505.

TANDEAU DE MARSAC, N. & COHEN-BAZIRE, G. (1977): Molecular composition of cyanobacterial phycobilisomes. – Proc. Natl. Acad. Sci. (Wash.) **74:** 1635–1639.

TAUSCHEL, H.-D. (1971): Der Geißelapparat von *Rhodopseudomonas palustris.* VI. Charakterisierung des Flagellins. – Arch. Mikrobiol. **76:** 91–102.

– (1985): ATPase and cytochrome oxidase activities at the polar organelle in flagellated bacteria: An ultrastructural study. – Arch. Microbiol. **141:** 303–308.

TAUSCHEL, H.-D. & DREWS, G. (1967): Thylakoid Morphogenese bei *Rhodopseudomonas palustris.* – Arch. Mikrobiol. **59:** 381–404.

– – (1969): Der Geißelapparat von *Rhodopseudomonas palustris.* I. Untersuchungen zur Feinstruktur des Polorganells. – Arch. Mikrobiol. **66:** 166–179.

– – (1970): Der Geißelapparat von *Rhodopseudomonas palustris.* III. Untersuchung zur Feinstruktur der Geißel. – Cytobiologie **2:** 87–107.

TAYLOR, B. L. & KOSHLAND JR., D. E. (1974): Reversal of flagellar rotation on monotrichous and peritrichous bacteria: Generation of changes in direction. – J. Bacteriol. **119:** 640–642.

TAYLOR, F. J. R. & HARRISON, P. J. (1983): Ecological aspects of intracellular symbiosis. – In: H. E. A. SCHENK & W. SCHWEMMLER (Eds.): Endocytobiology. vol. II, pp. 827–842. – de Gruyter, Berlin, New York.

TAYLOR, K. A., DEATHERAGE, J. F. & AMOS, L. A. (1982): Structure of the S-layer of *Sulfolobus acidocaldarius.* – Nature **299:** 840–842.

TERRY, K. R. & HOOPER, A. B. (1970): Polyphosphate and orthophosphate content of *Nitrosomonas europaea* as a function of growth. – J. Bacteriol. **103:** 199–206.

THOMPSON, B. G. & MURRAY, R. G. E. (1982): The association of the surface array and the outer membrane of *Deinococcus radiodurans.* – Canad. J. Microbiol. **28:** 1081–1088.

THOMPSON, L. M. M. & MACLEOD, R. A. (1974): Biochemical localization of alkaline phosphatase in the cell wall of a marine Pseudomonad. – J. Bacteriol. **117:** 819–825.

THUROW, H., NIEMANN, H., RUDOLPH, C. & STIRM, S. (1974): Host capsule depolymerase activity of bacteriophage particles active on *Klebsiella K 20* and *K 24* strains. – Virology **58:** 306–309.

THURSTON, E. L. & INGRAHAM, L. O. (1971): Morphology and fine structure of *Fischerella ambigua.* – J. Phycol. **7:** 203–210.

TIETZ, H. & HOPPE, W. (1984): Contrasted and non-contrasted regions in ribosomal particles. – Abstr. 8th Eur. Congr. Electron Microscopy. vol. II, pp. 1375–1376. Budapest.

TINGLU, G., GHOSH, A. & GHOSH, B. K. (1984): Subcellular localization of alkaline phosphatase in *Bacillus licheniformis 749/C* by immunoelectron microscopy with colloidal gold. – J. Bacteriol. **159:** 668–677.

TIPPER, D. J. & LINNETT, P. E. (1976): Distribution of peptidoglycan synthetase activities between sporangia and forespores in sporulating cells of *Bacillus sphaericus.* – J. Bacteriol. **126:** 213–221.

TOMOEDA, M., INUZUKA, M. & DATE, T. (1973): Bacterial sex pili. – Progr. Biophys. Molec. Biol. **38:** 23–56.

TORJESEN, P. A. & SLETTEN, K. (1972): C-phycocyanin from *Oscillatoria Agardhii.* I. Some molecular properties. – Biochim. Biophys. Acta **263:** 258–271.

TRAUB, P. & NOMURA, M. (1968): Structure and function of *E. coli* ribosomes. V. Reconstitution of functionally active 30S ribosomal particles from RNA and proteins. – Proc. Natl. Acad. Sci. USA **59:** 777–784.

TRUEBA, F. J. VAN SPRONSEN, E. A., TRAAS, J. & WOLDRINGH, C. L. (1982): Effects of temperature on the size and shape of *Escherichia coli* cells. – Arch. Microbiol. **131:** 235–240.

TRUEPER, H. G. & PFENNIG, N. (1981): Characterization and identification of the anoxygenic phototrophic bacteria. – In: M. P. STARR, H. STOLP, H. G. TRUEPER, A. BALOWS & H. G. SCHLEGEL (Eds.): The Prokaryotes. vol. I, pp. 299–312. – Springer, Berlin, Heidelberg, New York.

TSIEN, H. C. & HIGGINS, M. L. (1974): Effect of temperature on the distribution of membrane particles in *Streptococcus faecalis* as seen by the freeze-fracture technique. – J. Bacteriol. **118:** 725–734.

TULLY, J. G., TAYLOR-ROBINSON, D., ROSE, D. L., COLE, R. M. & BOVE, J. M. (1983): *Mycoplasma genitalium,* a new species from the human urogenital tract. – Internatl. J. Syst. Bacteriol. **33:** 387–396.

TULLY, J. G. & WHITECOMB, R. F. (1981): The genus *Spiroplasma.* – In: M. P. STARR, H. STOLP, H. G. TRUEPER, A. BALOWS & H. G. SCHLEGEL (Eds.): The Prokaryotes. vol. II, pp. 2271–2284. – Springer, Berlin, Heidelberg, New York.

TYLER, P. A. (1970): Hyphomicrobia and the oxidation of manganese in aquatic ecosystems. – Antonie van Leeuwenhoek J. Microbiol. Serol. **36:** 567–578.

TYLER, P. A. & MARSHALL, K. C. (1967a): Microbial oxidation of manganese in hydro-electric pipelines. – Antonie van Leeuwenhoek J. Microbiol. Serol. **33:** 171–183.

– – (1967b): Form and function in manganese-oxidizing bacteria. – Arch. Microbiol. **56:** 344–353.

– – (1967c): Hyphomicrobia – a significant factor in manganese problems. – J. Amer. Water Works Assoc. **59:** 1043–1048.

UNWIN, P. N. T. & HENDERSON, R. (1975): Molecular structure determination by electron microscopy of unstained crystalline specimens. – J. Molec. Biol. **94:** 425–440.

USUKURA, J., YAMADA, E., TOKUNAGA, F. & YOSHIZAWA, T. (1980): Ultrastructure of purple membrane and cell wall of *Halobacterium halobium.* – J. Ultrastruct. Res. **70:** 204–219.

VAARA, T. (1982): The outermost surface structures in chroococcacean cyanobacteria. – Canad. J. Microbiol. **28:** 929–941.

VALKENBURG, J. C. A. & WOLDRINGH, C. L. (1984): Phase separation between nucleoid and cytoplasm in *Escherichia coli* as defined by immersive refractometry. – J. Bacteriol. **160:** 1151–1157.

VALKIRS, G. E. & FEHER, G. (1982): Topography of reaction center subunits in the membrane of the photosynthetic bacterium *Rhodopseudomonas sphaeroides.* – J. Cell Biol. **95:** 179–188.

VARGA, A. R. & STAEHELIN, L. A. (1983): Spatial differentiation in photosynthetic and non-photosynthetic membranes of *Rhodopseudomonas palustris.* – J. Bacteriol. **154:** 1414–1430.

VARON, M., COHEN, S. & ROSENBERG, E. (1984): Autocides produced by *Myxococcus xanthus.* – J. Bacteriol. **160:** 1146–1150.

VARSHAVSKY, A. J., NEDOSPACOS, S. A., BAKAYEV, V. V., BAKAYEVA, T. B. & GEORGIEV, G. P. (1977): Histone-like proteins in the purified *Escherichia coli* deoxyribonucleoprotein. – Nucleic Acids Res. **4:** 2725–2745.

VERSCHOOR, A., FRANK, J. & WAGENKNECHT, T. (1984): Computer averaging of *E. coli* 70S monosomes. – Abstr. 8th Eur. Congr. Electron Microscopy. vol. II, pp. 1461–1462. Budapest.

VERWER, R. W. H. & NANNINGA, N. (1980): Pattern of meso-DL-2,6-diaminopimelic acid incorporation during the division cycle of *Escherichia coli.* – J. Bacteriol. **144:** 327–336.

DE VOE, I. W., COSTERTON, J. W. & MACLEOD, R. A. (1971): Demonstration by freeze-etching of a single cleavage plane in the cell wall of a Gram-negative bacterium. – J. Bacteriol. **106:** 659–671.

VOELZ, H. & REICHENBACH, H. (1969): Fine structure of fruiting bodies of *Stigmatella aurantiaca* (Myxobacterales). – J. Bacteriol. **99:** 856–866.

VOS-SCHEPERKEUTER, G.H., PAS, E., BRAKENHOFF, G. J., NANNINGA, N. & WITHOLT, B. (1984): Topography of the insertion of *LamB* protein into the outer membrane of *Escherichia coli* wildtype and *lac-lamB* cells. – J. Bacteriol. **159:** 440–447.

WABL, M. R. (1974): Electron microscopic localization of two proteins on the surface of the 50S ribosomal subunit of *Escherichia coli* using specific antibody markers. – J. Molec. Biol. **84:** 241–247.

WABL, M. R., BARENDS, P. J. & NANNINGA, N. (1973): Tilting experiments with negatively stained *E. coli* ribosomal subunits. An electron microscopic study. – Cytobiologie. Z. Exp. Zellforsch. **7:** 1–9

WAGENKNECHT, T., DE ROSIER, D., SHAPIRO, L. & WEISSBORN, A. (1981): Three-dimensional reconstruction of the flagellar hook from *Caulobacter crescentus*. – J. Molec. Biol. **151:** 439–465.

WAGENKNECHT, T., FRANK, J., BOUBLIK, M. & OFENGAND, J. (1984): Localization of the decoding region of the *Escherichia coli* small ribosomal subunit by electron microscopy and image averaging. – Abstr. 8th Eur. Congr. Electron Microscopy. vol. II, pp. 1357–1358. Budapest.

WAKIM, B., SCHRADER, M. & OELZE, J. (1979): Characterization of cell-envelope fractions of chemotrophically and phototrophically grown *Rhodospirillum tenue*. – Arch. Microbiol. **123:** 287–293.

WALKER, P. D. (1970): Cytology of spore formation and germination. – J. Appl. Bacteriol. **33:** 1–12.

WALKER, P. D., BAILLIE, A., THOMSON, R. O. & BATTY, I. (1966): The use of ferritin labelled antibodies in the location of spore and vegetative antigens in *Bacillus cereus*. – J. Appl. Bacteriol. **29:** 512–518.

WALL, F., PFISTER, R. M. & SOMERSON, N. L. (1983): Freeze-fracture confirmation of the presence of a core in the specialized tip structure of *Mycoplasma pneumoniae*. – J. Bacteriol. **154:** 924–929.

WALSBY, A. E. (1972): Structure and function of gas vacuoles. – Bacteriol. Rev. **36:** 1–32.

– (1980): A square bacterium. – Nature (Lond.) **283:** 69–71.

– (1981): Cyanobacteria: Planktonic gas-vacuolate forms. – In: M. P. STARR, H. STOLP, H. G. TRUEPER, A. BALOWS & H. G. SCHLEGEL (Eds.): The Prokaryotes. vol. I, pp. 224–235. – Springer, Berlin, Heidelberg, New York.

WALTHER-MAURUSCHAT, A., ARAGNO, M., MAYER, F. & SCHLEGEL, H. G. (1977): Micromorphology of Gram-negative hydrogen bacteria. II. Cell envelope, membranes and cytoplasmic inclusions. – Arch. Microbiol. **114:** 101–110.

WALTHER-MAURUSCHAT, A. & MAYER, F. (1978): Isolation and characterization of polysheaths, phage tail-like defective bacteriophages of *Alcaligenes eutrophus H 16*. – J. Gen. Virol. **41:** 239–254.

WANG, W. S. & LUNDGREN, D. G. (1969): Poly-β-hydroxybutyrate in the chemolithotrophic bacterium *Ferrobacillus ferrooxidans*. – J. Bacteriol. **97:** 947–950.

WARD, H. M. (1892): The ginger-beer plant, and the organisms composing it: A contribution to the study of fermentation-yeasts and bacteria. – Philos. Trans. Roy. Soc. London **183:** 125–197.

WARTH, A. D., OHYE, D. F. & MURRELL, W. G. (1963): Location and composition of spore mucopeptide in *Bacillus* species. – J. Cell Biol. **16:** 593–609.

WARTH, A. D. & STROMINGER, J. L. (1972): Structure of the peptidoglycan from spores of *Bacillus subtilis*. – Biochemistry **11:** 1389–1395.

WATSON, S. W., VALOIS, F. W. & WATERBURY, J. B. (1981): The family Nitrobacteraceae. – In: M. P. STARR, H. STOLP, H. G. TRUEPER, A. BALOWS & H. G. SCHLEGEL (Eds.): The Prokaryotes. vol. I, pp. 1005–1021. – Springer, Berlin, Heidelberg, New York.

WATSON, S. W. & WATERBURY, J. B. (1971): Characteristics of two marine nitrite-oxidizing bacteria, *Nitrospina gracilis* nov. gen. nov. sp. and *Nitrococcus mobilis* nov. gen. nov. sp. – Arch. Mikrobiol. **77:** 203–230.

WEBB, M. (1984): The action of lysozyme on heat-killed Gram-positive micro-organisms. – J. Gen. Microbiol. **2:** 260–274.

WEBER, H. J. & BOGOMOLNI, R. A. (1981): P_{588}, a second retinal-containing pigment in *Halobacterium halobium*. – Photochem. Photobiol. **33:** 601–608.

– – (1982): The isolation of *Halobacterium* mutant strains with defects in pigment synthesis. – Meth. Enzymol. **88:** 379–390.

WEIBULL, C. (1953): The isolation of protoplasts from *Bacillus megaterium* by controlled treatment with lysozyme. – J. Bacteriol. **66:** 688–695.

WEIBULL, C., CHRISTIANSSON, A. & CARLEMALM, E. (1983: Extraction of membrane lipids during fixation, dehydration and embedding of *Acholeplasma laidlawii* cells for electron microscopy. – J. Microsc. **129:** 201–207.

WEIDEL, W., FRANK, H. & MARTIN, H. H. (1960): The rigid layer of the cell wall of *Escherichia coli* strain *B*. – J. Gen. Microbiol. **22:** 158–166.

WEIDEL, W. & PELZER, H. (1964): Bagshaped macromolecules – a new outlook on bacterial cell walls. – Adv. Enzymol. **26:** 193–232.

WEISS, E. (1981): The family Rickettsiaceae: Human pathogenes. – In: M. P. STARR, H. STOLP, H. G. TRUEPER, A. BALOWS & H. G. SCHLEGEL (Eds.): The Prokaryotes. vol. II, pp. 2137–2160. – Springer, Berlin, Heidelberg, New York.

WEISS, E., SCHIEFER, H.-G. & KRAUSS, H. (1979): Ultrastructural visualization of *Klebsiella* capsules by polycationic ferritin. – FEMS Microbiol. Lett. **6:** 435–437.

WELLS, B. & HORNE, R. W. (1983): The ultrastructure of *Pseudomonas avenae*. II. Intracellular refractile (R-body) structure. – Micron **14:** 329–344.

WERNER, D., MELLOR, R. B., BASSARAB, S. & DITTRICH, W. (1983): *Glycine max* root response to symbiotic infection. – In: H. E. A. SCHENK & W. SCHWEMMLER (Eds): Endocytobiology. vol. II, pp. 585–586. – de Gruyter, Berlin, New York.

WETZEL, B. K., SPICER, S. S., DVORAK, H. F. & HEPPEL, L. A. (1970): Cytochemical localization of certain phosphatases in *Escherichia coli* – J. Bacteriol. **104:** 529–542.

WHITTENBURY, R. (1969): Microbial utilization of methane. – Process Biochem. **4 (1):** 51.

WHITTENBURY, R. & MCLEE, A. G. (1967): *Rhodopseudomonas palustris* and *Rh. viridis* – photosynthetic budding bacteria. – Arch. Mikrobiol. **59:** 324.

WHITTENBURY, R., PHILLIPS, K. C. & WILKINSON, J. F. (1970): Enrichment, isolation and some properties of methane-utilizing bacteria. – J. Gen. Microbiol. **61:** 205.

WILDGRUBER, G., THOMM, M., KÖNIG, H., OBER, K., RICCHIUTO, T. & STETTER, K. O. (1982): *Methanoplanus limicola*, a plate-shaped methanogen representing a novel family, the Methanoplanaceae. – Arch. Microbiol. **132:** 31–36.

WILKINSON, H. W. (1975): Immunochemistry of purified polysaccharide type antigens of group *B* streptococcal types *Ia, Ib,* and *Ic*. – Infect. Immun. **11:** 845–852.

WILKINSON, H. W. & JONES, W. L. (1976): Radioimmuno assay for measuring antibodies specific for group *B* streptococcal types *Ia, Ib, Ic, II,* and *III*. – J. Clin. Microbiol. **3:** 480–485.

WILKINSON, J. F. (1958): The extracellular polysaccharides of bacteria. – Bacteriol. Rev. **22:** 46–73.

WILLETTS, N. & SKURRAY, R. (1980): The conjugation system of F-like plasmids. – Annual Rev. Genet. **14:** 41–76.

WILLIAMS, R. C., GINGRICH, J. C. & GLAZER, A. N. (1980): Cyanobacterial phycobilisomes. Particles from *Synechocystis 6701* and two pigment mutants. – J. Cell Biol. **85:** 558–566.

WILSON, M. H. & COLLIER, A. M. (1976): Ultrastructural study of *Mycoplasma pneumoniae* in organ culture. – J. Bacteriol. **125:** 332–339.

WOESE, C. R. (1981): Archaebacteria. – Sci. Amer. **244:** 94–106.

WOLDRINGH, C. L. (1973): Effect of cations on the organization of the nucleoplasm in *Escherichia coli* prefixed with osmium tetroxide or glutaraldehyde. – Cytobiologie Z. Exp. Zellforsch. **8:** 97–111.

– (1976): Morphological analysis of nuclear septation and cell division during the life cycle of *Escherichia coli*. – J. Bacteriol. **125:** 248–257.

WOLDRINGH, C. L. & NANNINGA, N. (1976): Organization of the nucleoplasm in *Escherichia coli* visualized by phase-contrast light microscopy, freeze fracturing, and thin sectioning. – J. Bacteriol. **127:** 1455–1464.

– – (1985): Structure of nucleoid and cytoplasm in the intact cell. – In: N. NANNINGA (Ed.): Molecular Cytology of *Escherichia coli*. pp. 161–197. – Academic Press Inc., New York

WOLK, C. P. (1973): Physiology and cytological chemistry of blue-green algae. – Bacteriol. Rev. **37:** 32–101.

WOLLMAN, F. A. (1979): Ultrastructural comparison of *Cyanidium caldarium* wild type and *III-C* mutant lacking phycobilisomes. – Plant Physiol. **63:** 375–381.

WORCEL, A. & BURGI, E. (1972): On the structure of the folded chromosome of *Escherichia coli*. – J. Molec. Biol. **71:** 127–147.

WOROBEC, E. A., TANEJA, A. K., HODGES, R. S. & PARANCHYCH, W. (1983): Localization of the major antigenic determinant of *EDP 208*-pili at the pilus protein. – J. Bacteriol. **153:** 955–961.

WYSS, O., NEWMAN, M. G. & SOCOLOFSKY. M. D. (1961): Development and germination of the *Azotobacter* cyst. – J. Biophys. Biochem. Cytol. **10:** 555–565.

YAMAMOTO, T. (1967): Presence of rhapidosomes in various species of bacteria and their morphological characteristics. – J. Bacteriol. **94:** 1746.

YAMANAKA, G., GLAZER, A. N. & WILLIAMS, R. C. (1978): Cyanobacterial phycobilisomes. Characterization of the phycobilisomes of *Synechococcus* sp. 6301. – J. Biol. Chem. **253:** 8303–8310.

YOUNG, H. L., CHAO, F., TURNBILL, C. & PHILPOTT, D. E. (1972): Ultrastructure of *Pseudomonas saccharophila* at early and late log phase of growth. – J. Bacteriol. **109:** 862–868.

ZAVARZIN, G. A. (1960): The life cycle and nuclear apparatus in *Hyphomicrobium vulgare* Stutzer and Hartleb. – Microbiologiya **29:** 38–42.

ZEHNDER, A. J. B., HUSER, B. A., BROCK, T. D. & WUHRMANN, K. (1980): Characterization of an acetate-decarboxylating, non-hydrogen-oxidizing methane bacterium. – Arch. Microbiol. **124:** 1–11.

ZEIKUS, J. G. & BOWEN V. G. (1975): Comparative ultrastructure of methanogenic bacteria. – Canad. J. Microbiol. **21:** 121–129.

ZEIKUS, J. G. & WOLFE, R. S. (1972a): *Methanobacterium thermoautotrophicus* sp. n., an anaerobic, autotrophic, extreme thermophile. – J. Bacteriol. **109:** 707–713.

– – (1972b): *Methanobacterium thermoautotrophicus* should be *Methanobacterium thermoautotrophicum.* – Internatl. J. Syst. Bacteriol. **22:** 395.

– – (1973): Fine structure of *Methanobacterium thermoautotrophicum:* Effect of growth temperature on morphology and ultrastructure. – J. Bacteriol. **113:** 461–467.

ZENTGRAF, H., BERTHOLD, V. & GEIDER, K. (1977): Interaction of DNA with DNA-binding proteins. – Biochim. Biophys. Acta **474:** 629–638.

ZHILINA, T. (1971): The fine structure of *Methanosarcina.* – Microbiologiya **40:** 674–680.

ZILLIG, W., STETTER, K. O., SCHABEL, R., MADON, J. & GIERL, A. (1982): Transcription in Archaebacteria. – Zentralbl. Bakteriol. Hyg. 1. Abtl. Orig. C: 218–227.

ZIMMERMANN, T., GIFFHORN, F., SCHRAMM, H. J. & MAYER, F. (1982): Analysis of structure-function relationships in citrate lyase isolated from *Rhodopseudomonas gelatinosa* as revealed by cross-linking and immunoelectron microscopy. – Eur. J. Biochem. **126:** 49–46.

Subject Index

The "Handbuch der Pflanzenanatomie", founded by LINSBAUER and continued by A. PASCHER, G. TISCHLER, H. D. WULFF and W. ZIMMERMANN, has everywhere been recognized as the leading work on plant anatomy. It is reissued in a completely revised edition under the title

Handbuch der Pflanzenanatomie
Encyclopedia of Plant Anatomy
Traité d'Anatomie Végétale

2. völlig neubearbeitete Auflage / much revised edition. Herausgegeben von Prof. DR. H. J. BRAUN, Freiburg, Prof. Dr. S. CARLQUIST, Claremont (Calif./USA), Prof. Dr. P. OZENDA, Grenoble, und Prof. Dr. I. ROTH, Freiburg

DISPOSITION

Band I–III Zytologie
Band IV–V Anatomie einzelner Gewebe
Band VI–VII Anatomie niederer systematischer Gruppen
Band VIII–X Histologie der Kormophytae
Band XI–XII Anatomie phylogenetisch bedeutsamer Gruppen
Band XIII–XIV Ökologische, experimentelle und pathologische Anatomie

Band II: Allgemeine Pflanzenkaryologie, 2. Hälfte

Kernteilung und Kernverschmelzung
von Prof. Dr. GEORG TISCHLER. Großoktav. 472 Textabbildungen. VIII, 1040 Seiten. Fotomechanischer Nachdruck der 2. Auflage 1944. 1951. Ganzleinen. Vergriffen

Angewandte Pflanzenkaryologie *Band II: Ergänzungsband*
von Prof. Dr. GEORG TISCHLER, fortgeführt von Prof. Dr. HEINZ DIEDRICH WULFF. Großoktav. 127 Abb. X, 1227 Seiten. 1953–1963. Lfg. 1–6 mit Decke

The Plant Cell Wall (in English) *Band III, Teil 4*
3., completely revised edition by Dr. ALBERT FREY-WYSSLING, em. Prof. 17 × 24 cm. 193 figures, 27 tables in the text and on 20 plates. XI, 294 pages. 1976. Cloth

Fossile Wandstrukturen *Band III, Teil 5*
untersucht am Beispiel der Tracheidenwände paläozoischer Gefäßpflanzen von Dr. ERICH HENES. Großoktav. 132 Textabbildungen, davon 73 auf 9 Kunstdrucktafeln. VIII, 108 Seiten. 1959. Ganzleinen

Plant Hairs (in English and German) *Band IV, Teil 5*
by Prof. J. C. TH. UPHOF (in English). With an article
Die Verbreitung der Haartypen in den natürlichen Verwandtschaftsgruppen
von Dr. KARL HUMMEL und KARIN STAESCHE (in German). 17 × 24 cm. 96 figures, 10 photos on plates. XII, 292 pages. 1962. Cloth

Die mechanischen Elemente und das mechanische System *Band IV, Teil 6*
von Prof. Dr. FRIEDRICH TOBLER. Großoktav. 42 Textabbildungen. IV, 60 Seiten. 2. Auflage 1957. Broschiert

Das trophische Parenchym: A. Assimilationsgewebe *Band IV, Teil 7 A*
von Prof. Dr. FRITZ JÜRGEN MEYER. Großoktav. 119 Textabbildungen. X, 188 Seiten. 2. Auflage 1962. Ganzleinen

The Phloem (in English) *Band V, Teil 2*
by Prof. Dr. KATHERINE ESAU. 17×24 cm. 100 figures, 19 tables in the text and 48 plates. XI, 505 pages. 1969. Cloth

Bewegungsgewebe und Perzeptionsorgane *Allgemeiner Teil Band V, Teil 5*
von Prof. Dr. HERMANN VON GUTTENBERG †, Rostock. Großoktav. 231 Textabbildungen. VIII, 332 Seiten. 1971. Ganzleinen

Schizophyzeen *Band VI, Teil 1*
von Prof. Dr. LOTHAR GEITLER. Großoktav. 101 Textabbildungen. VIII, 131 Seiten. 2. Auflage 1960. Ganzleinen

Cytology of Bacteria (in English) *Band VI, Teil 2*
by Prof. Dr. F. MAYER, Göttingen. 17×24 cm. With 234 figures. IX, 290 pages. 1986. Cloth

Anatomie der Asco- und Basidiomyceten *Band VI, Teil 8*
von weil. Prof. Dr. HEINRICH LOHWAG. Photomechanischer Nachdruck der 1. Auflage 1941. Großoktav. 348 Abbildungen. XII, 572 Seiten. 1965. Ganzleinen

Anatomie des Lichens (en français) *Band VI, Teil 9*
par Prof. Dr. P. OZENDA. 17×24 cm. 123 figures et 8 planches. XII, 200 pages. 1963. Relié

Histogenese der Pteridophyten *Band VII, Teil 2*
von Prof. Dr. HERMANN VON GUTTENBERG †. Großoktav. 321 Textabbildungen. VIII, 312 Seiten. 1965. Ganzleinen

Band VII, Teil 3
Comparative Anatomy of Vegetative Organs of the Pteridophytes (in English)
by em. Prof. Dr. YUDZURU OGURA. 17×24 cm. 459 figures. VIII, 502 pages. 1972. Cloth

Anatomie des Blattes: I. Blattanatomie der Gymnospermen *Band VIII, Teil 1*
von Prof. Dr. KLAUS NAPP-ZINN. Großoktav. 86 Abbildungen. XIV, 369 Seiten, 1966. Ganzleinen

Anatomie des Blattes: II. Blattanatomie der Angiospermen *Band VIII, Teil 2*
A. Entwicklungsgeschichtliche und topographische Anatomie des Angiospermenblattes von Prof. Dr. KLAUS NAPP-ZINN. Großoktav.
1. Lieferung: 220 Textabbildungen und 46 Tabellen. XVI, 764 Seiten. 1973. Ganzleinen
2. Lieferung: 60 Textabbildungen und 22 Tabellen. VIII, 660 Seiten. 1974. Ganzleinen

B. Experimentelle und ökologische Anatomie des Angiospermenblattes von Prof. Dr. KLAUS NAPP-ZINN. Großoktav.
1. Lieferung: 129 Textabbildungen und 168 Tabellen. XV, 519 Seiten, 1984. Ganzleinen

Band VIII, Teil 3

Grundzüge der Histogenese höherer Pflanzen: I. Die Angiospermen
von Prof. Dr. HERMANN VON GUTTENBERG †. Großoktav. 328 Textabbildungen. VIII, 315 Seiten, 1960. Ganzleinen

Band VIII, Teil 4

Grundzüge der Histogenese höherer Pflanzen: II. Die Gymnospermen
von Prof. Dr. HERMANN VON GUTTENBERG † Großoktav. 178 Textabbildungen. VIII, 172 Seiten. 1961. Ganzleinen

Der primäre Bau der Angiospermenwurzel *Band VIII, Teil 5*
von Prof. Dr. HERMANN VON GUTTENBERG †. Großoktav. 372 Textabbildungen. VIII, 472 Seiten. 1968. Ganzleinen

Funktionelle Histologie der sekundären Sproßachse, I. Das Holz *Band IX, Teil 1*
von Prof. Dr. HELMUT J. BRAUN. Großoktav. 212 Textabbildungen. XI, 190 Seiten. 1970. Ganzleinen

Structural Patterns of Tropical Barks (in English) *Band IX, Teil 3*
by Prof. Dr. INGRID ROTH. 17×24 cm. 282 figures. XVI, 609 pages. 1981. Cloth

Fruits of Angiosperms (in English) *Band X, Teil 1*
by Prof. Dr. INGRID ROTH. 17×24 cm. 232 figures. XVI, 675 pages. 1977. Cloth

Embryology of Gymnosperms (in English) *Band X, Teil 2*
by Dr. HARDEV SINGH, Delhi. 17×24 cm. With 30 photos, 121 figures and 1 table in the text. XII, 302 pages. 1978. Cloth

Gnétophyta (en français) *Band XII, Teil 2*
par Prof. Dr. P. MARTENS, Louvain, Belgique. 17×24 cm. Avec 127 figures et 1 frontispice. XIV, 295 pages. 1971. Relié

Anatomie des Galles (en français) *Band XIII, Teil 1*
par Dr. JEAN MEYER et Prof. Dr. H. J. MARESQUELLE, Strasbourg.
17×24 cm. Avec 519 figures. XVI, 662 pages. 1983. Relié

Les Mycorhizes (en français) *Band XIII, Teil 2*
par Prof. Dr. Desiré Georges Strullu, Angers
17 × 24 cm. Avec 59 figures. IX, 198 pages. 1985. Relié

IN VORBEREITUNG — IN PREPARATION

Anatomy of the Asco- and Basidiomycetes
by Prof. Dr. E. MÜLLER, Zürich

Progymnospermae
von Prof. Dr. DIETER VOGELLEHNER, Freiburg-Kappel

Anatomie des Blattes: II. Blattanatomie der Angiospermen *Band VIII, Teil 2*
B. Experimentelle und ökologische Anatomie des Angiospermenblattes
von Prof. Dr. KLAUS NAPP-ZINN.
2. Lieferung

Gebrüder Borntraeger · Berlin · Stuttgart